Green Technologies and The Mobility Industry

Other SAE Books of Interest:

Particulate Emissions from Vehicles
(Product Code: R-389)

Diesel Emissions and Their Control
(Product Code: R-303)

Technologies for Near-Zero-Emission Gasoline-Powered Vehicles
(Product Code: R-359)

Combustion and Emission Control for SI Engines – Modeling and Experimental
Studies
(Product Code: PT-121)

For more information or to order a book, contact SAE International at

400 Commonwealth Drive, Warrendale, PA 15096-0001; phone 877-606-7323
(U.S. and Canada) **or 724-776-4970** *(outside U.S. and Canada);* **fax 724-776-0790;**
e-mail CustomerService@sae.org; website http://store.sae.org**.**

Green Technologies and the Mobility Industry

Edited by

Dr. Andrew Brown, Jr.

Published by
SAE International
400 Commonwealth Drive
Warrendale, PA 15096-0001 U.S.A.

Phone: (724) 776-4841
Fax: (724) 776-5760
www.sae.org
January 2011

PT-146

400 Commonwealth Drive
Warrendale, PA 15096-0001 USA

E-mail: CustomerService@sae.org
Phone: 877-606-7323 *(inside USA and Canada)*
724-776-4970 *(outside USA)*
Fax: 724-776-1615

ISBN 978-0-7680-3494-3
Library of Congress Catalog Number 2010937023
SAE Order No. PT-146

Information contained in this work has been obtained by SAE International from sources believed to be reliable. However, neither SAE International nor its authors guarantee the accuracy or completeness of any information published herein and neither SAE International nor its authors shall be responsible for any errors, omissions, or damages arising out of use of this information. This work is published with the understanding that SAE International and its authors are supplying information, but are not attempting to render engineering or other professional services. If such services are required, the assistance of an appropriate professional should be sought.

Applicable to SAE International technical papers: The Engineering Meetings Board has approved these papers for publication. They have successfully completed SAE's peer review process under the supervision of the session organizer. This process requires a minimum of three (3) reviews by industry experts. Positions and opinions advanced in these papers are those of the author(s) and not necessarily those of SAE. The author is solely responsible for the content of the paper.

To purchase bulk quantities, please contact:
SAE Customer Service
E-mail: CustomerService@sae.org
Phone: 877-606-7323 *(inside USA and Canada)*
724-776-4970 *(outside USA)*
Fax: 724-776-1615

Visit the SAE Bookstore at
http://store.sae.org

TABLE OF CONTENTS

Introduction

Vehicle Electrification

Fuels & Emissions

Sustainable Mobility

New Technologies

INTRODUCTION

INTRODUCTION

The Quest for a Green and Sustainable Mobility Industry

The new "green paradigm", which considers the environment as a vital aspect of doing business, is now a reality that touches all industries, with its technological challenges and victories.

The mobility industry has responded to this new state of affairs with a gradual yet unwavering commitment to be less polluting, more sustainable and efficient in the long run.

The future of transportation rests on the steady pillars of environmental care, safety at all levels, and the ability to communicate with other vehicles, and with the infrastructure we rely upon when we drive.

The economic growth and rising incomes we have enjoyed in the past four decades brought about increased urbanization and more demand for motorized mobility. For many years, though, it seemed that we as consumers and engineers had to choose between vehicles that were safe as opposed to vehicles that were fuel efficient. That is no longer the case.

From the development of new high-capacity energy storage systems for hybrid and fully electric vehicles, to biofuels and the use of innovative lightweight materials, we now witness the emergence of technologies we did not even dream of ten years ago.

Innovation is coming to the market place with real-life products and solutions. We see its results in emissions and waste reduction, the expanded use of recyclable and renewable materials and a wider focus on sustainability. This teaches us that a cleaner present is possible, and a much greener future a goal within reach.

In the U.S. alone, cars and light trucks consume an average of 8.2 million barrels of oil each day, translating to emissions of more than 300 million metric tons of carbon every year. The foregone conclusion is that significant changes in how we develop new products for the transportation needs of a growing world population are required.

Overall we consider the main drivers of this new predicament to be:

- The environment, on which we all depend, including the consequences of climate change.
- The price of oil and its future availability, including energy security.
- Economic concerns for the next decade or so.

Advances in electric vehicle development and fuel cells are an excellent example of how engineering ingenuity is responding to the challenges ahead.

At the same time, gasoline and diesel engines, under extreme pressure to become ever less polluting, are now being transformed into new power-efficient machines.

Alternative fuels, such as biodiesel, DME, E-85, natural gas, biomass-generated gas, and hydrogen are also top priorities on the list of possible solutions. Together with next-generation lubricants, we can expect more efficiency in overall engine performance coming from these technologies.

This all leads us to a path in search of sustainable transportation not only for mature markets but for all markets. The global synergies of production and the new players such as China and India amplify the scope and impact of any decision made by OEMs and suppliers alike. The quest for environmentally-minded designs will require intense collaboration among international engineering teams. New products have to meet the technical and customer requirements of global markets at launch. The costs and risks of not doing so are becoming simply prohibitive.

Additionally, consumers themselves will be also drivers of change by demanding more eco-friendly means of transportation, both at the mass and individual levels. After all, it is for them that the industry works.

New Solutions in Sight: The Case for Electrification

Although no industry can ever turn its back to cost efficiencies, the equation now also includes the monetary and non-monetary price of environmental damage, the lasting impact of carbon footprint, and the value of being socially responsible. Around the world, we see governments increasingly regulating and restricting GHG emissions, and incentivizing the creation and use of non-polluting technologies.

Over the next decade, industry forecasts show a strong growth in production of electric vehicles in Asia. Most analysts are conservative showing penetration rates under 5%, but some of the industry predictions actually foresee up to 15% market share by 2020. There are more than 40 vehicle development programs taking place in Asia as this book goes to press.

The Chinese government has made its intentions clear on how important it considers the development and consumer purchase of hybrid and electric vehicles. The mandate is that by year 2012, vehicle makers produce at least 500,000 units (or 5%) per year of their total output as hybrid and/or electric. All Chinese vehicle manufacturers must have at least one HEV or EV model in the market by the same year. As of today, the country has invested over US$3.5 billion to stimulate the production of electric vehicles and the necessary infrastructure to support them.

In Japan, by 2020, it is forecasted that the HEV market share will be as much as 20%-30%, with pure EV market share getting to be 15%-20%. Korea and India, with aggressive incentives for HEVs and EVs both for OEMs and consumers, are helping to shape Asia as a leader in this technology.

In the United States, new emissions standards have been established and the focus is growing on pollution prevention instead of pollution treatment.

With the stated objective to achieve 35.5 mpg and 250 g/mi CO_2 by 2016, the American government is also placing emphasis on accelerating new standards adoption and, for the first time, is putting effort into ruling emissions from medium and heavy-duty vehicles.

The European Community, on the other hand, is strongly focused on regulating and controlling CO_2 emissions from passenger vehicles, which should achieve a standard of 95g of CO_2 per kilometer by 2020. Super credits are offered for ultra-low carbon vehicles (<50 g CO_2/km), and penalties imposed on those which fall below par. From 2012 to 2018, the penalties range from 5 Euros for the first gram of CO_2 above the target, 15 Euros for the second gram, 25 Euros for the third and up to 95 Euros per gram over the target after that multiplied by number of vehicles sold. From 2019 onwards, 95 Euros per gram of CO2 will be charged to above-target emissions multiplied by number of vehicles sold. Non-compliance could become a very expensive proposition.

Technology for the Future and the Environment

These scenarios show us the importance of the commitment to sustainable and green transportation. Electrification of powertrain will play a vital role in decades to come, together with strict pollution regulation, government incentives, rebates and reduced tax fees to consumers, and industry investments.

Along that line, the world has its eyes set on how battery technology will develop and which architecture will eventually win the race. The competition between voltage and current capacity might be won by a brand new topology still to be fully designed. Yet, no matter which, the final technological answer will have to meet some basic requirements pertaining to energy consumption, power generation potential, cost, lifetime, and safety.

Mobility engineers are tasked to offer solutions that support better vehicle range as well as torque and acceleration.

How much markets are willing to pay for battery power is one of the most strategic questions at hand. The targeted cost established by the US Department of Energy, for PHEVs, is between <US$300/kWh by 2014. (1)

According to a study by the Boston Consulting Group, the cost of electric batteries is estimated to decrease between 60-65% from 2009 to 2020. (2)

So far, the lithium-ion chemistry seems to hold increasing promise for a commercial-level solution, but significant research and testing still needs to be done to make it a final choice.

Besides the batteries themselves, the use of ultra-capacitors will become more prevalent as they offer almost unlimited lifecycle, buffering the battery from extreme discharge (acceleration) and charge (regenerative braking) events, thus extending the battery life.

Additionally, powertrain electrification also means electric motors, with two primary types of machines which are currently used in hybrid and electric vehicles: induction and permanent magnets. In simple terms, induction engines generate their magnetic fields from an electric current flowing through the copper windings wrapped around the rotor's iron core. The rotors in permanent magnet machines use magnets to generate the magnetic field exclusively without the need for current. It is not yet clear which design shall be favored.

Every hybrid and electric vehicle has at least one electric machine, and some have multiple motors, depending on their layout and intended use. Motor technology development is increasing at OEMs and suppliers, buoyed by millions in R & D funding from the public sector. And, with growth, also comes the need to decide between making it and buying it. The jury is still out on which model will work more efficiently.

From a sourcing trend standpoint, it is likely that the early movers in the hybrid markets will build their own motors and/or inverters, transitioning to relying on suppliers as the markets stabilize. The growth of the EV market could be accompanied by the creation of niches geared towards full integration of e-motors and the transmission system. This would be expanded by the use of inverters with the purpose of EMI reduction and additional weight saving from optimized wiring and battery/ultra-capacitor technology.

Another trend that will claim an even bigger space in the picture of hybridization is the increased use of power electronics, which can make up 20% of hybrid vehicles' material costs.

According to Yole Developpement, located in Lyon, France, in the power modules arena there are two current dominant technologies for low-voltage and high-voltage applications: MOSFET and IGBT respectively. This is a market expected to grow at the aggressive rate of close to 30% p.a. to reach US$5 billion in 2020. With power modules representing about 50% of the inverter and converter cost, a strong effort in its reduction is expected in the coming years. (3)

Still according to Yole Developpement, new technologies are being worked on in this sphere to include prototypes based on SiC diodes and switches. This could lead to significant reduction in the size and cost of silicon devices.

The possible success of this new endeavor would have meaningful impact on HEV/EV applications as the availability of SiC switches would allow reduction of the overall cooling systems cost.

An alternative to SiC, in case the targeted cost reductions cannot be obtained, would be GaN, with its better performance-to-cost ratio.

Green is for All Sectors of the Industry

Although we seem to see constant new "green" development in the automotive sector, the aviation, commercial and off-highway vehicle sectors are also jumping on the bandwagon and showing strong commitment to the adoption of a new, more environmentally-friendly mindset.

Air travel continues to grow, raising issues of fuel and energy consumption, noise and their impact on the environment. New engines are being designed and tested to use more biofuels, batteries and be more efficient altogether. This is supported by enhanced airport design and aircraft operations logistics in order to minimize fuel burn.

The heavy-duty vehicle group is in a prime position to see the most improvement in terms of fuel consumption reduction and a more permanent transition to hybridization. In this arena too, the future scarcity of oil supplies, ever more stringent emissions regulations and noise limitations are the main reasons for the increased investment in R&D. The efforts are beginning to pay off, with meaningful focus on the possibilities of energy recovery through regenerative braking systems.

The Quest for a Green and Sustainable Mobility Industry

With the publication of "Green Technologies and the Mobility Industry," SAE Intl. brings to the reader a unique, handpicked collection of twenty of its technical papers. They are solid examples of how the mobility industry, in different parts of the world, is developing greener products, and at the same time staying responsive, if not ahead, of new standards and legal requirements.

Available in print and eBook formats, it covers the areas of vehicle electrification, fuels and emissions, sustainable mobility and new technologies.

The authors come from both from industry and academia, bringing fresh and creative ideas to the discussion forum.

On the subject of **electrification**, we have chosen the following papers:

- **"Bridging the Gap from Gasoline to Hybrids: Using Systems Engineering to Deliver Advanced Powertrain Technologies,"** by Robert Brincheck, 2010, who writes about how technologically aggressive OEMs are becoming in order to meet new fuel economy standards, lower manufacturing costs and consumers' demands for performance.
- **"A Survey on Electric/Hybrid Vehicles,"** 2010, by B. Ribeiro, F. Brito and J. Martins, presenting us with an interesting historical perspective on the subject.
- **"Technology Improvement Pathways to Cost-effective Vehicle Electrification,"** 2010, authored by A. Brooker, M. Thornton and J. Rugh, introduces the evaluation of how to make plug-in electric vehicles and plug-in hybrid electric vehicles cost effective.
- **"Hybrid Drive Systems for Industrial Applications,"** 2009, by F. Böhler, P. Thiebes, M. Geimer, J. Santoire and R. Zahoransky who discuss the use of hybrid drives for industrial and off-highway applications.
- **"Diesel Hybrids- The Logical Path towards Hybridisation,"** 2009, by A. Srinivas, T. Kumar Prasad, T. Satish, S. Dhande and C. Nandagopalan, approaches the idea of diesel/hybrid electric powertrains as a way to an ideal platform to maximize the benefits of hybridization.

On the subject of **fuels and emissions**, we selected:

- **"Demonstration of Power Improvements on a Diesel Engine Operating on Multiple Fuels,"** 2010, written by E. Vance, D. Giordano, J. Rogers and J. Stewart. The paper describes the results of their work in showing improved power density of a diesel engine, with a wide range of fuels, through the combination of flexible, high-pressure fuel injection technology and advanced engine controls with cylinder pressure feedback.
- **"Automotive Materials Engineering Challenges and Solutions for the Use of Ethanol and Methanol Blended Fuels,"** 2010, by P.K. Yuen, J. Beckett and W. Villaire, presents a comprehensive overview of the materials selection and engineering challenges facing metals, plastics and elastomers when engine components are exposed to the chemistry and different quality levels of ethanol and/or methanol fuel blends.
- **"A Study on Refrigerant Irregular Emission from China Mobil Air Conditioning Vehicles Based on JD Power Result,"** 2010, authored by B. Li and W. Hill, presents the study of current refrigerant emission levels in China, drawing attention to irregular refrigerant emissions related to system design and reliability.
- **"Review of CO_2 Emissions and Technologies in the Road Transportation Sector,"** 2010, by T. V. Johnson, provides an introductory, high-level of the current status of mobile CO2 regulations and technologies to address them.
- **"Fuel Economy Impact Evaluation of Hybrid Vehicles in the Brazilian Fleet,"** 2009, written by D. Queiroz Luz and G. de Paula, Jr., discusses the negative effects of excessive fuel consumption of Brazil's current fleet, proposing an estimation of the possible fuel economy with the future availability of hybric technology.
- **"Solution for India Towards Clean and CO_2 Efficient Mobility,"** 2009, by R. Kishore, P. Leteinturier and W.S. Long, discusses the possible technologies to be used in CO_2 emissions reduction, in a country whose number of on-road vehicles could surpass 40 million by 2012.

The challenges of **sustainable mobility** are addressed by the following papers:

- **"FCV's for a More Sustainable Mobility in 2050,"** 2010, by E. Velasco, is an investigation to support the vision of global sustainable mobility well into the future, providing some insights of how current technologies may survive or evolve up to year 2050.
- **"Is Mobility as We Know It Sustainable?"** 2009, written by P.G. Gott, explores various aspects of energy demand and GHG emissions, while identifying opportunities to minimize their negative environmental impact.
- **"Sustainable Green Design and Manufacturing Requirements and Risk Analysis Within a Statistical Framework,"** 2009, by P.G. Ranky, emphasizes the integration of advanced process modeling, customer requirements analysis, statistical methods and risk analysis. The goal was to develop a generic and systematic, sustainable green design and manufacture architecture.
- **"Sustainable (Green) Aviation: Challenges and Opportunities,"** 2009, authored by R. K. Agarwal, expands on how the growth of air travel could impact environmental issues such as noise, emissions and fuel burn, including possible mitigation strategies.

Finally, in the arena of **new technologies**, the ensuing papers were picked:

- **"In-Vehicle Networking Technology for 2010 and Beyond,"** 2010, by C. A. Lupini, is an overview of the current state of in-vehicle multiplexing and the emerging technologies in this field, also presenting usage and trends of in-vehicle networking protocols.
- **"Development of Injector for the Direct Injection Homogeneous Market Using Design for Six Sigma,"** 2010, authored by E.A. Rivera, N. Mastro, J. Zizelman, J. Kirwan and R. Ooyama, describes how the Design for Six Sigma innovation methodology was used to develop a new injector for the homogeneous direct injection market. Market needs and drivers were understood and led into functional requirements and concepts.
- **"Boosted HCCI for High Power Without Engine Knock and With Ultra-Low NOx Emissions - Using Conventional Gasoline,"** 2010, by J.E. Dec and Y. Yang, talks about how well-controlled boosted HCCI has a strong potential for achieving power levels close to those of turbo-charged diesel engines.
- **"Measuring Near Zero Automotive Exhaust Emissions – Zero Is a Very Small Precise Number,"** 2010, written by W. Thiel, D. Eason and R. Woegerbauer, discusses how modern engine and pollution control technology has moved so quickly towards zero pollutant emissions that the testing technology is no longer able to accurately measure the trace levels of pollutants.
- **"Nanotechnology Applications in Future Automobiles,"** 2010, a collaborative effort by E. Wallner, D.H.R. Sarma, B. Myers, S. Shah, D. Ihms, S. Chengalva, R. Parker, G. Eesley and C. Dykstra, addresses the potential of nanotechnology to be a game changer in redefining the methods used for developing lighter, stronger, and high-performance structures, focusing on automotive applications.

In support of this endeavor, Delphi and SAE International have teamed up to offer the readers of "Green Technologies and the Mobility Industry" unique companion booklets, **Worldwide Emissions Standards,** developed by Delphi Corporation.

First published in 1992, these brochures summarize the emissions regulations from around the world, helping industry professionals keep abreast of the newest emissions standards and related requirements. They also contain information on fuels, evaporative standards emissions and on-board diagnostics, and material related to the proposed worldwide driving cycle for motorcycles and mopeds.

We hope the reader will find "Green Technologies and the Mobility Industry" useful, compelling and, above all, an affirmation of the positive outcomes that are possible when technology is committed to finding the right solutions for the ever-changing mobility realities of our times.

Dr. Andrew Brown Jr., P.E., FESD, NAE
President, SAE International, 2010
Executive Director & Chief Technology Officer, Delphi Corporation

References:

1. U.S. Department of Energy, Office of Vehicle Technologies.
2. Martin, Ripley, Xavier Mosquet, Maximilian Rabl, Dimitrios Rizoulis, Maximo Russo and Georg Sticher. "Batteries for Electric Cars: Challenges, Opportunities, and the Outlook to 2020." *The Boston Consulting Group*. The Boston Consulting Group, January, 2010. Web. 3 Sep 2010.
3. Yole Developpement, Lyon, France (www.yole.fr)

VEHICLE ELECTRIFICATION

Bridging the Gap from Gasoline to Hybrids: Using Systems Engineering to Deliver Advanced Powertrain Technologies	2010-01-0012 Published 04/12/2010

Bob Brincheck
Dassault Systèmes America

Copyright © 2010 SAE International

ABSTRACT

In order to meet the new and aggressive fuel economy standards, rapid development of advanced engine technologies relying upon hybrid powertrains will be critical. However, it won't be enough for manufacturers to just meet emission regulations; they will also need to address reduced fuel consumption, decreased manufacturing costs, consumers' desire for performance such as power and torque, as well as maximization of reliability and quality.

Hybrid technologies that can meet EPA demands will involve increased cost, complexity, cooling requirements and battery weight. Add to that the fact that hybrids are fairly new, so that the existing foundation of knowledge is not as strong as for the veteran gasoline engine.

Reliable vehicle operation of hybrids will depend upon successful integration and verification of all drivetrain component interactions under varying operational and environmental conditions, such as cold-weather testing of battery capacity. Traditional powertrain testing methodologies have proven the validity of various parts of the system in the virtual world, but total system testing has typically relied upon physical prototypes.

With the increased complexity of the hybrid engine involving the integration of mechanical, electronic and software components, it is crucial to develop the system in a virtual environment where 'what if' scenarios can be quickly evaluated to make up for the lack of existing experience. There will neither be enough time or resources to physically build and test all the numerous potential scenarios needed to ensure optimal performance of a complete hybrid powertrain system.

To manage such increasing complexity, systems engineering has emerged as a collective, integrated multi-disciplinary approach to product development that is easily understood from the product planning to engineering to design to manufacturing perspectives. This paper will explore the modeling, simulation, and analysis capabilities that are available to improve system performance, reduce cost, and maximize reliability of these critical systems through implementation of a Virtual Systems Engineering approach in a timely manner. It provides a comprehensive, collaborative definition across a product's different views (requirements, functional, logical, physical), which allows for a full spectrum of virtual design and simulation capabilities across the enterprise and far beyond the traditional CAD design and core engineering users.

INTRODUCTION

The auto industry is facing twin, yet potentially diametrically opposed challenges as it struggles with economic viability at the same time as it is being mandated to meet stringent fuel economy standards. The 2007 CAFÉ (corporate average fuel efficiency) energy bill set a target of 35 MPG for new vehicle sales to be phased in beginning with model year 2011. The Obama administration has added to this mandate by moving up compliance dates for all vehicles to meet this standard from 2020 to 2016.. Although there are environmental and national security benefits to be gained as we reduce our dependence on gasoline and petroleum products controlled by other regions of the world, there is also a higher cost that will be imposed on all automakers.

In order to meet the new fuel economy and greenhouse targets, automakers will look to develop technologies that can offer the lowest cost while improving fuel economy. These technologies may come from a variety of powertrain and vehicle body solutions - such as improvements to the internal combustion engine, more advanced transmissions, reduction of vehicle weight, and various advanced engine management options. Designed to offer 20% better fuel economy and 15% lower emissions, Ford Motor Company's new Eco-boost direct injection, turbo-charged engine is a good example of what is coming. Diesel engines will also have their place, but may remain challenging due to pollution regulations and customer acceptance.

However, the most excitement and potential exists among the next-generation technologies, such as hybrid powertrains and fuel cells. Although hybrids offer better mileage and smoother acceleration at lower speeds, they also currently come with a very high price tag - 30 to 40% more than conventional vehicles. Additionally, as hybrids rely upon multiple power systems, including an internal combustion engine, electric motor, and battery, these additional sources of power naturally add complexity to the system - not only in how the two work together, but in areas such as cooling system effectiveness as well as body structure dynamics.

These new requirements, which of themselves are enough to challenge any automaker, have to be taken into consideration along with the already daunting task that the industry meets daily just in keeping up with consumer preferences and demands. Consider the number of electronics introduced into vehicles over the past few years as automakers strive to provide consumers an immersive multi-media experience that leverages cell phones, computers, emails, video, etc. Consumers are not going to waive their demands for innovation as the auto companies take time to address the CAFÉ mandates.

This challenges automakers on numerous fronts - increased vehicle complexity of multiple systems, reduced time to market, quick reaction to accommodate consumer demands, best standard practices that make it easy to ensure each vehicle footprint meets its mandate, and doing all of this in a manner that provides cost-efficiency in order to deliver a reasonably affordable vehicle. It is believed that to meet these demands that the rate of new technology introduced will triple1. Today's engineering processes and technologies are not equipped to manage the plethora of change that will be required to meet all of these requirements. There is need for a robust engineering methodology that can efficiently manage innovation that comes at a break-taking speed.

A LESSON FROM HISTORY

In conventional auto industry wisdom, platforms evolved slowly over time. The changes that occurred from model-to-model were closely linked to the company's development process and its ability to manage the change and predict the outcome of that change with a high degree of confidence.

However, this methodology received a shake-up in the late 1970s and early 1980s. At this point in time, automakers were facing a similar situation to what is currently occurring - they had to comply with a slew of safety, fuel economy and CAFÉ regulations coming at them all at once. This demanded more vehicle change than any current development process could handle. Hence, all the product development resources went into meeting the government mandates, leaving little effort focused on customer satisfaction. In the 1990s, much of the regulation backed off, allowing companies to go back to concentrating on the consumer experience and making vehicle changes in a manner that aligned with the process and its inherent ability to successfully deliver that change.

With today's government mandates leading to advanced powertrain technologies in conjunction with the importance of customer experience, we are again faced with this situation of too many revolutionary changes needing to be implemented all at once.

When Toyota introduced the first generation Prius, it was a hybridized version of its existing Echo. It was launched only in Japan and only in small quantities until it performed to satisfaction. The second generation was then released and, although it had many changes, they were changes implemented upon the capability of the development process to execute successfully.

However, the new regulations impose an external force that will drive change beyond what any of today's automakers are comfortable with. In addition to this is the fact that, given the state of the industry and a recent lack of investment in new vehicle design, there is also a huge need on many automaker's part to re-invent their vehicle lineup.

So, take an already big task and make it bigger. It now becomes a situation that insists upon more change than what can be managed within the existing product development processes.

CURRENT STATE OF THE PROCESS

All product development begins with requirements from which the technical specifications are then derived. Traditional approaches keep each domain - mechanical, electronic, controls, etc. - separated in its own silo focused on its own perspective regarding how to deliver on the requirement. Each sub-system group is developing its solution (Figure 1) to fulfill the requirement with no ability to know completely how it influences the other sub-system until the systems are brought together in a physical prototype. At this point, the cost to change the design is exponentially higher than if design integration issues had been identified and dealt with earlier in the design process.

Traditional development approach

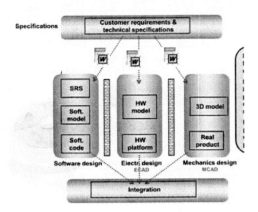

Figure 1. traditional engineering approach

To combat this issue, systems engineering was developed as a methodology that provides a collaborative business process to manage all these requirements from a holistic point of view The product definition is managed according to the different views - logical, functional, requirements - in a way that provides a unified foundation to cover each of the different engineering domains (Figure 2). In a nutshell, this approach enables a common view on what needs to be accomplished and a common model (Figure 3) to develop it in, providing a better understanding of how the system will operate as a whole.

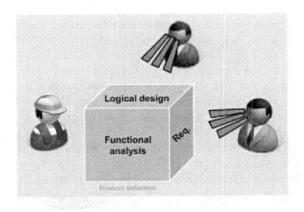

Figure 2.

Systems Engineering approach

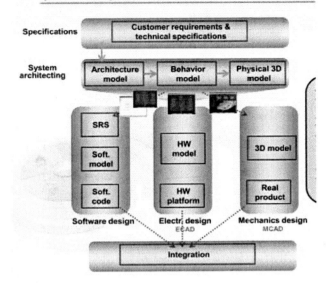

Figure 3.

Systems engineering provides a tremendous amount of downstream value, and value moving forward by capturing best practices and helping to minimize downstream costly design changes. Through systems engineering, you can improve the quality of the outcome. Most systems engineering approaches, which are technology neutral, rely upon taking all the data from the various disciplines and their disparate systems and uniting them within a common product data management system (PDM). This allows all revisions to be managed and viewed once they have occurred to see how any change impacts the other disciplines.

However, this is really a reactive change process with little to no traceability back to the requirements among the typically multiple heterogenous software tools (Figure 4), and there is disconnected validation of the sub-systems. Finding regulations regarding any design conflicts depends largely on the knowledge-base of the staff; the more seasoned the engineer, the more likely he knows where to go for the information. Requirements information could be stored in any of multiple databases with few companies possessing the necessary tools to integrate the design data of all the product elements. And, although PDM solutions have improved this process, the system still often results in poor communication of requirements though the design process, errors in design, and limited traceability between the design and the requirements.

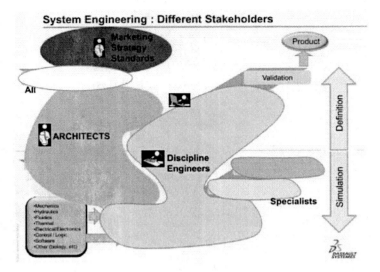

Figure 4.

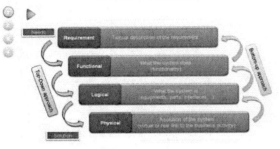

Figure 5.

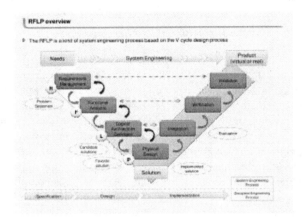

Figure 6.

To meet the necessary changes required within today's automotive industry there is no time to wait while a change gets passed around to other disciplines via a process that simply manages the change. That solves yesterday's problem. There is an absolute must to have a system capable of managing change pro-actively and to understand the impact of any change PRIOR to it being made through virtual design and validation.

RFLP AND THE V6 ENVIRONMENT

Dassault Systèmes addresses this challenge through a unique process model called RFLP (Requirement /Functional/ Logical/Physical) (Figure 5) to be applied within its V6 environment, based upon a single, open, and scalable service oriented architecture platform. RFLP is an enhanced model-based approach to applying the systems engineering methodology.. It is based upon the V-Cycle design process (Figure 6), with the premise that the shortest way to get to the opposite side is not always a straight line, enabling concurrent engineering. Each aspect of the RFLP functions is matched to various aspects of the V-Cycle.

To begin the process, "R" requirements - whether corporate, end-user, financial, design manuals, guides, etc. - are defined within the ENOVIA V6 authoring environment. The system architects can then pull from these requirements to define the functional and logical aspects of a product and link them to the physical definition of a product, ensuring full traceability from product specifications to the actual 3D design.

For example, the requirements indicate the need for an engine that produces a certain amount of torque - it can be any kind of engine. The requirements also state specific mass, size, and other requirements. Once the requirements are defined, system architects begin the "F" functional decomposition of the product in dedicated editors -This describes "what" functions must be performed to satisfy the requirements. This is developed as a reuseable template against which the program requirements are linked. This then enables the impact of any change to be seen and aids in the decision making process.

At this level, users understand that they have to produce power and torque, but that there are different ways to fulfill the requirement. If they choose to produce the power via a combustion engine, this then leads to a need for air intake, fuel intake, compression, ignition, power transfer and exhaust. But, the engine could still be a diesel, traditional gas, direct injection, etc. At this functional level, it will also be discovered that in producing the power and torque, heat will be created, and that heat has to be managed. This then leads to fulfilling a cooling requirement. Additionally, there is the noise that the power and torque will produce; which has to be designed to meet the NVH targets.

The next step - the "L" logical piece defines "how" the functional requirement will be achieved, and there can be multiple ways to achieve the requirements. For example, in order to produce the combustion, there is then a need for a combustion system, which consists of features that allow air and fuel to enter the combustion chamber, the exhaust system to deal with waste, etc.. If a gasoline direct-injection engine is the best solution to meet the requirements, this would indicate the need for a compression chamber, cylinder, air intake manifold, cylinder head, etc. At this stage, users can use a first level of 3D representation of the system in order to perform space reservation and check specific requirements on pathways, for instance.

The V6R2010 technology embeds specialized modeling and simulation tools (dynamic behavior modeling based on the Modelica language and Logic & Control for reactive systems). So, as each logical entity is defined and identified, a dynamic behavior is assigned by attaching a Modelica model to each of these behaviors allowing for dynamic simulation of the behavior associated with a logical entity. Modelica's non-proprietary, object-oriented, equation-based language allows users to conveniently model complex physical systems containing mechanical, electrical, electronic, hydraulic, thermal,, electric power or process-oriented sub-components. With this capability, the model now has the physics built into it. At this logical model juncture, users can also define the control strategy. Logic & Control Modeling provides users with a formal model suited to manage parallel systems through a comprehensive set of editors (grafcet, statechart, dataflow…). Both dynamic behaviors and Logic & Control can be co-simulated on a virtual execution platform.

Last, is the physical "P" level where actual parts are specified to execute the logical model. There will be a need for a valve train to let the air fuel mixture come in and exhaust go out. But, how many valves are needed? Two, four, five, six? Overhead cam? Underneath?-Pushrod? Or, perhaps a brand new technology that is being developed. Whatever it is must be defined.

Those decisions should be based upon meeting the requirements in an optimal fashion. If the primary requirement is fuel economy, then the solution with the least amount of mass and greatest air flow through the combustion cycle may be the best choice.

At this point, all engineering domains and solutions are linked together in a common and dynamic engineering template, which allows for virtual simulation and validation at any system level. Components from multiple disciplines, which may include their 3D representation, as well as the numerous interactions between them are modeled in the authoring environment of CATIA to enable dynamic simulation of the complete system via a virtual prototype. So, in the case of a hybrid engine, the functions being produced - power and torque - will be accomplished through two different power sources, but shown in one single model that everyone is working from. The behavior of the product in operation is assessed while various design alternatives can be tested very early on.

This is where the approach can make huge gains for automakers. Numerous 'what-if' scenarios can be run either from a bottom-up approach - changing the real product and tracing it back to how this affects the specifications, or from a top-down approach - changing a specification and seeing how this impacts the physical product. And, most importantly, with any change, how are all the different functions/requirements impacted?

This becomes extremely important in a dual-powertrain system as there will almost always be a trade-off in engine development. A powerful engine may be producing a lot of noise, so it can't go in the luxury car according to noise level requirements. To make it quieter, the power needs to be reduced,. With this approach, changing the product and/or requirements is completely traceable as to how it impacts the other systems. The requirements are directly linked to the design decision - performance, fuel economy, or cost will all directly influence the design choice. The logic that drove the decision to select that part is transparent to all users in the system.

RFLP is the Dassault Systèmes V6 solution to model-based system engineering (Figure 7), providing a collaborative system engineering methodology that can capture, manage and track product requirements with full traceability, all from one engineering desktop window. Users can easily access the requirements of the product, its functional decomposition, its logical architecture and its physical definition (Figure 4) Companies can model complete systems with the CAD-solution being just one component (geometric representation) of the process.

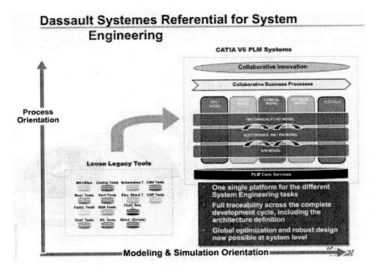

Dassault Systemes Referential for System Engineering

Figure 7.

PRACTICAL APPLICATION TO HYBRID POWERTRAIN DEVELOPMENT

Although this solution is new, the benefits should be apparent, especially as vehicle complexity continues to grow. The primary design tasks for development hybrid powertrains are the packaging of the various subsystems. Developing purpose-built sub-systems to meet a specific set of requirements is relatively straight-forward. However, developing each of these sub-systems to work in conjunction with each other is a new frontier for the industry, given the infancy of the hybrid technologies and complicated by the compressed need to deliver product to meet CAFÉ standards.

Testing the performance of a given engine and electric motor combination to ascertain real-world performance and fuel economy will be critical in delivering successful hybrid design. Understanding how the hardware components of the driveline - engine, motor; and an energy storage device interact with each other as well as with the rest of the vehicle including the torque converter, transmission, fuel storage system, HVAC system, high level vehicle controller, accessory loads, etc. requires a common engineering process that can link changes back to requirements. The only way to manage this level of complexity and interaction is through a model-based pro-active RFLP systems engineering approach.

The new Chevy Volt is an example of the challenges presented when there is too much change occurring within a demanding time frame. Given the time constraints General Motors is facing with product launch, it made a decision to design the vehicle with redundant HVAC systems. Batteries must be heated up when cold, and cooled down when warm. This is being accomplished by two separate HVAC systems.

Relying upon its existing product development methodologies and its ability to manage the complexity of change, the company knew it could deliver the vehicle by optimizing and validating these solutions separately. The process went right from requirements to delivering on function and parts; totally skipping over the logical aspect of the design. This was a safe approach that the company knew they could execute, but there will be a cost factor.

If an RFLP model-based systems engineering approach had been implemented, the company would have first looked at the HVAC requirements and the features associated with those. By allowing the team to look at that basic level of data, there is a much better opportunity to combine all the features into one solution, or to at least understand the conflicts in a more definitive, quicker manner.

Referring back to the impact of gasoline prices and consumer demand for continual innovation in vehicles, changes in market conditions may demand quick reaction in the product design. For example, an engine that is being developed for use in a truck may find two years into the process that gas prices are too prohibitive to create a demand for this vehicle. Now, there is a quick need to re-design that engine to go in a sedan that has a different set of requirements. To meet these requirements, perhaps you need to remove 15 - 20 mm off the engine height. With the RFLP process, the impact of the reduced height can quickly be traced to the requirements to see if it adversely affects the other requirements such as NVH or torque. You need a very robust engineering process to understand the impact of changes like this quickly.

With hybrid powertrain development, there is no stored base of knowledge upon which to build. And development of test procedures is very difficult as HEV designs are extremely diverse with limited access to real HEVs to test new concepts. With no wiggle room in meeting the CAFÉ standards, a systems engineering multi-disciplinary approach can help encourage visibility into the overall product design as teams methodically update and interact with each other, flagging potential problems prior to physical build. Reliance upon simulation to help identify system level problems is also helpful in running virtual tests earlier in the design cycle.

And, it is worth noting that in hybrids it is not only the powertrain that will require massive engineering change. Additionally, the body structure requirements will be severely impacted. Specifically, the battery requirements are forcing changes in the body structure dynamics that will cause engineers to redesign aspects of the vehicle that significantly affect the performance and feel of the vehicle.

Model-based engineering as produced by the RFLP process puts more rigor into the process, providing a link back to requirements for each logical, functional and physical change. (Figure 8) Features in the logical definition have equivalent features in the CAD model, whether electrical, software or mechanical. Any change is not only quickly connected back to the requirements, but also to the impact it has on all the other disciplines, allowing for quick analyzation of various modifications.

<figure 8 here>

SUMMARY/CONCLUSIONS

When a company works in a reactive manner, there will always be a requirement that will get compromised. Most often, it is the cost requirement, which can then spiral out of control. Given the precarious state of many automakers, controlling cost is of utmost importance. To meet the current demands placed upon the automakers, they must implement a holistic approach that links all disciplines together during product development, providing a virtual environment that dynamically tests all the systems together. Additionally, there is need for a robust system that can handle the complexity and rate of change that is required providing transparency of requirement data throughout the process.

CONTACT INFORMATION

Bob Brincheck
Dassault Systemes
900 N. Squirrel Road
Auburn Hills, MI 48326
Bob.brincheck@3ds.com
248-267-9696

ACKNOWLEDGMENTS

Sohair Varnhagen, Dassault Systemes America. Khursid Qureshi, Dassault Systemes America.

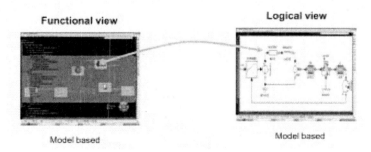

Figure 8. Relation of Logical Definition features to the CAD Model.

A Survey on Electric/Hybrid Vehicles

2010-01-0856

Published
04/12/2010

Bernardo Ribeiro, Francisco Brito and Jorge Martins
Univ. do Minho

ABSTRACT

Since the late 19th century until recently several electric vehicles have been designed, manufactured and used throughout the world. Some were just prototypes, others were concept cars, others were just special purpose vehicles and lately, a considerable number of general purpose cars has been produced and commercialized.

Since the mid nineties the transportation sector emissions are being increasingly regulated and the dependency on oil and its price fluctuations originated an increasing interest on electric vehicles (EV).

A wide research was made on existing electric/hybrid vehicle models. Some of these vehicles were just in the design phase, but most reached the prototype or full market production. They were divided into several types, such as NEVs, prototypes, concept cars, and full homologated production cars. For each type of vehicle model a technical historic analysis was made. Data related to the vehicle configuration as well as the embedded systems were collected and compared. Based on these data future prospect of evolution was subsequently made.

The main focus was put on city vehicles and long range vehicles. For city vehicles the market approach normally consists in the use of full electric configuration while for the latter, the hybrid configuration is commonly used. The electrical systems and combustion engines found in these vehicles are compared in order to forecast the evolution trend in terms of specifications and performance of the whole vehicle and of each system.

INTRODUCTION

Environment policy measures are being implemented worldwide aiming at the reduction of emissions of greenhouse gases and other pollutants from the transportation sector. This sector is one of the major CO_2 emitter and road vehicles constitute the main pollutant sources in cities. Furthermore, vehicles with internal combustion engines are noisy, therefore increasing the level of noise pollution. With this in mind, several locations in the world (e.g. London [1]) are introducing policies for the establishment of electric fleets, by offering electrical connection (recharging) for electric vehicles and other advantages such as free parking and no access restriction to inner city areas. These measures resulted in the increasing trade and use of electric vehicles. At the moment it is possible to buy a variety of vehicles, from small NEV (neighborhood electric vehicles) to large delivery vans, obviously including sedan vehicles and even sports cars. Electric vehicles are produced in various locations in the world, with emphasis to North America, Europe, Japan and recently in China ([2]).

Following the nomenclature used by SAE, [3,4] an electric vehicle is a vehicle in which its propulsion is accomplished entirely by electric motors only, regardless for the means of obtaining that electric energy. Therefore, what previously was known as a series hybrid vehicle, is now referred to as an electric vehicle, and the expression "hybrid car" is only used for parallel hybrid systems. At the same time, the internal combustion engine used by "series hybrid" to produce electricity within the car is now known by the term "range extender". In this category lie vehicles using internal combustion engines (turbines included) and fuel cells, as these produce electricity from a fuel such as hydrogen.

Electric cars rely on lead-acid batteries as the energy source for locomotion since the end of the 19th century. These batteries were outdated by the NiMH and Li-ion types, but in terms of the price/energy ratio, lead batteries still rule. Therefore it is common to find lead-acid batteries on the lower priced electric cars. Another conventional disadvantage of batteries is their low power, which is not sufficient to allow for proper brake recovery. These extremely high powers cannot be supplied to batteries but they may be given to supercapacitors or ultracapacitors. Therefore, the use of supercapacitors combined with batteries would allow a strong increase of the potential for braking energy recovery, while additionally extending the life of the batteries, filtering the electric peaks. Lately, innovative nano-materials applied to lithium based batteries [5,6] and carbon based supercapacitors [7] are surfacing, having properties that extend significantly the range of operation of common Li-ion batteries and supercapacitors, enabling huge power for the batteries and significant energy storage for the capacitors. But on the whole, storage is still the major hurdle in the mass production and use of electric cars, as the energy density of batteries is orders of magnitude away from that of liquid fuels.

Electricity, within the vehicle can also be produced by thermal engines (gasoline or Diesel IC engines, gas turbines or other types) or by any type of fuel cells. In such cases the range of the vehicle is achieved by the batteries plus the range provided by the use of the fuel stored in the car.

The energy sources for the thermal engines or fuel cells may vary, from fossil fuels (gasoline, diesel, LPG, NG) to biofuels (biodiesel, alcohols, biogas), or synthetic fuels (Fischer Tropsch fuels), and includes the energy carrier hydrogen. Batteries are a form of chemical storage, but it is possible to use mechanical storage devices, such as flywheels [8]. Recently, Formula 1 cars started using energy storage devices known as KERS (kinetic energy recovery system) that enable cars to recover part of the braking energy and use it during acceleration [9].

Cars with internal combustion engines always use transmission systems to connect the engine to the wheels and a differential in order to allow both wheels to have different speeds when cornering. Electric cars seldom use these devices, as each power wheel may be connected to one electric motor. Furthermore, the speed of each motor can be that of the corresponding wheel, therefore eliminating the need for transmission and gears. When this happens the rotation of the motor is the same of the revolving wheel, which usually is somehow small (1 500 rpm max), leading to slow rotating electric motors. However, electric motors can have a much better power/weight ratio than internal combustion engines, mainly if they achieve high speed (~15 000 rpm). For lower speeds, electric motors have a lower power/weight ratio but can be installed inside the driving wheel with obvious advantages of packaging [10]. Some cars have the motors connected to the wheels through driveshafts and gears, while others use just one electric motor connected to a differential in the conventional layout.

Electric cars and hybrids with internal combustion engines tend to use higher efficiency cycles, such as over-expanded cycles for spark ignition combustion type. This cycle, also known as Miller or Atkinson, has the particularity of having the expansion longer that the compression, leading to a higher expansion of the burned gases, therefore achieving a higher efficiency. [11]. Hybrid vehicles rely on the IC engine as the main propulsion system, being the electric motor just an assistance in terms of locomotion. These engines use a large load/speed spectrum during normal operation in the running cars. On the contrary, newer engines developed as range extenders for electric vehicles are developed to work at a more limited range of load and speed, sometimes only at one or two conditions. [12, 13]. These engines are much more efficient than the usual "accelerating" engines found in conventional cars, as they always run at specified conditions and are designed only for running at these conditions. In Europe, where the Diesel engine reigns, some of the hybrid cars have Diesel engines. These cars have a price disadvantage, as Diesel engines are more expensive than spark ignition engines. Hybridizing a SI engine car is also more effective than hybridizing a Diesel engine car, as SI engines have a higher reduction in efficiency when driven at light loads.

Fuel cells are available in various types but vehicles tend to use just PEM (polymer exchange membrane) operating on plain hydrogen. In some cases it is possible to feed natural gas to a reformer, with the produced hydrogen being supplied to the fuel cell. The advantages of fuel cells are their inherent high efficiency, silent operation and the sole production of steam as a byproduct. On the down side, they are extremely expensive and they are easily contaminated, requiring an expensive rebuilt.

INFORMATION STRUCTURE

A database was built using MS Access structure and Visual Basic programming. This data base can store data related to EVs and their powertrain systems/components such as energy storage systems, motors, controllers, IC engines (in the case of hybrid vehicles), fuel cells, solar PV panels (Figure 1).

<figure 1 here>

Electric Vehicle models were categorized using several criteria, the vehicle type, the vehicle powertrain and the vehicle status.

From the first classification criteria vehicles were divided into the following types:
1. Conventional design (Sedan, MPV, SUV, Pick-up)
2. Sport EV

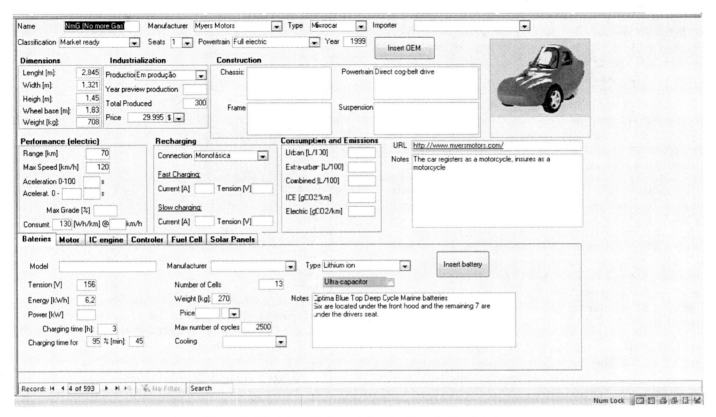

Figure 1. Electric vehicle model main form in the database.

3. Microcar / Tricycle

4. NEV / Quadricycle

5. Van & Utility

6. Bus

7. Truck

8. Competition, Golf cart, ATV, etc

It is worth noting that the difference between an NEV and a personal and microcar is not the size but the vehicle performance, once a NEV has its top speed limited by regulations, while a microcar differs from other cars due to its size and its passenger capacity of usually 2 or less. Two-wheel vehicles were not included in this survey.

A second vehicle classification criterion is the powertrain configuration. In the present work the following powertrain types were considered:

1. Full electric

2. Parallel hybrid

3. Parallel hybrid "plug-in"

4. Electric with range extender or series hybrid

5. Electric with fuel cell

6. Electric with turbine generator

Concerning the vehicle development status field, it is intended to classify each vehicle in an industrial/marketing status, allowing the following categories:

• Design concept - Vehicles that have been publicly presented and still exist only in drawings or CAD models. No physical prototype exists yet. Vehicles under this status of development may not go further than this.

• Concept car (converted) - Vehicles presented to the public but still not tested, being a conversion from previously existing models and in which just the powertrain has been converted into hybrid/electric. Vehicles under this status of development may not go further than this. These converted vehicles may be presented by the original vehicle manufacturer (OEM) or by a 3rd party company which performs the conversion.

• Concept car- Vehicles presented to the public but not tested, which are intentionally designed to be hybrid/electric vehicles. Vehicles under this status of development may not go further than this.

• Demonstration car (converted) - Vehicles presented to the public and tested, which are a conversion from previously existing models and in which just the powertrain has been converted into hybrid/electric. Vehicles under this status of development are very often produced on very small series and tested on specially selected regions or markets. These converted vehicles may be presented by the original vehicle manufacturer (OEM) or by a 3rd party company which performs the conversion.

• Demonstration car- Vehicles presented to the public and tested, which are intentionally designed to be hybrid/electric vehicles. Vehicles under this status of development are very often produced in very small series and tested on specially selected regions or markets.

• Competition - Single unit (or very small series) vehicles designed and built for competition purposes or for record performance.

• Market ready (converted) - Vehicles designed and produced by OEMs from an original model with an IC engine powertrain, but now having hybrid or full electric powertrain. These vehicles are already on the market. Some models of these class of vehicles may have been already taken out of production.

• Market ready - Vehicles intentionally designed to be hybrid/ electric vehicles. These vehicles are already on the market. Some models of these class of vehicles may have been already taken out of production.

The information concerning a total amount of 593 vehicle models was stored and analyzed, although not all specifications/features were available or released for each of these vehicles.

Considering the vehicle type classification presented above, the first four vehicle types include almost all vehicle models present in the database (Figure 2). The most common vehicle types being released with an electric or hybrid powertrain are Sedans and SUVs. The following is the NEV group. Microcars (including tricycles) and sport vehicles are also used very often. Frequently sport vehicles are also built to prove the effectiveness of the electric powertrain. From all these sport vehicles a significant amount of them are concept cars, as it will be seen below.

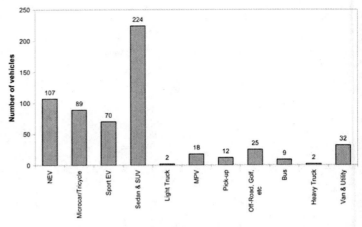

Figure 2. Number of vehicle models surveyed by vehicle type.

On Table 1 all the vehicle models from the database are divided into to the several possible powertrain configurations, from the full electric to hybrids and electric with on-vehicle power generation devices.

Table 1. Number of vehicle models by powertrain configuration.

Powertrain type	Number of vehicles
Full electric	381
Parallel hybrid	92
Electric with fuel cell	52
Electric with range extender ICE	30
Plug-in hybrid	18
Electric with turbine	1

Nearly 66% of all surveyed vehicle models were full electric. These exist since over a hundred years. Hybrid vehicles and electric vehicles with on-board power generation (either ICE range extender or hydrogen fuel cells) are more recent and their technology is still under development. That is why only a reduced number of vehicles has been released.

The development status of the released models is also a good information source to understand how this technology stands in terms of maturity. The total amount of vehicle models on this survey and their status of development is shown in figure 3.

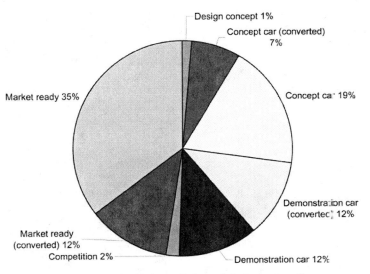

Figure 3. Vehicle models and their status of development.

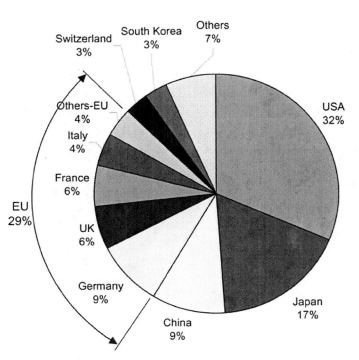

Figure 4. Percent of new model vehicles per manufacturing country.

As it can be seen from figure 3, nearly 50% of all vehicle models have already hit the market. This fact shows that the technology involved on this type of vehicles is still under hard development. Another fact that also corroborates this is the amount of vehicles which are converted from existing conventional powertrain platforms. Nearly 31% of all vehicles surveyed were built using existing platforms, reducing costs and development time for OEMs.

It is also important to note that from the vehicle models that have been already released to the market, and from those labeled as market ready, 45% are NEV, 21% are Sedan or SUV and 14 % are microcars. This reveals the great amount of NEV models on the market and the early stage of development of the major OEMs in terms of releasing full electric vehicles. From another point of view, of the converted market-ready vehicles 46% of them are sedans and SUVs and 19% of them are commercial vans and utility vehicles. Again, from these figures it is possible to understand the market strategy followed by OEMs in terms of electric mobility. The main focus is put on the development of powertrains using previously developed platforms, leaving the development of completely new designed vehicles for a second stage.

Analyzing the vehicle models from a manufacturing country point of view, it can be seen (Figure 4) that the USA is the leader, with the highest number of vehicles released to date, followed by Japan and China. Only after these three come some European countries. However, if all European Union countries are summed, EU has a higher number of new model vehicles than Japan and China, becoming the second world economy with higher number of new released electric/hybrid vehicles.

Looking with more detail to the hybrid and electric vehicle manufacturing, some other results may be found. In terms of NEVs, for example, and considering that some USA labels have their production in China, 78% of the models come from these two countries, corresponding 43 to USA and 35 to China. If just hybrid vehicles are considered, then a different composition exists. Japan is the country with more hybrid and plug-in hybrid model vehicles released (42%), while USA comes in second place with 24% and Germany with 17%.

The distribution of new model vehicles which hit the market is presented on figure 5.

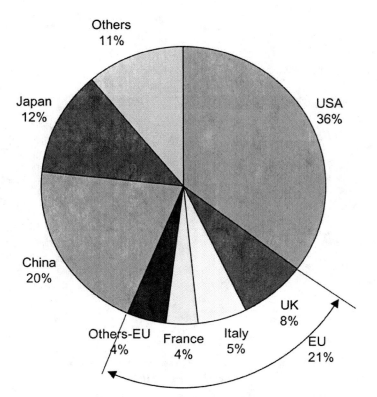

Figure 5. Distribution of new model vehicles released to the market.

On Figure 5 it can be seen that the role of USA is quite the same but EU comes to the second place in terms of new vehicles released to the market. The role of China is more relevant on this analysis, being at the same level as the total EU countries. This means that China is mainly a manufacturer, once the number of released models is higher than Japan. Comparing the number of models that reached the market to the total number of models released, China displays the best ratio, 89%, followed by UK and Italy with 62% and USA with 53%. These figures represent the amount spent on R&D for each country and the complexity of the technology used. In fact, vehicles from China are mostly full electric and from these most are NEVs. These vehicles have a very simple structure and powertrain using lead-acid batteries and DC motors.

VEHICLE TYPES

NEV

Neighborhood electric vehicles (NEVs) are the most popular electric vehicles on the market. From a total 107 models analyzed 89% are on production now, with 7% without any information on the market situation.

NEVs have their top speed limited as well as their curb weight. The configuration must be a four-wheel vehicle, excluding from this category tricycles. In terms of passengers

capacity, NEVs have a distribution such as the one presented on Figure 6. As it can be seen 55% of all these vehicles have two seating places while other 26% have four seat capacity. This kind of vehicles are mostly used as golf cars and city cars where maneuverability and space are much more important issues than speed.

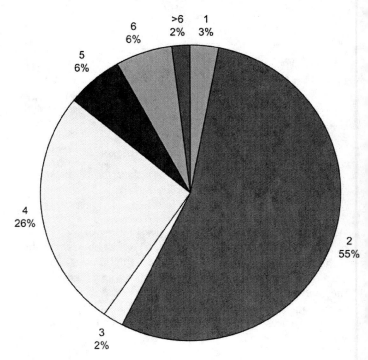

Figure 6. Number of seating places on NEVs.

On this type of vehicles the most used motor is the DC type (used on 67% of the surveyed models), with average power around 5.5 kW and an average peak power of 13 kW. The AC motors are used with less frequency with an average power of 8.4 kW. These motors are used in vehicles which have an average curb weight of 742 kg.

Concerning batteries, NEVs use mostly lead acid batteries (90%) followed by Li-ion batteries. Unfortunately from the information gathered on these type of vehicles, there is none concerning battery capacity or voltage. The battery packs from NEVs provide an average range of 90 km.

MICROCAR / TRICYCLE

These vehicles are a kind of upgraded NEVs, since they have a size or capacity of a conventional NEV, but display a higher performance namely concerning their top speed, which is not limited. Such vehicles may assume configurations very different from what is the "norm" on the automotive industry design, such as only three wheels (tricycles), in-line sitting (tandem), just one passenger (the driver), etc. In fact from all personal cars surveyed, near 53% are already in the market and have original design plus 17% are in a prototype phase heading for the market. Modified vehicles using existing platforms represent just 5% of all released models.

In some countries legislation includes tricycles on the motorcycle class, having thus some tax benefits in relation to four wheeled vehicles. The conception of microcars as tricycles has that benefit and that is the reason for its inclusion under this class of vehicles on this survey.

Microcars and tricycles are almost always commuter vehicles, which are designed to accomplish some high efficiency requirements such as low space/volume occupation, reduced number of places (once on most of daily travels made are on a vehicle, either in US or Europe, the driver is the only passenger), reduced energy consumption. From the vehicles surveyed, the area occupied, on average for this type of vehicles, is 4.4 square meters, while on the conventional sedan vehicles this value goes to 7.4 square meters. This same tendency exists with the number of sitting places on board, where the great majority of the models has 2 or less sitting places. The full electric powertrain is the most widely used (on 92% of the models released), and the rest, which are just 6 models, have hybrid or range extended powertrains.

As for the top speed obtained with this type of vehicles, it can go as high as 240 km/h on the Tango T600 from Commuter Cars, already on the market. In terms of range, the capability offered by batteries can go up to 450 km on the two Obvio models from Brazil the 828E and the 012E. However no specifications are presented on this vehicle battery except that is of Li-ion technology and a capacity of 39 kWh. In fact on this class of vehicles these two models are those which have a higher battery capacity. Batteries are, on a great majority, of the Li-ion type. But their capacity is on average of 11 kWh for the Li-ion type batteries and 8 kWh for the lead-acid type.

On this type of vehicles AC motors, either induction or permanent magnet synchronous, are the most widely used. Their power is on average 8.5 kW continuous and the average peak power is approximately 23 kW. DC motors were identified on just 17% of the vehicles of this type, with power very similar to the power of the AC motors used.

SPORT EV

70 vehicle models have been included on the type of sport vehicles. The criteria were the performance as well as the design, usually two seat vehicles. A significant part of these vehicles are concept cars or prototypes. Some are already in the market, however their price is significantly higher than the average car. The average price of a sport electric car is around 116,000 $.

Their performance can be evaluated based on the top speed and vehicle acceleration. The top speed can go up to 330 km/

h and the acceleration from 0 to 100 km/h (or approximately 60 mph) can reach 2.5 s. All these best performance marks belong to the Shelby Ultimate Aero EV.

In terms of powertrains, two thirds of all sport model vehicles are full electric using one, two or even 4 motors, one for each wheel. These can be either motor-in-wheel or directly linked to the wheel via a transmission axle. Almost all motors identified are AC type. Just one vehicle was identified using DC motors, the Advanced Mechanical Products conversion of the Saturn Sky, which uses two brushless DC motors, one driving each of the two rear wheels. Although the information released about the motors of sport vehicles is too little, the power of the complete powertrain of full electric vehicles goes from 122 kW continuous to 258 kW of peak power on average. Looking at the motors individually, their average power goes to 105 kW and 211 kW respectively.

On sport vehicles the main energy storage medium is the Li-ion batteries, with an average capacity of 26 kWh. Several other types of batteries are used but only one or two models were identified for each of these other types.

SEDAN & SUV & MPV & PICK-UP

On this vehicle group a total of 254 models are included. In terms of powertrain they are divided as presented on Table 2.

Table 2. Number of vehicle models by powertrain configuration.

Powertrain type	Number of vehicles
Full electric	108
Paralel hybrid	72
Electric with fuel cell	40
Electric with range extender ICE	17
Plug-in hybrid	12
Electric with turbine	1

As can be seen, there is a significant effort of the auto industry on the full electric vehicles and on parallel hybrids. Analyzing these types of vehicles from the models development stand point (Table 3) it can be seen that a significant amount of these type of vehicles are in fact modifications from previously developed platforms but with different powertrains, made by the original manufacturer or other companies specialized on powertrain conversions. There is still also a significant amount of concept cars meaning that industry is still on an early phase of development of such vehicles, once the amount of vehicles that do not hit the market is significant.

Table 3. Number of vehicle models by development status.

Development status	Vehicles
Concept car	54
Demonstration car (converted)	47
Market ready	46
Market ready (converted)	42
Demonstration car	33
Concept car (converted)	29
Design concept	1

In terms of batteries on this type of vehicles, Li-ion battery type is the most used, on 45% of the models surveyed. The second most used battery type is the NiMH type, specially used on hybrid vehicles, where a small energy storage capacity is needed. In the case of the Li-ion batteries the average energy storage capacity is 20 kWh, while in the case of the NiMH batteries this value drops to an average 9.2 kWh.

AC induction motors are the most powerfull and those with the highest torque among all motor types used in this type of vehicles. Continuous power averages 62 kW while the peak power is 103 kW. For this motors the average torque is 349 Nm. The motor most widely used is the permanent magnet synchronous AC motor, with less power than the previous ones (32 kW) and torque of 212 Nm. DC motors are also used however with much lower frequency and with much lower power. The average power of these motors is 19 kW with torque averaging 178 Nm.

POWERTRAIN

FULL ELECTRIC VEHICLES

Full electric vehicles have their range limited by battery capacity. The R&D efforts on this technology are nowadays under a great pressure aiming at the increase of power density and of battery life cycle. From the vehicles analyzed it can be concluded that since 1989 the maximum range increased five times from 80 km to 400 km in 2009 announced for the BYD E6. However, the great majority of new released models have their range within the lower values, as presented on Figure 7.

The average range of the vehicles released each year shows a clear tendency for an increase as a result of the development of the batteries industry.

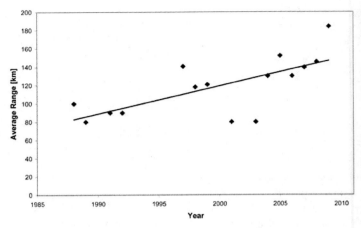

Figure 7. Average range of full electric vehicles.

Concerning the electric motors used in full electric vehicles, 59 % of new models opted for the use of AC motors and from these the most used type was the AC induction motor with an average continuous power of 40kW. Electric vehicles using DC motors mostly use those of separate excitation, with an average power of 8.5 kW. Clearly these are typically used in NEVs, while the AC motors are preferred in vehicles with higher power demand. In fact, the peak power of DC motors averages at 29 kW, while for AC motors the average peak power goes to 84 kW.

On this type of powertrains, it is possible to use the motor-in-wheel, or motor-in-hub technology and the motors used for this type of applications are usually different from the previous, since their torque requirement is higher. In fact, the motors used in this application have an average torque of 700 Nm in the case of the AC induction motors and 772 Nm in the case of the AC permanent magnet motors.

Considering the batteries used in full electric vehicles, from a total 227 new models, 45 % use Li-ion batteries and 27% use lead-acid. Other types of batteries represent less than 10% of new models. The lead-acid battery vehicles display the smallest range, with a fleet average of 96 km, while the Li-ion battery vehicles have an average range that goes up to 227 km. In terms of battery pack voltage, again the lead-acid vehicles have much lower voltages (79 V) than Li-ion with an average fleet voltage of 208 V.

The significant increase of the use of Li-ion batteries since 2007 must be referred. The release rate of new models with each type of battery has been around 3 per year. Except for the Li-ion battery type, which had 10 new vehicles released in 2008 and 9 in 2009, showing the increased interest and development put on this type of battery.

PARALLEL HYBRID VEHICLES

Parallel hybrid vehicles include vehicles with powertrains that use both electric power and ICE power. Also included in this category are parallel hybrid vehicles which can be plugged in for battery recharge, also known as Plug-in Hybrid Electric Vehicles (PHEV). This type of powertrain can use both electric-only mode, ICE- only mode or both power sources simultaneously. Although widely known as parallel hybrids the positioning of the electric motor in relation to the ICE can assume several possible assemblies. In some cases, the electric motor is assembled in parallel with the ICE and both are linked with a planetary geartrain to have only one output shaft normally linked to the gear box. This is the case of the Toyota Prius [14]. Another positioning is the one presented by Honda with their Integrated Motor Assist (IMA) system [15] where the motor is placed in the position of the flywheel in conventional ICE engines. This motor is used as motor and generator. In the case of Peugeot 3008 Hybrid4 system, a Diesel engine is used to power the front wheels, while an electric motor powers the rear wheels. Ford presented its concept Reflex which has a diesel hybrid powertrain on the front axle and an additional electric motor to power the rear axle.

OEMs are releasing a significant number of vehicle models with this type of powertrain configuration. From Figure 8 it can be seen that from 2003 the rate of new models release is increasing until now.

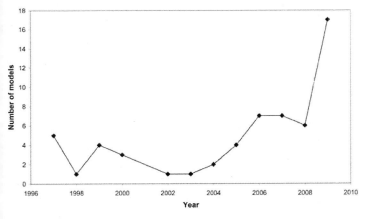

Figure 8. Rate of parallel hybrid new model vehicles released per year.

The engines used in hybrid vehicles usually are conventional ICE with some improvements in terms of fuel economy such as the adoption of the Atkinson cycle, cylinder deactivation and shutting off of part or all cylinders during speed reduction or breaking. Some of these vehicles have curb weights over 2000 kg and their IC engines have powers over

200 kW. In the case of the GMC hybrids the engine is the 6.0L Vortec V8 VVT, LIVC with Active Fuel Management, with 248 kW (332 hp) power.

Electric motors mostly used in this type of powertrains are AC permanent magnet synchronous. About 50% of new parallel hybrids use this type of motor with an average power of 30 kW and a peak average power of 97 kW. Other types of AC motors (induction) are also used but with much less frequency. DC motors are used in 23% of new models when reduced power level is required, with rated average power of 7.5 kW.

For this type of vehicles, 54% of new models released have NiMH battery packs and 21% have Li-ion batteries. The battery cost may be the main factor for the great predominance of the NiMH battery type. For this type of powertrain the required battery energy storage capacity is not high (just enough to perform a few vehicle accelerations), thus the problem of weight and volume of the battery pack is not an issue.

RANGE EXTENDED ELECTRIC VEHICLES

Extended range electric vehicles are vehicles with a battery pack displaying a capacity which is similar to that of full electric vehicles and motor(s) with a power which is equivalent to that of full electric vehicles. The only difference is that they have an on-board electric generator powered by an internal combustion engine.

In this type of powertrain the main engine configuration is the naturally aspirated spark ignition engine. The spark ignition engine was used in 64% of them, followed by the compression ignition present in 3 models (see Table 4). Two vehicles were identified as having a supercharged direct injection spark ignition engine as range extender, the Fisker Karma and a converted vehicle from Raser Technologies which uses a Hummer SUV. And again only one vehicle displayed a supercharged spark ignition engine, the Opel Ampera.

Generally, the IC engine/generator power for range extenders is bellow 70 kW, providing additional vehicle ranges of more than 500 km. The exception is again the Fisker Karma, which has an engine of 198 kW.

Most of the engines used for range extenders use the Atkinson cycle having a longer expansion stroke than the compression, with a significant increase of the engine thermal efficiency and thus fuel consumption reduction.

Table 4. Engine types used for range extenders.

Engine type	Vehicles	[%]
CI	3	13.6
CI - turbo	2	9.1
SI	14	63.6
SI - DI	1	4.5
SI - turbo - DI	2	9.1
SI - turbo	1	4.5

In this type of vehicles Li-ion batteries represent almost 76% of the cases, followed by other types of batteries with much less frequency. In the case of the Li-ion batteries the range provided by the battery pack had an average value of 72 km.

FUEL CELL VEHICLES

52 fuel cell (FC) powered vehicle models have been surveyed and from these only two are already in the market, the Honda FCX Clarity and the Mercedes-Benz Class B F Cell. The former was released at the end of 2008, while the latter was announced for the beginning of 2010. The number of fuel cell vehicles released annually is presented in Figure 9. As it can be seen, the increase of new model vehicles is significant, showing the interest of almost all OEMs in this technology.

The main cause for the lack of market penetration of this automotive technology is the hydrogen supply infrastructure which is still at its early stages and the extremely high price of the fuel cell stack. To overcome these issues OEMs propose different approaches: For instance, Honda proposes and a domestic natural gas reformer to fuel their Fuel Cell vehicle FCX Clarity [16].

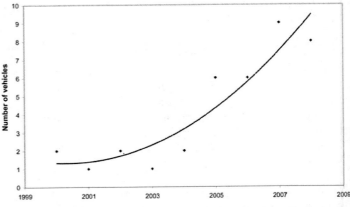

Figure 9. Rate of fuel cell new model vehicles released per year.

The batteries used in FC vehicles are mainly Li-ion and NiMH. When using Li-ion batteries, the battery pack is able to supply 12.5 kWh in average, while if NiMH type is used, the energy capacity of the battery pack falls to an average 6.2 kWh. Fuel cell vehicles display an average range of 92 km when relying solely in battery power. The fuel cell increases this range usually by 200 to 800 km. The fuel cell starts working when further range is required beyond the capacity of the batteries.

From the data collected only proton exchange membrane (PEM) fuel cells are used in automotive applications. They are fed with hydrogen either reformed on board [17], or stored in tanks on board, using pressures from 350 to 700 bar and with volumes that usually can go from 75 to 150 liters. A relation obviously exists between the fuel cell power and vehicle weight as shown in Figure 10.

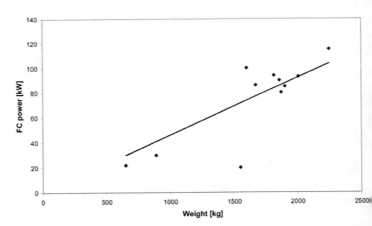

Figure 10. Relation between the vehicle weight and the fuel cell power.

As it can be seen, fuel cell power has an approximately linear relation with vehicle weight. This relation can be established at 0.05 kW/kg of vehicle mass. It is worth noting that a significant number of vehicles with fuel cells have a curb weight higher than 1500 kg. This is justified by the volume needed for the hydrogen storage, which demands a bigger body and thus a heavier vehicle. For this reason it is not expected that a wide variety of commuters with microcar size with a fuel cell powertrain will exist.

Fuel cell vehicles are full electric using mainly AC motors. In fact just one model was found using DC motors, the Pac-Car from Esoro, which is a small competition vehicle. The motors from FC vehicles display an average power of 81 kW. Most often these vehicles deploy a single motor linked with a differential either to the front or rear wheels. When several motors are used, the solution of the motor-in-wheel is preferred over the direct link using a cardan joint.

BATTERIES

Vehicles with ICE range extenders, fuel cells, and mild and full hybrid parallel powertrains all have battery packs but usually with smaller available range. By gathering the information from all these vehicle types (including full electric vehicles) and by relating the vehicle weight with the corresponding energy consumption it is possible to find a relation between these two variables as shown on Figure 11.

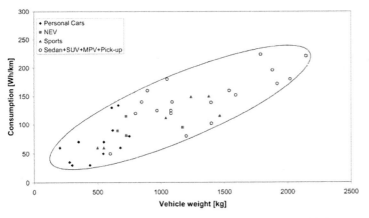

Figure 11. Energy consumption as a function of vehicle weight.

As it can be seen, a clear relation exists between the vehicle weight and its consumption. This means that a significant effort must be put on vehicle weight reduction so that an improvement on energy saving may be attained. From Figure 11 a relation representing on average 0.1 Wh/km for each kg of vehicle mass can be drawn.

Concerning batteries and the vehicle range they provide, a relation between these two variables may be estimated, as presented on Figure 12. Despite the dispersion of the results, it is possible to apprehend a relation of roughly 5.5 km of range per each kWh of energy capacity/stored in the battery pack. It can also be seen from Figure 12 that Li-ion battery

packs are mainly used for longer vehicle ranges while lead acid, Ni-Cd and Na-Ni-Cl battery types are used for shorter range vehicles. The main reason for this is clearly the battery pack weight.

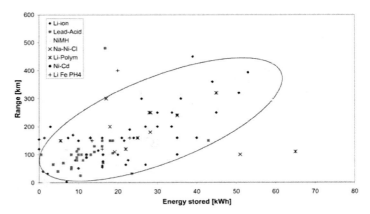

Figure 12. Vehicle range as a function of battery energy stored and battery type.

The comparison of the battery pack weight with the corresponding energy stored (shown in Figure 13), clearly highlights the potential of the Li-ion battery type. In Figure 13 two zones are shown representing the Li-ion and lead-acid battery packs in terms of relation between weight and energy stored. From the information gathered, Li-ion battery packs have three times more energy density than lead-acid battery packs. The other advantage is the battery life, that in the case of the lead-acid is reported by vehicles manufacturers to be around 500 cycles while for the Li-ion the battery average life is 2700 cycles. All these advantages have put a strong focus on Li-ion battery development, with car manufacturers starting up new industrial projects for mass production of these battery packs [18]. The results concerning the NiMH batteries presented in Figure 13 may be divided into two zones, which are approximately between the Lead-Acid and Li-ion battery packs zones. This means that their energy density is higher than that of Lead-based batteries but lower than Lithium-based ones. A first zone of reduced energy stored corresponds to the mild hybrid vehicles. A second zone exists, with higher energy stored than the lead battery packs.

Some vehicles report the use of ultra-capacitors. These have higher power than any battery pack and simultaneously have reduced energy density, with higher life duration than battery packs. However, no detailed information is released about specifications of the ultra-capacitors used on these vehicles. The use of these technologies will allow for higher rates of regenerative braking energy recovery, due to its higher power.

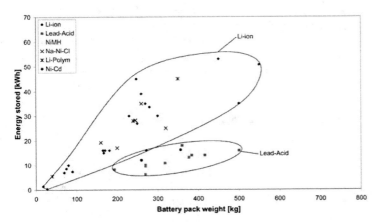

Figure 13. Energy stored on the battery pack as a function of battery pack weight and battery type.

ELECTRIC MOTORS

Electric motors are mostly described on vehicle specifications with their peak power. Some times the continuous power is also presented. From these data was possible to compare this two motor characteristics. This comparison is shown on Figure 14. It can be stated that on average the peak power is 1.5 times the continuous power.

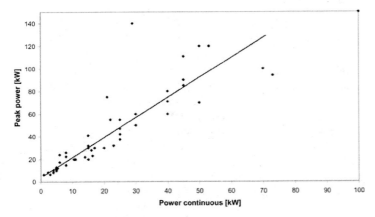

Figure 14. Relation between the continuous and peak power of vehicle motors.

Peak power on electric motors is available during short working periods from a few seconds until 1 minute, although is not common a so long period of time to work under peak power conditions. Usually this power is used for quick vehicle accelerations. A relation can also be made between these two variables as presented on Figure 15. That relation can be estimated based on the data collected from motors and vehicle acceleration times from 0 to 100 km/h under full electric working conditions. The relation between the peak power and the acceleration was preview to be 19 kW to reduce 1 s on the 0 to 100 km/h time.

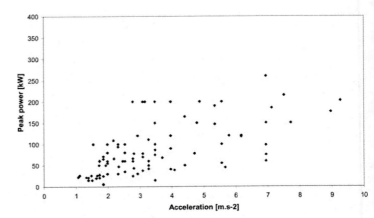

Figure 15. Relation between the peak power and vehicle acceleration.

The selection and dimensioning of electric motors to electric/ hybrid vehicles is also dependent on the vehicle mass as shown on Figure 16. From Figure 16 the relation between these two variables is calculated as 0.025 kW of motor power per kg of vehicle mass, bit the data is scattered.

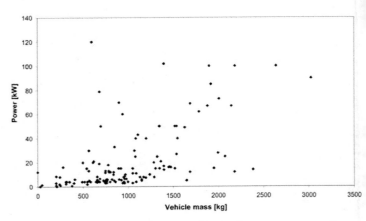

Figure 16. Relation between the continuous power and vehicle mass.

In terms of motoring, a new trend exists on full electric vehicles, which is the integration of the motor on the wheel hub. Just 28 vehicles with electric traction have their motors "in-the-wheel", either on two or on the four wheels, and one example was found with eight wheels with its one motor, the Eliica from Japan. This type of power transmission configuration has a significant benefit on saving mechanical losses from the electric motor to the wheel. On this kind of devices, the motor is built in the wheel and on this module may also be integrated the suspension and steering. This design solution also allows design teams to rethink the vehicle layout, as more space is free within the vehicle, once all moving parts are placed on or near the wheel.

SUMMARY/CONCLUSIONS

A computer database was created using MS Access and information of technical specifications of electric and hybrid vehicle models was stored. Vehicle models were classified in categories concerning several criteria, namely vehicle type, powertrain configuration and status of development.

This database of vehicle model information allowed for the definition of several queries relating data from the vehicle and its powertrain components such as batteries, electric motors and internal combustion engines.

The data collected revealed that on the electric mobility industry the most common vehicle model configuration is the conventional sedan and SUV, followed by NEVs and microcars. In terms of status of development nearly 50% of all vehicles launched did not get into the market revealing a lack of maturity of the electric/hybrid technology.

For each vehicle type an analysis was made describing the average powertrain characteristics such as battery pack type and capacity, motor type and power and internal combustion engine type and other features. The same type of analysis was made for each powertrain type. A deeper analysis was made to the battery packs and motors. This analysis allowed to determine some average relations between battery power, vehicle weight and range, as well as relations between motors power, vehicle mass and acceleration.

REFERENCES

1. Mayor of London "An Electric Vehicle Delivery Plan for London", May 2009, http://www.london.gov.uk/mayor/publications/2009/docs/electric-vehicles-plan.pdf, last access 16 of September 2009.

2. "Powertrain 2020 China's Ambition to Become Market Leader in E-Vehicles", Roland Berger Strategy Consultants, Munich/Shanghai, April, 2009.

3. SAE J1715 Information Report, Hybrid Vehicle (HEV) and Electric Vehicle Terminology, July 2007.

4. Tate, E. D., Harpster, M. O., Savagian, P. J., "The Electrification of the Automobile: From the Conventional Hybrid, to Plug-in Hybrids, to Extended Range Electric Vehicles", *SAE Int. J. of Pass. Cars - Electron. Electr. Syst.* 1(1):156-166, 2008.

5. Chu, A., "The Use of A123 Systems Technology in PHEV Applications", PHEV2007 Conference, Winnipeg, Canada, November 1-2, 2007.

6. Burke., A., Miller, M., "Performance Characteristics of Lithium-ion Batteries of Various Chemistries for Plug-in Hybrid Vehicles", EVS24, Stavanger, Norway, May 13-16, 2009.

7. Signorelli, R., Schindall, J., Kassakian, J., "Carbon Nanotube Enhanced Double Layer Capacitor", Proceedings of the 14th intern. Seminar on Double Layer Capacitors and Hybrid Energy Storage Devices, FL, USA, p. 49-61, December 6-8, 2004.

8. Chau, K. T., Wong, Y. S., Chan, C. C., "An Overview of Energy Sources for Electric Vehicles", Energy Conversion & Management 40 (1999) 1021-1039.

9. Brockbank, C., Cross, D., "Mechanical Hybrid System Comprising a Flywheel and CVT for Motorsport & Mainstream Automotive Applications," SAE Technical Paper 2009-01-1312, 2009.

10. Rahman, K.M., Patel, N. R., Ward, T. G., Nagashima, J. M., Caricchi, F., Crescimbini, F., "Application of Direct-Drive Wheel Motor for Fuel Cell Electric and Hybrid Electric Vehicle Propulsion System", IEEE Transactions on Industry Applications, Vol. 42, NO. 5, September/October 2006.

11. Ribeiro, B., Martins, J., "Direct Comparison of an Engine Working under Otto, Miller and Diesel cycles: Thermodynamic Analysis and Real Engine Performance," SAE Technical Paper 2007-01-0261, 2007.

12. "AVL Range Extender", http://www.avl.com/emag/Range_Extender/ last access 21 of September 2009.

13. "Lotus Range Extender Engine Revealed", Lotus Press release, 07 September 2009.

14. "Toyota Hybrid System THS II", Toyota Motor Corporation, Tokyo, May 2003.

15. Abe, S., Murata, M., "Development of IMA Motor for 2006 Civic Hybrid", SAE 2006-01-1505, 2006.

16. http://automobiles.honda.com/fcx-clarity/ last access 21 of September 2009.

17. Bowers, B. J., Zhao, J., Ruffo, M., Dattatraya, D., Khan, R., Quet, P-F., Sweetland, V., Darby, E., Shi, Y., Dorfman, Y., Dushman, N., Toro, A., Alberti, I., Conti, A., Beziat, J-C., Boudjemaa, F., "Multi-Fuel Fuel Processor and PEM Fuel Cell System for Vehicles," SAE Technical Paper 2007-01-0692, 2007.

18. "Renault-Nissan Alliance Takes Initial Zero Emission Mobility Move in Europe", Nissan Press release, 20 July 2009.

CONTACT INFORMATION

Bernardo Ribeiro is a Post-Doctoral researcher at the Universidade do Minho, Portugal. He can be contacted through bernardor@dem.uminho.pt.

Francisco P. Brito is a Post-Doctoral researcher at the Universidade do Minho, Portugal. He can be contacted through francisco@dem.uminho.pt

Jorge Martins is an Associate Professor at the Universidade do Minho at Guimaraes, Portugal, where he is head of the I.C. Engines Laboratory. He can be contacted through jmartins@dem.uminho.pt

ACKNOWLEDGMENTS

Bernardo Ribeiro thanks the FCT and MIT Portugal Program for the financial support given for his research activities (SFRH / BPD / 48189 / 2008).

This research project was supported by MIT-Pt/EDAM-SMS/ 0030/2008.

DEFINITIONS/ABBREVIATIONS

EV
Electric Vehicle

ICE
Internal Combustion Engine

LIVC
Late Intake Valve Closure

MPV
Multi-Purpose Vehicle

NEV
Neighborhood Electric Vehicle

OEM
Original Equipment Manufacturer

PEM
Proton Exchange Membrane

PV
PhotoVoltaic

SUV
Sport Utility Vehicle

VVT
Variable Valve Timing

Technology Improvement Pathways to Cost-effective Vehicle Electrification	2010-01-0824
	Published 04/12/2010

Aaron Brooker, Matthew Thornton and John Rugh
National Renewable Energy Laboratory

ABSTRACT

Electrifying transportation can reduce or eliminate dependence on foreign fuels, emission of green house gases, and emission of pollutants. One challenge is finding a pathway for vehicles that gains wide market acceptance to achieve a meaningful benefit. This paper evaluates several approaches aimed at making plug-in electric vehicles (EV) and plug-in hybrid electric vehicles (PHEVs) cost-effective including opportunity charging, replacing the battery over the vehicle life, improving battery life, reducing battery cost, and providing electric power directly to the vehicle during a portion of its travel. Many combinations of PHEV electric range and battery power are included. For each case, the model accounts for battery cycle life and the national distribution of driving distances to size the battery optimally. Using the current estimates of battery life and cost, only the dynamically plugged-in pathway was cost-effective to the consumer. Significant improvements in battery life and battery cost also made PHEVs more cost-effective than today's hybrid electric vehicles (HEVs) and conventional internal combustion engine vehicles (CVs).

INTRODUCTION

It has been well documented that the United States (U.S.) is faced with a transportation energy problem. The transportation sector is almost entirely dependent on a single fuel - petroleum. The future of the petroleum supply and its use as the primary transportation fuel threatens both personal mobility and economic stability. The U.S. currently imports nearly 60% of the petroleum it consumes and dedicates more than 60% of its petroleum consumption to transportation [1]. As domestic production of petroleum steadily declines and U.S. consumption continues to climb, imports will continue to increase. Internationally, the growing economies of China and India continue to consume petroleum at rapidly increasing rates. Many experts are now predicting that world petroleum production will peak within the next 5-10 years [2]. The combination of these factors will place great strain on the supply and demand balance of petroleum in the near future.

Hybrid electric vehicle (HEV) technology presents an excellent way to reduce petroleum consumption through efficiency improvements. HEVs use energy storage systems combined with electric motors to improve vehicle efficiency by enabling engine downsizing and by recapturing energy normally lost during braking events. A typical HEV will reduce gasoline consumption by about 30% over a comparable conventional vehicle. This number could approach 45% with additional improvements in aerodynamics and engine technology. Since their introduction in the U.S., HEV sales have grown at an average rate of more than 60% per year [3]. However, after 10 years of availability, they represent less than 1% of the total U.S. vehicle fleet [3]. There are 237 million vehicles on the road today and more than 16 million new vehicles sold each year [4]. Each new vehicle (the vast majority of which are non-hybrids) will likely be in use for more than 15 years [5]. With continued growth in the vehicle fleet and in average vehicle miles traveled (VMT), even aggressive introduction rates of efficient HEVs to the market will only slow the increase in petroleum demand. Reducing U.S. petroleum dependence below present levels requires vehicle innovations beyond current HEV technology.

Plug-in electric vehicle (EV) and plug-in hybrid electric vehicle (PHEV) technologies are options with the potential to displace a significant portion of transportation petroleum consumption by using electricity for all or portions of given trips. Plug-in electric vehicles use an electric motor powered by an energy storage system and only use electricity from the utility grid. A plug-in hybrid electric vehicle is an HEV with the ability to recharge its energy storage system with electricity from the utility grid. With a fully-charged energy storage system, a PHEV will bias toward using electricity over liquid fuels. A key benefit of plug-in electric and plug-in hybrid electric technologies is that the vehicle is no longer dependent on a single fuel source. The primary energy carrier would be electricity generated using a diverse mix of domestic resources including coal, natural gas, wind, hydro, and solar energy. In the PHEV case the secondary energy carrier would be a chemical fuel stored on the vehicle (i.e., gasoline, diesel, ethanol, or even hydrogen).

EV and PHEV technologies are not without their own technical challenges. Energy storage system cost, volume, and life are the major obstacles that must be overcome for these vehicles to succeed. Nonetheless, these technologies provide a relatively near-term possibility for achieving petroleum displacement. One of the key factors in assessing the potential fuel use reductions of EVs and PHEVs is to assess their fuel use relative to specific configurations and component sizes (energy storage trade-offs) and how they compete with both conventional vehicles and other advanced technology vehicles, such as HEVs, in terms of cost, performance, and petroleum displacement potential. By doing this relative comparison, cost-effective pathways to vehicle sector electrification can be identified.

APPROACH

There are many possible pathways to cost-effective vehicle electrification. This study evaluates a variety of scenarios and technology improvements. Prior to the analysis the vehicle performance, cost, and battery life models were checked to match today's technologies and cost. Next, a variety of vehicle electrification scenarios were run. One scenario sized the battery to last for the life of the vehicle. A second assumed battery replacement: that the battery will be replaced during the life of the vehicle. A third scenario assumed opportunity charging: that the vehicle will be able to recharge after every trip rather than just at the end of the day. A fourth assumed both battery replacement and opportunity charging. These scenarios were then all rerun with improvements in battery cost or battery life. In each case they are all compared to conventional vehicles and HEVs.

ESTIMATING COST

A large share of the market needs to switch to electric vehicles to realize the national and global benefits of vehicle electrification. According to the J.D. Power and Associates 2008 Alternative Powertrain Study, most people will purchase a fuel saving vehicle if the fuel savings will pay back the extra upfront cost [6]. Alternatively, most would not be willing to purchase a fuel saving vehicle if it didn't provide payback [7]. Therefore, this study uses the cost-effectiveness as a metric to reflect the potential to successfully achieve the individual, national, and global goals.

The cost-effectiveness is estimated by comparing the net present vehicle and fuel cost of each electric vehicle against today's options. Since insurance, maintenance, and repairs have not been consistently higher or lower for advanced vehicles such as HEVs [8], they were not included.

Component costs were based on previous study estimates [1, 9] as shown in Table 1. The exception is the $700/kWh [10] battery energy cost coefficient. This was calibrated to match estimates of a range of today's HEV, PHEV, and EV vehicles, as seen in Figure 1.

Table 1. Component manufacturing cost and markup factor applied to calculate price to consumer.

Battery	$22/kW + $700/kWh + $680
Motor and controller	$21.7/kW + $425
Engine	$14.5/kW + $531
Markup factor	1.75

<figure 1 here>

The last three vehicle prices are much higher than the others. The Tesla Roadster is listed at $109,000 [33]. The Scion EV, known as the E-Box, was a conversion of a roughly $15,000 Scion by AC Propulsion for $55,000 [34]. The estimated Volt manufacturer suggested retail price (MSRP) estimate of $48,000 is based on the cost that would be required to make it profitable today [11], not the $40,000 ($32,500 after tax incentive) it is expected to sell for [11].

The conventional vehicle costs are used to estimate the HEV, PHEV, and EV costs. The engine cost is subtracted from the conventional vehicle price. Then the advanced vehicle component costs are added. This approach matched closely for a range of advanced vehicles with different component sizes.

UNIQUE IMPROVEMENTS

This study expands on previous efforts. As in previous studies, it accounts for the impact of larger batteries on cost, weight, and performance using a vehicle model. In addition, it improves on other aspects including the driving distance assumption, battery life, battery sizing, battery use strategy, and the method for estimating fuel economy. It also looks at another method of plugging-in, connecting electrically along the roadway while driving.

Distribution of Driving Distances

This study's assumption for driving distance between recharge expanded the constant distance assumption from other studies [12, 13] to a distribution of distances. This had important impacts on battery life, control strategies, and fuel economy. A constant distance is often used to represent a consistent commuting distance. Commuting, however, only represents one third of the miles driven [14]. Therefore, most driving may not be a consistent distance. To improve this assumption, this study uses a distribution of daily driving distances based on national statistics [14]. Figure 2 was generated using the 2001 National Household Travel Survey (NHTS) DAYPUB database and filtering consistent with SAE J2841. The frequency of occurrence assumed 2-mile bins with a total of 600 bins, which was required to capture the maximum daily driving distance of 1200 miles. While long trips are infrequent, they are important because their length can make them a significant portion of the total miles traveled.

<figure 2 here>

The long trips reduce the average PHEV fuel economy. Therefore, it could be argued that PHEVs shouldn't be used for long trips. However, PHEVs still have high efficiency after the charge depleting range, similar to an HEV. Therefore, using them on long trips saves fuel relative to conventional vehicles to help capitalize on the higher initial cost.

EVs cannot travel many of the long distances without recharging. For daily driving distances greater than the electric range, this study assumes that the vehicle recharges each time it reaches its maximum range. This increases the frequency of daily driving distances at the maximum EV range, as seen in Figure 3. This is an optimistic assumption for EVs because it assumes greater use than is likely, and thus higher fuel cost savings, for EVs. It assumes greater use than likely would occur because people may use a different vehicle for long trips to avoid having to stop along the way and take the time to recharge.

<figure 3 here>

The opportunity charging scenario assumed charging after each trip instead of daily charging resulting in a different distribution of driving between recharges, as seen in Figure 4. The shift increases the amount of driving done electrically, especially for shorter range PHEVs.

<figure 4 here>

Battery Life and Sizing

The driving distribution has important implications on battery life and sizing. For PHEVs and EVs, the trip length is used to

estimate the level of discharge to the battery based on the vehicle's charge depleting efficiency. Each discharge causes a specific level of battery wear based on data from Johnson Controls [15], as seen in Figure 5. Using the trip driving distribution data, battery discharge efficiency, and battery cycle life data, the average charge depleting wear per mile was calculated. The acceleration and regenerative breaking cycle wear per mile based on the drive cycle simulations, which can account for as much as 5% of the wear for low range PHEVs, was then added to calculate the total wear per mile.

<figure 5 here>

The original battery life curve in Figure 5 represents the published data. Since this data does not consider calendar, temperature, or power level effects for the current technology case, the trend was adjusted to match published Nissan Leaf [16] and Chevy Volt [17, 18] battery life expectations. The future case was adjusted to match the 7,000 cycle life published by A123 [19], which is similar to the U.S. Department of Energy's (DOE) target [20]. It is used for the future improved case because again the published data does not include the calendar, temperature, or power level effects that would occur for a vehicle application.

The life estimates are used to size the battery for the different scenarios. As the battery cycle life curves show, decreasing the depth of discharge will increase the number of cycles that battery can sustain. Therefore, to increase the battery life to match the vehicle life, or to match the vehicle life with one replacement, the model iterates on depth of discharge to find the smallest battery that will last the required amount. Finding the smallest battery to meet the requirements minimizes the battery size and thus total vehicle cost.

Battery Use Strategy

The driving distribution assumption impacts not only battery life but also the PHEV battery use strategy. Assuming a constant driving distance may suggest that the best control strategy very selectively depletes the battery over the entire distance [21]. However, assuming that people drive a distribution of distances and that consumers don't enter in information about their trip every time they get in the vehicle, a better control strategy displaces gasoline as quickly as possible to minimize gasoline use before the trip ends.

Fuel Economy

A vehicle model is used to predict fuel economy. To gain confidence in the model, component sizes and vehicle characteristics were entered into the model for a variety of vehicles. As seen in Figure 6, the model predicted the fuel economy within 10% except for the Hymotion Prius. This overestimation was accepted to account for the non-ideal implementation of an aftermarket conversion vehicle.

<figure 6 here>

The method to estimate PHEV fuel economy also builds from previous studies. A recent paper discusses a few methods on how to estimate in-use PHEV fuel economy [22]. The approaches involve repeating a drive cycle enough times to deplete the battery and then running one charge sustaining cycle. The charge depleting and charge sustaining fuel economies can then be calculated, adjusted to better represent in-use fuel economy, and combined based on the utility factor (UF), or the percent of driving that would likely occur in charge depleting mode. This study used a slight variation from the approach described in the paper to remove fuel economy variations caused by the test approach.

Dynamic Plug-in

This study also expands on the type of plug-in hybrid vehicle evaluated. It assesses a vehicle that plugs-in dynamically, connecting electrically along the roadway while moving, similar to the way trolley buses or streetcars currently do, although research to improve the connection approach would be required. Since the vehicle is connected while driving, it doesn't need a large battery to gain PHEV fuel economy benefits, although it does need infrastructure along a small fraction of roadway. The fraction of infrastructure is small because most travel occurs on just a few roads. The interstate, for example, makes up 1% of the miles of roadway but carries 22% of the vehicle miles traveled [28, 29]. This scenario assumes that 50% of the distance driven is connected dynamically. It also assumes an additional $1,000 cost to the consumer for the dynamic connection, the same fuel cost per mile as an HEV when not connected dynamically, and the charge depleting fuel cost per mile of a PHEV when connected.

ADDITIONAL ASSUMPTIONS

Additional assumptions used in this study are listed in Table 2.

Table 2. Additional assumptions used in the study.

Vehicle miles traveled per year [23]	12375
Vehicle type	Compact Car
Vehicle life (years) [24]	15
Gasoline price (average 2008 price) [25]	$3.21
Electricity price ($/kWh) [26]	$0.10
Sales tax [27]	7.8%
Discount rate	8%
Battery cost reduction per year	3%
Battery salvage value (percent of future manufacturing cost)	20%
Percent of distance dynamically plugged-in	50%
Dynamic connection cost to the consumer	$1,000
Low power PHEV battery power (kW)	21
High power PHEV battery power (kW)	45

RESULTS

CURRENT BATTERY TECHNOLOGY

The improvements in assumptions and approach led to unique results. Using today's battery assumptions, while the gasoline consumption decreases significantly, no electrification pathways were cost-effective compared to HEVs or CVs except one, as indicated by the red line in Figure 7. The vehicles listed on the figure follow the naming convention of vehicle type, charge depleting range, and then battery power level. For example, the PHEV10 Low Power stands for plug-in hybrid electric vehicle with 10 miles of charge depleting range using a low power battery. Increasing battery power had little effect on fuel consumption results because in both cases the battery power can provide most of the driving on the test cycles, so the fuel economy only differs slightly. For the electric powered vehicles, the electricity cost is relatively low, reflecting the low cost of electricity and the high efficiency of batteries and motors. The gasoline, on the other hand, is a large expense, especially for the conventional vehicle. Even so, the extra battery costs in PHEVs and EVs outweighed the gasoline cost savings.

<figure 7 here>

Battery replacement had minor overall improvements in cost-effectiveness. These cases reduced the size of the battery but used it more aggressively to reduce upfront cost and weight and take advantage of lower future battery costs. The advantages, however, were mostly balanced out by the increase in battery wear. For a smaller battery to provide the same electric range and regenerative braking, it must use a greater portion of the battery energy, and thus have greater depths of discharge. Since battery wear increases non-linearly with depth of discharge, each battery has to be larger than half of the single battery case. For example, in the high power PHEV10 case, a 5.9 kWh battery would last the life of the vehicle using 34% of the energy. Having one replacement, however, required more than half of a 5.9 kWh battery. It required purchasing two 3.7 kWh batteries using 54% of the energy to meet the life requirement. Although it was assumed that future batteries cost less and that there is a time value of money advantage to purchasing the second battery, these advantages did not make up for the total added cost of buying more total battery energy. The nonlinear wear trend balanced out the advantages for little overall gain.

Opportunity charging further decreased the gasoline consumption, and thus gasoline cost, of PHEVs, but at a greater increase in battery cost. Opportunity charging reduced gasoline consumption for the PHEV10 by 35%. A 35% reduction amounted to roughly a $2,400 reduction in present fuel cost. Although the fuel cost went down, opportunity charging increased the use of the battery. In order to sustain the additional use and wear, the battery energy had to be increased from 5.9 kWh to 10 kWh. This added more than $5,500 to the vehicle cost. Including the additional electricity cost, opportunity charging increased the total by $4,400.

Opportunity charging decreased the EV cost. Unlike the PHEV, which drives more on the battery with opportunity charging, the EV has to cover all of its distance on the battery with or without opportunity charging. Opportunity charging increased the frequency of recharging, reducing the depth of discharges and the amount of wear, and thus reducing the amount that the battery has to be oversized to last the required life. Specifically, it reduced the battery size from 47 kWh to 32 kWh. This reduced the battery cost and the vehicle cost overall, but the EV still exceeded the cost of all the other vehicle types.

Combining battery replacement and opportunity charging increased the use of the high cost battery to better leverage the investment. With the current battery life assumption, however, little to nothing was gained by adding battery replacement to the opportunity charging cases.

One case may warrant further investigation because it reduced total cost to the consumer and it reduced fuel use. This is labeled EHEV, for electrified HEV. This case assumes that an HEV could connect to an external source of energy along some roadways while moving, similar to the way trolley buses or streetcars do in some cities such as Boston, Cambridge, Philadelphia, and San Francisco [35], though it would require research to improve the connection. Although it would require infrastructure along a small percentage of heavily traveled roadways [28,29], if the design can be flexible for both mass transit and private vehicles, then cities may install it for the mass transit benefit alone [35]. On the consumer side, the EHEV is cost-effective even with the extra $1,000 cost to the consumer for the connection mechanism. The cost is low because it gains the low cost electric mode operation similar to a battery PHEV without the cost, wear, efficiency losses, and weight of a large battery.

REDUCED BATTERY COST

An additional pathway to cost-effective vehicle electrification is to reduce battery cost. As seen in Figure 8, if the battery energy cost comes down around DOE targets [20] to $300/kWh, PHEVs get close to breaking even with today's vehicles. Battery replacement didn't add any further advantages for the PHEVs. Opportunity charging, however, significantly reduced gasoline consumption for the PHEVs for little additional cost.

<figure 8 here>

IMPROVED BATTERY LIFE

A third pathway to cost-effective vehicle electrification is to improve battery life. As seen in Figure 9, PHEVs became cost-effective by improving the battery life as illustrated previously in Figure 5. Unlike reducing battery cost, improving battery life makes opportunity charging slightly more cost-effective, providing more potential to reduce gasoline consumption further.

<figure 9 here>

CONCLUSIONS

Electrifying transportation can reduce or eliminate dependence on foreign fuels, emission of green house gases, and emission of pollutants. However, finding a cost-effective pathway to gain widespread adoption and provide a significant impact is challenging. Three possible pathways include improving battery life, reducing battery cost, or connecting to the grid more directly.

Using current battery cost and life, PHEVs and EVs were not cost-effective for many different configurations. PHEVs with 10, 20, or 40 miles of electric range, with low or high electric power, with or without battery replacement, and with or without opportunity charging were all less cost-effective than conventional vehicles and HEVs. EVs' cost-effectiveness improved with battery replacement and opportunity charging, but not enough.

One approach with current battery technology could be cost-effective. If an acceptable method for plugging in while traveling along the roadway can be devised, it may provide a cost-effective pathway to vehicle electrification. This approach benefits from the low electric fuel cost of a large battery without the high cost, cycling wear, weight, and efficiency loss. Even with assuming a $1,000 price for the connection device, the cost to the consumer was still lower than for today's conventional and hybrid vehicles. This pathway requires infrastructure, but only along a small fraction of heavily traveled roadways to gain the same gasoline saving benefits as battery PHEVs.

Significant battery improvements can also provide cost-effective pathways to vehicle electrification. If today's battery energy cost component goes down from $700/kWh to $300/kWh, PHEVs start becoming cost-effective. PHEVs also become cost-effective if battery life improves by a factor of 10.

REFERENCES

1. U.S. Energy Information Administration, www.eia.doe.gov, accessed July 10, 2006.

2. Hirsch, R., Bezdek, R., and Wendling, R., "Peaking of World Oil Production: Impacts, Risks, and Mitigation," U.S. Department of Energy, http://www.netl.doe.gov/energy-analyses/pubs/Oil_Peaking_NETL.pdf, February 2005.

3. HybridCARS Auto Alternatives for the 21st Century, "September 2009 Dashboard: End of Clunkers Hurts Hybrids," http://www.hybridcars.com/hybrid-sales-dashboard/september-2009-dashboard.html, posted Oct. 6, 2009, retrieved Oct. 19, 2009.

4. U.S. Department of Transportation, Federal Highway Administration, "Highway Statistics 2004," http://www.fhwa.dot.gov/policy/ohim/hs04/index.htm, updated March 2006.

5. Davis, S. and Diegel, S., "Transportation Energy Databook: Edition 24," Oak Ridge National Laboratory Center for Transportation Analysis Report, ORNL-6973, December 2004.

6. Power J.D. and Associates, "J.D. Power and Associates 2008 Alternative Powertrain Study," July 2008.

7. Power J.D. and Associates, "J.D. Power and Associates 2002 Alternative Powertrain Study," 2002.

8. Mello, T., "The Real Costs of Owning a Hybrid," http://www.edmunds.com/advice/hybridcars/articles/103708/article.html, July 23, 2008, retrieved Oct. 12, 2008.

9. Simpson, A., "Cost-Benefit Analysis of Plug-In Hybrid Electric Vehicle Technology," 22nd International Battery, Hybrid and Fuel Cell Electric Vehicle Symposium and Exhibition (EVS-22), Yokohama, Japan, October 23-28, 2006.

10. Green, J. and Ohnsman, A., "Obama Battery Grants May Help GM Market Cheaper Electric Cars," http://www.bloomberg.com/apps/news?pid=20601103&sid=aHxVR7B_zrtk, posted Oct. 20, 2009, retrieved Oct. 20, 2009.

11. "At $40,000 the Volt Would Result in No Profit for GM," http://gm-volt.com/2008/03/25/at-40000-the-volt-would-result-in-no-profit-for-gm/, posted March 25, 2008, retrieved Oct. 13, 2009.

12. Shiau, C-S. N., Samaras, C., Hauffe, R., and Michalek, J. J., "Impact of battery weight and charging patterns on the economic and environmental benefits of plug-in hybrid vehicles," *Energy Policy* 37(7):2653-63, 2009.

13. Elgowainy, A., Burnham, A., Wang, M., Molburg, J., and Rousseau, A., "Well-to-Wheels Energy Use and Greenhouse Gas Emissions Analysis of Plug-in Hybrid Electric Vehicles," Argonne National Laboratory Report ANL/ESD/09-2, February 2009.

14. Hu, P., and Reuscher, T., "2001 National Household Travel Survey, Summary of Travel Trends," U.S. Department of Transportation Federal Highway Administration, December 2004.

15. Duvall, M., "Batteries for Plug-In Hybrid Electric Vehicles," presented at The Seattle Electric Vehicle to Grid (V2G) Forum, June 6, 2005.

16. Nissan USA Q&A, http://www.nissanusa.com/leaf-electric-car/?dcp=ppn.39666654.&dcc=0.216878497#/car/index, posted Aug. 10, 2009, retrieved Oct. 13, 2009.

17. Lienert, A., "GM Hustles to Bring Chevrolet Volt to Market by November 2010," http://www.edmunds.com/insideline/do/News/articleId=125465, posted Apr. 4, 2008, retrieved Oct. 13, 2009.

18. Interview with Chevrolet Volt Vehicle Chief Engineer Andrew Farah, http://www.coveritlive.com/mobile.php?option=com_mobile&task=viewaltcast&altcast_code=073e2147ec, retrieved Oct. 12, 2009.

19. "Life, Thousands of Low Rate Cycles. Chart 1/4," http://www.a123systems.com/a123/technology/life, retrieved Oct. 13, 2009.

20. Howell, D., "FY 2008 Progress Report for Energy Storage Research and Development", U.S. Department of Energy, January 2009 (page 3).

21. Rousseau, A., Pagerit, S., Gao, D., "Plug-in Hybrid Electric Vehicle Control Strategy Parameter Optimization," Argonne National Laboratory, December 2007.

22. Gonder, J., Brooker, A., Carlson, R., Smart, J., "Deriving In-Use PHEV Fuel Economy Predictions from Standardized

Test Cycle Results," presented at the 5th IEEE Vehicle Power and Propulsion Conference, Dearborn, Michigan, 2009 (NREL/CP-540-46251).

23. U.S. Department of Transportation Federal Highway Administration Highway Statistics 2005, "Annual Vehicle Distance Traveled in Miles and Related Data," http://www.fhwa.dot.gov/policy/ohim/hs05/htm/vm1.htm, posted November 2006, retrieved Oct. 14, 2009.

24. Research and Innovative Technology Administration Bureau of Transportation Statistics, "Table 1-11. Number of U.S. Aircraft, Vehicles, Vessels, and Other Conveyances," http://www.bts.gov/publications/national_transportation_statistics/2002/html/table_01_11.html, retrieved Oct. 14, 2009.

25. International Energy Agency (IEA), "Weekly U.S. Regular Conventional Retail Gasoline Prices," http://www.eia.doe.gov/oil_gas/petroleum/data_publications/wrgp/mogas_history.html, retrieved Jan. 13, 2009.

26. International Energy Agency (IEA), Official Energy Statistics from the U.S. Government, "Average Retail Price of Electricity to Ultimate Customers: Total by End-Use Sector," http://www.eia.doe.gov/cneaf/electricity/epm/table5_6_a.html, posted Sept. 11, 2009, retrieved Oct. 14, 2009.

27. Consumer Reports, "High Cost of Hybrid Vehicles, Ownership Cost Comparison," http://www.consumerreports.org/cro/cars/new-cars/high-cost-of-hybrid-vehicles-406/ownership-cost-comparison/0609_ownership-cost-comparison_ov.htm, retrieved Oct. 14, 2009.

28. Federal Highway Administration, "Federal - Aid Highway Length - 2004 Miles by System," http://www.fhwa.dot.gov/policy/ohim/hs04/xls/hm15.xls, posted October 2005, retrieved Oct. 15, 2009.

29. Federal Highway Administration, "Annual Vehicle Distance Traveled in Miles and Related Data - 2004 1/By Highway Category and Vehicle Type," http://www.fhwa.dot.gov/policy/ohim/hs04/xls/vm1.xls, retrieved Oct. 25, 2009.

30. Carlson, R.W., Duoba, M.J., Bohn, T.P., and Vyas, A.D., "Testing and Analysis of Three Plug-in Hybrid Electric Vehicles," SAE Technical Paper 2007-01-0283, 2007.

31. Electrical Power Research Institute (EPRI), "Comparing the benefits and impacts of hybrid vehicle options," Report 1000349, EPRI, Palo Alto, California, July 2001.

32. Gonder, J. and Simpson, A., "Measuring and reporting fuel economy of plug-in hybrid electric vehicles," presented at the 22nd International Battery, Hybrid and Fuel Cell Electric Vehicle Symposium and Exhibition (EVS-22), Yokohama, Japan, 2006 (NREL/CP-540-40377).

33. Tesla Motors, http://www.teslamotors.com/buy/buyshowroom.php, retrieved Oct. 22, 2009.

34. Wikipedia, "AC Propulsion eBox," http://en.wikipedia.org/wiki/AC_Propulsion_eBox, retrieved Oct. 22, 2009.

35. Wikipedia, "Trolleybus," http://en.wikipedia.org/wiki/Trolleybus#United_States_of_America, retrieved Oct. 22, 2009.

CONTACT INFORMATION

Aaron Brooker
Senior Research Engineer
National Renewable Energy Laboratory (NREL)
Tel: 303-275-4392
Aaron.Brooker@nrel.gov

Aaron holds a B.S. in mechanical engineering from Michigan Technological University (1998) and an M.S. in mechanical engineering from the University of Colorado (2000).

Matthew Thornton, Senior Engineer, task leader for the Vehicle Systems Analysis team.

Matthew holds a Ph.D. in civil and environmental engineering from the Georgia Institute of Technology, a master's degree from Michigan State University, and a bachelor's degree from the University of Oregon.

John Rugh, Senior Engineer, task leader for the Vehicle Ancillary Load Reduction Project.

John holds a B.S. in mechanical engineering from Colorado State University and an M.S. in mechanical engineering from Purdue University.

ACKNOWLEDGMENTS

The authors gratefully acknowledge the support for this work provided by Lee Slezak and Patrick Davis in the Vehicle Technologies Program of the U.S. Department of Energy's (DOE) Office of Energy Efficiency and Renewable Energy.

DEFINITIONS/ABBREVIATIONS

CD

Charge depleting - The PHEV mode of operation where electricity from the grid is being used by the battery.

CVs

Conventional vehicles

DOE

Department of Energy

EHEV

Electrified hybrid electric vehicles

EVs

Electric vehicles

HEVs

Hybrid electric vehicles

kW

Kilowatts

kWh

Kilowatt hours

MPG

Miles per gallon

MSRP

Manufacturer Suggested Retail Price

NHTS

National Household Travel Survey

PHEVS

Plug-in hybrid electric vehicles - This vehicle is plugged in dynamically. It is similar to an HEV but it connects to an outside source of electricity while moving.

UF

Utility Factor - The percent of travel done in charge depleting mode based on driving distance statistics and the charge depleting range.

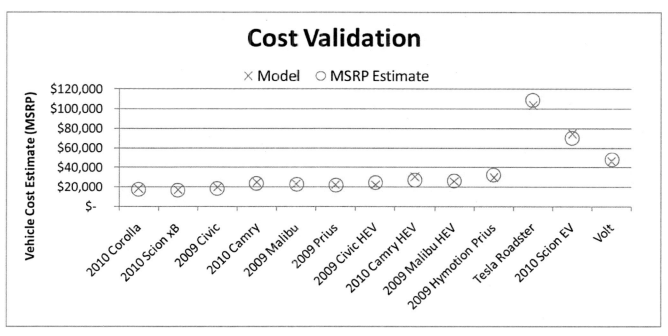

Figure 1. Vehicle cost validation.

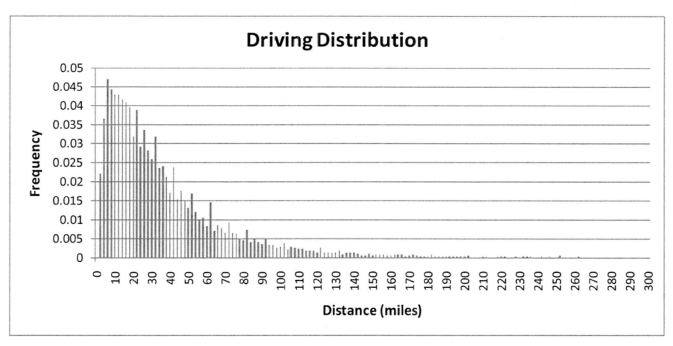

Figure 2. Distribution of daily driving distances.

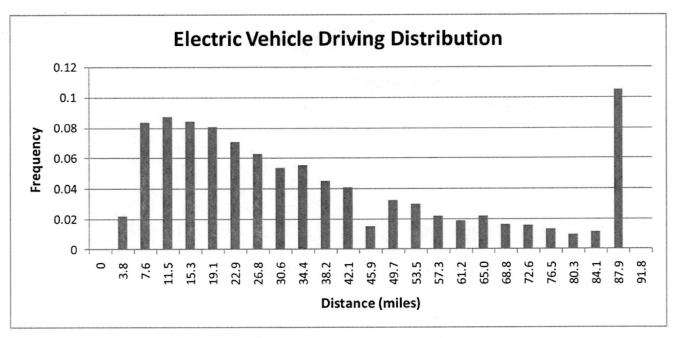

Figure 3. EV driving distribution between recharge events.

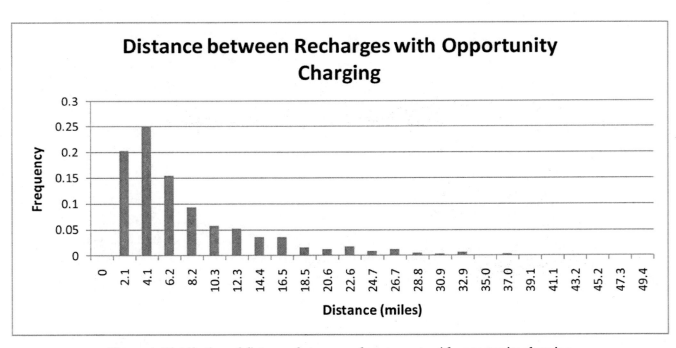

Figure 4. Distribution of distances between recharge events with opportunity charging.

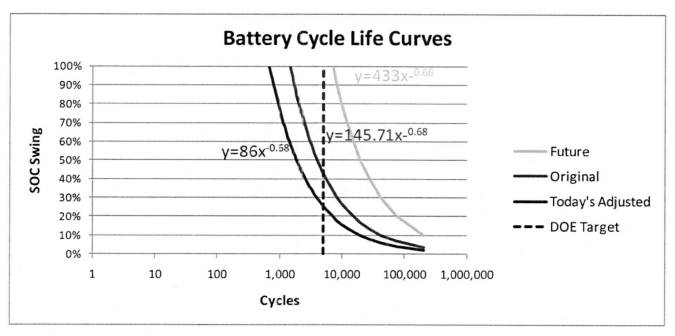

Figure 5. Original and modified battery cycle life curves.

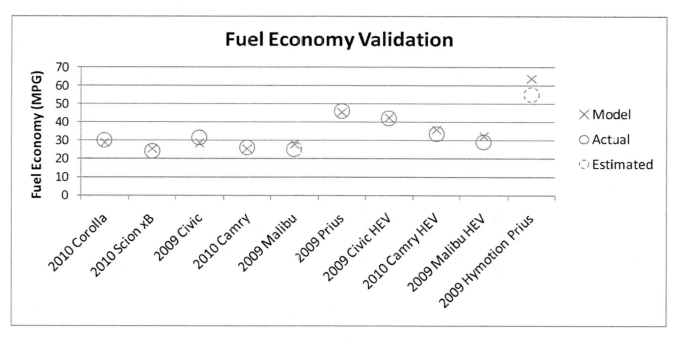

Figure 6. Vehicle model fuel economy validation.

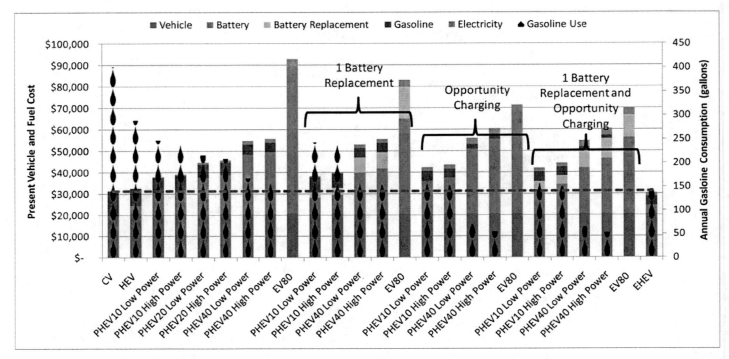

Figure 7. Cost-effectiveness of vehicle electrification using today's assumptions.

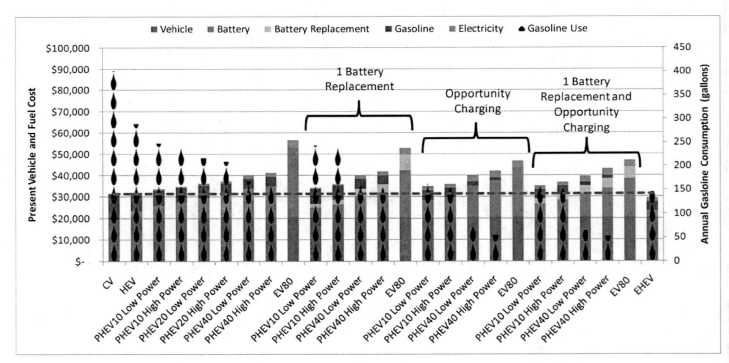

Figure 8. Cost-effectiveness of vehicle electrification with battery energy cost at $300/kWh.

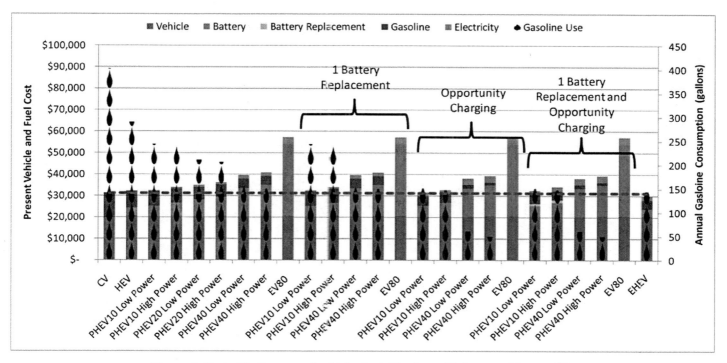

Figure 9. Cost-effectiveness of vehicle electrification with improved battery life.

2009-24-0061

Hybrid Drive Systems for Industrial Applications

Dipl.-Ing. (FH) Frank Böhler
Heinzmann GmbH & Co. KG, Am Haselbach 1, D-79677 Schönau

Dipl.-Ing. Phillip Thiebes
Prof. Dr.-Ing. Marcus Geimer
Cand.-Ing. Julien Santoire
Institute of Vehicle Science and Mobile Machines, Universität Karlsruhe (TH), D - 76131 Karlsruhe

Prof. Dr.-Ing. Richard Zahoransky
Heinzmann GmbH & Co. KG, Am Haselbach 1, D-79677 Schönau

ABSTRACT

Hybrid drives for automotive application are extensively discussed and have been available for several years. In contrast, hybrid drives for industrial and off highway applications are less developed. This is even though the fuel savings potential and expected other benefits are much more promising than in automotive applications.

The company HEINZMANN and the University of Karlsruhe concentrate on hybrid drives for these applications. HEINZMANN develops and produces control systems and electrical motors which can be used in hybrid drives. The Institute of Mobile Machines at the University of Karlsruhe is experienced in simulating off highway machines. In a joined project the two partners are applying an electrical mild hybrid drive to a municipal multi-purpose vehicle.

In this paper the diesel-electric hybrid technology is presented as well as the different applications in construction equipment, fork lifts. A closer analysis is performed with a municipal multi-purpose vehicle equipped with the hybrid drive. Measurements of a conventional vehicle were used for the validation of a simulation model. Results of the simulation are being presented. More than 20% savings in fuel consumption have been determined.

1 BASIS

HEINZMANN GmbH & Co. KG is a system supplier for off-highway diesel and CNG/LPG engines. The core business are control units, common rail systems and air-fuel ratio solutions up to electric drives and hybrid drive systems.

The Institute of Vehicle Science and Mobile Machines, University Karlsruhe is specialized in advanced drive systems for mobile machines, i.e. hybrid drives of different combinations.

Hybrid systems incorporate different engineering disciplines like vehicle technology, combustion engines, electric motors, inverters, control technology, battery systems where the two project partners have a sound experience.

2 MOTIVATION

Hybrid drive systems for passenger vehicles gained public interest worldwide. This triggered also the idea to use hybrid drives for industrial applications. The driving forces for the development have been:
- Scarcity of oil supplies and rising fuel prices.
- Stricter emissions regulations (TIER 4, EURO 5).
- Noise limitations.

Hybrid drive systems represent a potential solution for this problem. They represent the future of today's combustion-based drive systems and will also act as an intermediate step and link to solely electric drive systems.

3 PREMISES

Hybrid drive systems for industrial applications must prove cost-effectiveness. This depends directly on the particular application. Forklift trucks (figure 1) and certain construction machines with a working profile made up of many short load peaks are ideal candidates [1].

The additional components required for a hybrid drive system result in greater initial cost. These additional costs must be compensated by the benefits of the hybrid drive system. These include:

- Savings through lower fuel consumption.
- Reduced exhaust after-treatment requirements through the use of smaller combustion engines.
- Additional functionality/application areas due to better dynamics of hybrid drive system.
- Noise reduction.
- Further potential for electrically operated auxiliary units.
- Potential in house operation if all electric mode is implemented.

Figure 1   Forklift application [3]

The most important cost factor is currently the battery. The latest developments in the automotive sector give cause for hope. Almost every automotive manufacturer is working to develop a hybrid vehicle. Accordingly, battery technology is sure to develop quickly.

4   MILD HYBRID FOR INDUSTRIAL APPLICATION

"Mild hybrid" technology - based on the following concept - is most promising for industrial applications. An electric motor is combined with a combustion engine. The electric motor has 3 functions:

- Boost: The electric machine is switched on during load peaks to provide additional torque.
- Generator/regeneration: The electric machine switches to generator mode when the full output of the combustion engine is not required for the particular application. The electrical energy is then recharged to the battery. The system switches to generator mode during braking ("regeneration") feeding the braking energy into the battery.
- Starter: The relatively large electric machine can start the combustion engine very quickly, typically between 150 and 300 ms. This makes it possible to implement a start-stop function whereby the combustion engine is switched off when idling but is made immediately available when it is required for operation.

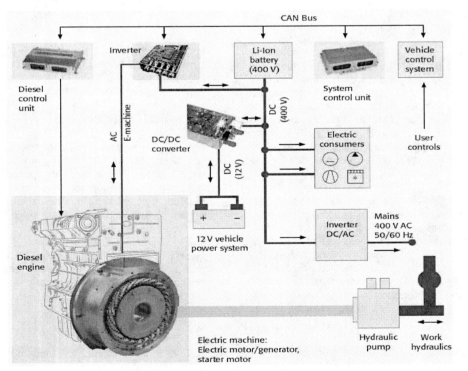

Figure 2   System overview

4.1 SYSTEM OVERVIEW

Figure 2 shows an example of the mild hybrid system. The functions of the individual components are:
- Electric motor/generator or starter
- Hybrid system control unit: To control individual units. Activates boost, charge or start function.
- Diesel control unit: Governor of diesel engine
- Inverter: Speed/torque control for the electric machine
- DC/DC converter: Converts the 400V DC bus into the on-board electrical system voltage

4.2 TORQUE ADDITION

The additional electric machine of the hybrid drive system provides a much higher torque. This makes it possible to use a smaller combustion engine, known as "downsizing" or "rightsizing" the diesel engine (

Figure 3). By combining the torque provided by the smaller combustion engine and the electric machine, it is possible to obtain the same or a slightly better performance than that of the original combustion engine, at least for short periods of time. The typical torque curve of a PSM (permanent magnet synchronous motor) provides advantages at low rpm. In practice, the aim is to select a smaller combustion engine with reduced exhaust aftertreatment requirements.

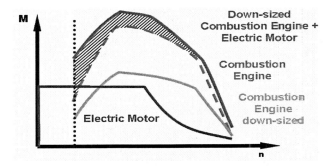

Figure 3   Torque addition of a mild hybrid system [1]

4.3 POWER OUTPUT VERSUS TIME

The diagram of power output versus time visualizes the mild hybrid strategy (see Figure 4).

The start-stop function reduces fuel consumption drastically. For example, the vehicle equipped with a mild hybrid system will not be left anymore idling.

Load peaks - which the downsized combustion engine can no longer cover - are covered by the electric machine. In addition, the electric motor can be engaged right at the start of the load peak, providing a high level of torque, even at low rpm. This is the "dynamic boost", improving the dynamics, reducing the "roaring noise" of the diesel engine and the transient smoke associated.

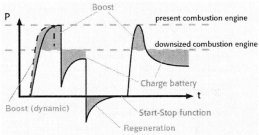

Figure 4   Mild hybrid power output over time

If the particular load demand does not require the full output of the diesel engine, the electric motor charges the battery. The electric motor also recuperates the braking energy before the mechanical brake engages.

Figure 4 visualizes another advantage of this principle: The load pattern of the combustion engine is smoothened out when the system is tuned correctly (phlegmatisation). The load setup should be tuned to achieve either the lowest fuel consumption or the lowest level of emissions. The maximum output level of the diesel engine is exploited more, which should improve the service life of the combustion engine.

4.4 HYBRID CONTROL STRATEGY

The hybrid control unit basically maps the electrical machine's operation. According to the actual speed of the combustion engine and the actual combustion engine's torque, a value for the torque of the electric machine is assigned. The basic control hierarchy and the basic CPU functionalities are outlined in Figure 5 and 6, respectively.

The values "engine torque" and "engine speed" typically are transmitted from the diesel control unit via the SAE J1939 protocol. The torque of the motor can be either positive for boost-mode or negative for charging the battery. This torque of the motor must be limited by several parameters e.g. expected cycle and cycle frequencies, actual state of charge battery, temperatures of battery, inverter and motor.

A basic hybrid-map is shown in figure 7. It takes into account the characteristics of the motor and the combustion engine. As the motor can be adapted to practically all combustion engines, the torque of the diesel engine is plotted in percentage of the available torque at the corresponding speed of revolution. If the driver demands maximum load, both, the motor and the combustion engine have to deliver the maximum torque. If there is a lower load requested, the combustion engine can take over gradually all the load. At a highly reduced load demand, the torque of the motor can be chosen to be negative in order to charge the battery.

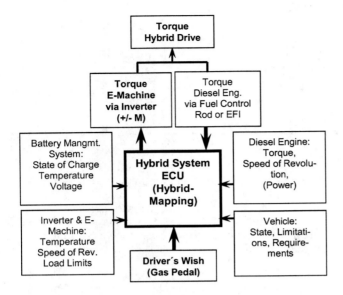

Figure 5   Basic hybrid system control hierarchy

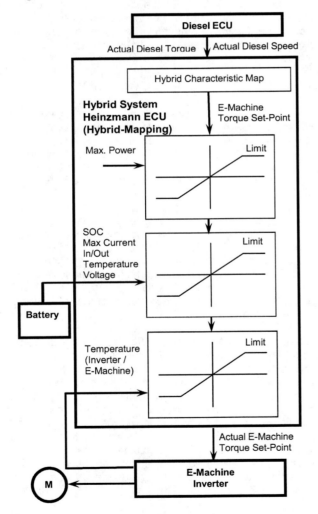

Figure 6   Basic hybrid system CPU functionalities

Electric braking or regeneration is in figure 7 at negative torque of the motor and zero torque of the combustion engine. Again, the regeneration is limited to the electric/electronic design of the inverter so that there is only a limited electric braking power possible; the rest of the braking power must be taken over by the mechanical brake.

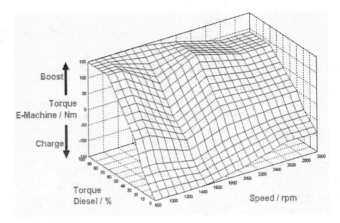

Figure 7   Hybrid characteristic map

The hybrid characteristic map has to be adjusted due to demands of the individual application. This strategy leads to a torque split between combustion engine and motor showed in figure 8. This picture is slightly different to the idealized schematic diagram figure 4. The blue line shows the diesel torque: It can be seen, that there is a "phlegmatisation" of the combustion engine.

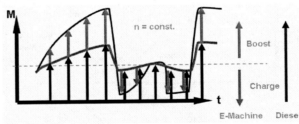

Figure 8   Torque split between combustion engine and motor

4.5 DESIGN OF ELECTRIC MACHINE

Electric motors are usually categorized according to their rated output, which corresponds to allowed rotor/stator temperature. However, electric machines used in mild hybrid applications do not operate continuously - the short boost or regeneration phases generally last only seconds. Therefore, only the short-term power capacity of the electric motor matters for mild hybrid systems.

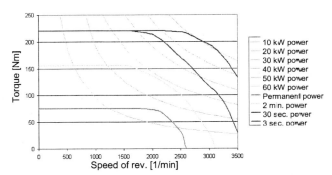

Figure 9: Torque characteristic 15kW electric machine

4.6 MECHANICAL INTEGRATION

On the HEINZMANN mild hybrid, the rotor of the electric machine replaces the flywheel of the combustion engine and is bolted directly to the crankshaft. As a result, the electric machine uses the crankshaft bearing of the combustion engine (figure 10). The advantage of this arrangement is that the motor can be kept very compact and the overall drive train remains as short as possible (see figures 11 & 12). The manufacturer of the combustion engine must give its approval for the additional load placed on the crankshaft bearing by the electric motor.

Figure 10   Electric machine

Figure 11   HEINZMANN electric machine on DEUTZ 2011 diesel engine

Figure 12   HEINZMANN's electric motor on VW SDI diesel engine with hydraulic pump

4.7 MAGNETIC PULL

On a PSM, a magnetic force is generated which attempts to pull the rotor towards the stator. If the rotor is centered relative to the stator, these forces balance each other (see Figure 13, left).

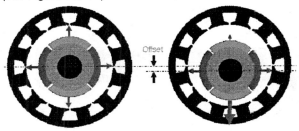

Figure 13   Magnetic pull of rotor

In a combustion engine, the ignition of the fuel/air mixture applies a force to the crankshaft. This causes the crankshaft to bend slightly. Since the rotor is directly bolted to the crankshaft, any bending of the shaft causes the rotor to wobble. The rotor is then no longer centered and the forces become unbalanced. This produces a force called "magnetic pull" (see Figure 1, right).

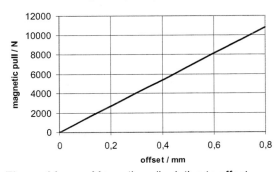

Figure 14   Magnetic pull relative to offset

This force also acts on the crankshaft bearing. For this reason, it is important to clarify the following points during the development process:

- What is the deflection of the crankshaft?

- Is the crankshaft bearing able to cope with the additional load caused by the magnetic pull?
- Is the clearance between rotor and stator sufficient?

The magnitude of the force depends on the stator offset and the electric machine used. Figure 14 shows an example of the force for a 30 kW (peak) electric machine.

Crank shafts of modern diesel engines have a low eccentricity below 0.3 mm and bearings with reasonable safety margins. This is sufficient for a safe operation of the motor, directly mounted on the crank shaft without own bearings. However, the air gap between stator and rotor of the motor must be sufficiently large, e.g. 1 mm. Also, the active rotor length is limited to approx. 60 mm to minimize the lever effect. Longer, more powerful motors would need own bearings.

4.8 BATTERY SIZE / BATTERY TYPE

The minimum number of cells required depends on the system voltage of the DC bus. For example, a DC bus voltage of 400 V needs:
- 110 Li-Ion cells, or
- 125 LiFePO$_4$ cells

Figure 15   Battery for a mild hybrid [5]

The minimum capacity of the battery is determined by the capacity of the individual cells. The battery capacity required depends on the application. E.g., the red colored area in figure 4 represents the energy consumed when the battery is used for a boost procedure. For this reason, the load profile for the application must be known in order to estimate the battery size. The selection of the battery size is based on the following conditions:
- Required capacity according to load profile.
- SOC window within the battery is operated. A small SOC window (low discharge depth) is necessary if a battery a long service life is demanded.
- Required battery current peaks (charge/discharge).
- Space requirements.
- Cooling system.
- Cost of battery.

Figure 1 depicts a battery for a mild hybrid system with a rated voltage of 400 V and a capacity of 4.5 Ah.

4.9 BATTERY MANAGEMENT SYSTEM

Hybrid drive systems need a Battery Management System (BMS) to ensure that the battery remains in safe conditions. The BMS performs the following tasks:
- Monitoring of individual cells (voltage, temperature)
- Charge balancing
- SOC determination
- Insulation monitoring
- Safety shutoff

The BMS communicates with the system control unit over a suitable bus. For example, a CAN bus with SAE J1939 protocol can be used. Figure 16 illustrates schematically the BMS.

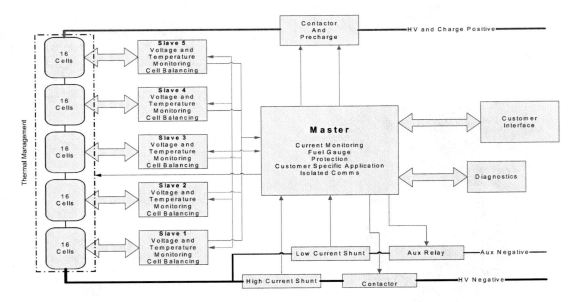

Figure 16   Schematic illustration of a BMS (Battery Management System) [5]

5 SIMULATION OF THE REDUCTION IN FUEL CONSUMPTION

5.1 OBJECTIVE

Objective of the simulation approach is the estimation of potential fuel savings through hybridization. The first example which is investigated is the drive train of a municipal multi-purpose vehicle, figure 18. The original vehicle is equipped with a 72 kW diesel engine, a variable hy-drostatic transmission inline with a mechanical transmission with fixed ratio. The hybridized vehicle has the additional electrical motor and the engine Power is varied.

5.2 TEST RUNS

During the test runs various values were measured like vehicle speed, engine speed, instantaneous fuel consumption and the pressure in the hydrostatic transmission. Figure 17 displays some recorded cycles.

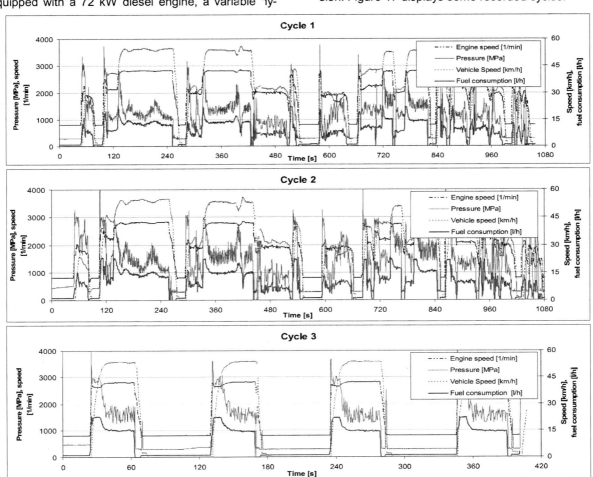

Figure 17   Measured values for 3 different cycles

5.3 SIMULATION MODELS

The simulation models for vehicles with and without hybrid drive were created in Matlab Simulink. Static efficiency maps ($b_e = f(M, n)$, $\eta = f(M, n)$) are used for the internal combustion engine and the electric motor. With b_e being the specific fuel consumption in g/kWh, M being the torque in Nm and n being the speed in min^{-1}. As a simplification the battery was implemented with constant efficiency. The transmission efficiency as well as losses in the tire and the tire-ground contact was modeled as a black box. The parameters of this black box were set within the validation process. The increase of weight due to battery and electric motor was not modeled.

Description	Unit	Value
Vehicle weight	[kg]	2.950 (cycle 1)
		4.600 (cycle 2)
		2.800 (cycle 3)
Engine power (conventional)	[kW]	72
Engine power (right sized)	[kW]	42
Motor power	[kW]	30
Top speed	[km/h]	53,5

Table 1   Vehicle data

Different drive train configurations and operation strategies (hereinafter called configurations) were investigated throughout the simulations. As a first step, the combustion engine and the electric motor were modeled via static efficiency maps ($M_{el} = f(M_{vkm}, n_{vkm})$)[1] and the engine was operated only along the line of minimal specific fuel consumption. In a consecutive step, the combustion engine was turned off during idling phases (start-stop). The third step modeled the possibility to recuperate breaking energy. Simulations were conducted with both, the conventional 72 kW combustion engine and with the smaller, right sized[2] 42 kW engine. The electrical motor remained unchanged in all configurations. Hydraulically driven auxiliaries were modeled as constant load. An overview of the fuel consumption with different configurations and cycles is presented in table 2. The driven cycles are shown in figure 17.

Figure 18   Test unit: Municipal multi-purpose vehicle

5.4 SIMULATION RESULTS

The battery's state of charge and the fuel consumption for each cycle and each configuration were simulated. Whereas the fuel consumption was calculated with the engine's efficiency map. The state of charge (SOC) resulted from the charged and discharged amounts of energy.

Configuration	Engine =72 kW	Engine =42 kW	Efficiency map	Start-Stop	Recuperation	Fuel economy [%]		
						Cylce1	Cycle 2	Cycle 3
Conventional (measured)	x					0,0	0,0	0,0
Hybrid 1	x		x			0,2	0,3	0,4
Hybrid 2	x		x	x		3,1	3,4	9,8
Hybrid 3	x		x	x	x	5,7	5,3	13,4
Hybrid 4		x	x			16,7	14,6	15,7
Hybrid 5		x	x	x		19,0	15,5	19,0
Hybrid 6		x	x	x	x	21,6	17,6	22,8

Table 2   Simulation results

---

[1] M_{el}: Torque of electric motor; M_{vkm}: Torque of combustion engine

[2] The term "rightsizing" is used as defined in [6]: Matching the installed engine power to the actual need.

Figures 19 and 20 show examples of the battery's SOC.

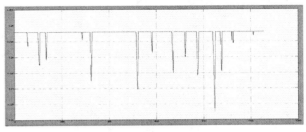

Figure 19   SOC for hybrid 1 cycle 1

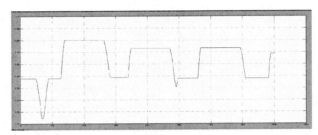

Figure 20   SOC for hybrid 3 cycle 3

6   CONCLUSION

Construction equipment, fork lifts, multi-purpose vehicles and other industrial equipment like lifting platforms are ideal candidates for hybrid drives. Their frequent load cycles promise high fuel savings applying a hybrid drive. The savings depend, of course, strongly on the frequency and amplitudes of the load cycles.

Under the various diesel-electric hybrid concepts, the mild hybrid drives emerges for above mentioned applications as the favorite due to its simple configuration. The compact design allows to integrate this hybrid drive into existing vehicles without major redesigns. Lithium-ion batteries or super caps are highly flexible energy storages. Compared to other hybrid configurations, e.g. hydraulic ones, the high price of the electric storage unit is actually the main disadvantage of the diesel-electric hybrid. The economy of using a diesel-electric drive must be shown to potential applicants in advance. Therefore, simulation algorithms are mandatory to prove the economy and to support the distribution of hybrids in suitable applications.

The present application, the municipal multi-purpose vehicle, provides an enormous savings potential of over 20% under the assumed conditions, as the summarized simulation results in table 2 show. It also reveals, however, that certain configurations lead only to insignificant savings. The main increase in efficiency derives from rightsizing the engine, while start-stop and recuperation have less influence. This gives an idea of how different effects enabled through hybridization have different results in fuel saving. This can lead to prioritizing development goals in ongoing and further projects. The size of the electric machine can be adjusted to electric consumers. If the energy consumers of the vehicle (e.g. the air blower for engine cooling, the cooling water pump, even the hydraulic pump) could be electrified, the savings would be much higher. These drives could then be controlled to work according to need.

It must be pointed out that the municipal multi-purpose vehicle is not an ideal candidate for hybridization. Appreciably higher savings are expected from wheel loaders and fork lifts.

The low eccentricity of the crank shaft movement of most diesel engines is suitable to support the rotor of the brushless motor without own bearings. The air gap between rotor and stator must be sufficiently large, approx. 1 mm to avoid contact. The length of the rotor is, however, limited to approx. 60 mm.

7   OUTLOOK

The next step in the investigations will be the electrification of energy consumers on test vehicles. Fuel savings of over 30 % are expected.

The mild hybrid system described in chapter 4 can not be operated on a solely electrical basis. A full hybrid system is obtained by a disengageable clutch between combustion engine and electric machine. When the clutch is disengaged, the combustion engine can be stopped to allow solely electrical operation. When the clutch is engaged, the system can either be operated with the combustion engine alone or in combination with the electric machine, as on a mild hybrid.

A full hybrid has some advantages over a mild hybrid:
- Solely electrical operation feasible (usage in halls, tunnels etc.).
- Greater potential for fuel savings.
- Combustion engine can be downsized further.

The disadvantages compared to a mild hybrid are:
- A much larger battery is required.
- Larger electric machine and inverter are required.
- An additional clutch is required.
- The electric motor must have separate bearings.
- The mild hybrid is more compact.
- Higher system costs.

In the short to medium-term, mild hybrid systems will be a beneficial addition to the drivetrain for industrial applications. Mild hybrids have already been proven to offer positive benefits. On a long-term basis, full hybrids will eventually become cost-effective when battery prices fall as a result of the series production of hybrid vehicles in the automotive industry.

It is conceivable that these developments will lead to a battery-operated vehicle with a small, optimised genset diesel-generator which is used to extend the range of the vehicle and/or to supply base load energy.

8   REFERENCES

[1]   R. Prandi, A True Heavy-Duty Hybrid – Deutz, HEINZMANN, Atlas-Weyhausen team up to develop prototype hybrid wheel loader, Diesel Progress Int. Edition, Sept.-Oct., 2007
[2]   T. van der Tuuk, Dieselhybridantriebe als alternative Antriebslösungen, Informationstagung des VDMA und der Univ. Karlsruhe "Hybridantriebe für mobile Arbeitsmaschinen", Karlsruhe 2007
[3]   LINDE Material Handling GmbH, picture, 2008
[4]   F. Böhler, R. Zahoransky (Interview), Industrial Engines Meet Hybrid Technology, Diesel Progress Int. Edition, March-April, 2008
[5]   Axeon Power Ltd., picture credit, 2008
[6]   Thiebes, P., M. Geimer and G. Jansen. Hybridantriebe abseits der Straße - Methodisches Vorgehen zur Bestimmung von Effizienzsteigerungspotentialen. In: 2. Fachtagung Hybridantriebe für mobile Arbeitsmaschinen. 18.02.2009 in Karlsruhe, S. 125-135.

9   CONTACT

Prof. Dr. Richard Zahoransky,
Heinzmann GmbH & Co., Am Haselbach 1,
D-79677 Schönau, Germany
Phone: 49-(0)7673 8208 -0
e-mail: r.zahoransky@heinzmann.de

Diesel Hybrids – The Logical Path towards Hybridisation

A. Srinivas, T. Kumar Prasad, T. Satish, Suhas Dhande, C. Nandagopalan

R&D Centre, Mahindra & Mahindra

ABSTRACT

Concerns on emissions and their effects on climate changes are currently the drivers behind automotive technology. Diesels have been faring better with CO_2 and fuel economy norms. However the state of art diesel engines emerging from various OEMs and trend as far as marginal improvements in the emissions and fuel economy potentials indicate that diesels are at the limits. In this scenario, diesel hybrid electric powertrains have shown promise to push the limits even further and offer the ideal platform to maximize benefits of hybridization. This paper shows that the diesel hybrids are the logical way to bridge the gap between conventional vehicles and electric vehicles.

INTRODUCTION

Based on the configuration of the powertrain the hybrids are mainly classified as series and parallel. In series electric hybrid, traction power is delivered by the electric motor. The engine via a generator produces the electric energy to drive the motor [1], [2]. The schematic of a series hybrid is shown in figure 1.

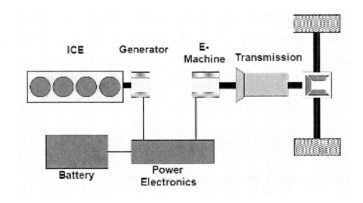

Figure 1. Series Hybrid Configuration

In case of a parallel electric hybrid, the engine is mechanically connected to the wheels, thus able to supply power to the wheels. Motor supplements the engine torque in parallel. It also acts as a generator to charge the battery [4]. The schematic of a parallel hybrid is shown in figure 2.

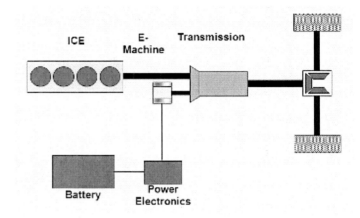

Figure 2. Parallel Hybrid Configuration

The Scorpio Hybrid, a demonstrator vehicle, is a parallel diesel-electric hybrid system i.e., the diesel engine and the electric motor are providing power/torque in parallel to the transmission. The vehicle is a mild hybrid with no pure electric drive mode. The aim of the program was to demonstrate that a hybrid vehicle can be "fun to drive" and be fuel efficient at the same time. The engine was not down-sized as compared to the conventional vehicle. This paper explains the overall architecture of the diesel hybrid electric vehicle developed, integration challenges tackled, safety requirements considered, simulation studies and fuel economy improvement realized in real world and various emission cycles. This paper discusses the influence of selection of various hybrid components on fuel economy and performance.

MAIN SECTION

BASIC CONFIGURATION - The Scorpio hybrid is a parallel diesel-electric hybrid. As mentioned earlier, the engine on the hybrid is same as the conventional vehicle. The main components are

- 4 cylinder, 2.2 litre, 87 kW common rail diesel engine

- 39 kW Permanent Magnet Synchronous Motor (PMSM)

- 6 speed automatic transmission

- 288V Ni-MH battery

The schematic of the hybrid powertrain is shown in figure 3. The major additional components due to hybridization are

<u>Hybrid Controller Unit (HCU)</u> - This acts as a master supervisory controller arbitrating the torque request between the engine and the motor. A rapid prototyping unit is used as the controller.

<u>Integrated Power Unit (IPU)</u> - This consists of a DC/DC converter, DC/AC inverter and also the Motor Control Unit (MCU).

<u>High Voltage Battery</u> - This supplies the energy to the motor when engine needs boost from the electrical source and also stores the energy from the generator (motor in the generating mode) during recuperation/generation. The battery details are discussed in the succeeding section.

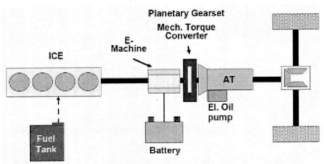

Figure 3: The Scorpio Hybrid Powertrain Schematic

The various hybrid functionalities implemented on the vehicle are

- Start-stop: This ensures the engine switches off during idling conditions. This feature is active when the gear lever is either in "D" (drive) or "N" (neutral) or "P" (park)

- Recuperation: This feature enables the battery to be charged during either braking or during coast down conditions. The motor runs in the generator mode to charge the battery utilizing the engine and transmission torque

- Efficient Engine Operation: One of the benefits of having a secondary power source (like a motor) is that it enables the engine to be run in a more efficient region i.e., at better B.S.F.C points.

- Boost: Motor provides additional boost (torque) during hard accelerations, thus lowering the load on the engine.

TECHNICAL CHALLENGES - Quite a few technical challenges were faced due to the fact that it is one of the first diesel electric hybrids. The main challenges were integrating the powertrain, ensuring smooth start-stop operation, packaging of the hybrid components, etc...

<u>Powertrain Integration</u>- This was one of the prime challenges encountered. One option was to use the existing torque converter along with additional elements (like automatic clutch) to ensure all the hybrid functions could be implemented. Since this is an automatic, the start-stop functionality would occur when the vehicle is in "D". While starting the vehicle in "D" it had to be ensured that the vehicle doesn't move. With a conventional torque converter, by itself, this would not be possible. Using an automatic clutch along with the torque converter would lead to loss in the launch feel of the automatic transmission.

Torque converter was replaced with a planetary gear set (PGS). The PGS emulates the functions of a torque converter (with lesser losses) and also facilitates start-stop function in "D". In case of the PGS, the three elements (engine, motor & transmission) sit on the three gears of the PGS and the three elements rotate at different speeds. This ability to rotate at different speeds is used to implement the start-stop functionality in "D". The motor is on the sun gear, the engine on the planetary carrier and transmission on the ring gear. During the start-stop operation in "D", the ring gear is locked to ensure that the vehicle doesn't move. The PGS is a double pinion system i.e., there are two pinion gears between the planet and the ring.

The speed and torque equations between the motor, engine and transmission are listed below.

$$\omega_{trm\_in} = I_1\omega_{mg} + I_2\omega_{ice}$$

$$T_{ice} = I_2 T_{trm\_in}$$

$$T_{mg} = I_1 T_{trm\_in}$$

ω_{trm_in} - rotational speed of the input transmission (ring)

I_1 - gear ratio between the sun and the ring gear

ω_{mg} - rotational speed of the motor/generator (on sun)

I_2 - gear ratio between the planet carrier and the ring gear

ω_{ice} - rotational speed of the engine (planet)

T_{ice} - engine torque

T_{trm_in} - input transmission torque

T_{mg} - motor torque

Transmission Oil Circulation during Stop-Start - During start-stop, in "D" mode, the ring gear (transmission) is locked. Oil does not circulate within the transmission due to the locking. When engine cranks in "D" and the vehicle starts to move, the main oil pump fails to provide the necessary oil pressure to the transmission. This might be detrimental to the transmission. There is a loss in driver comfort due to staggered oil circulation. An additional electric oil pump is used to ensure oil circulation during the stop in "D" and at slow speeds.

Lock-Up Clutch - When the engine and the motor are running in the PGS configuration, the speeds are different as they are on different gears. In this mode, the speed and torque of the motor have to be controlled to ensure the transmission is running at the required speed and producing the requisite torque. It would be simpler if only motor/engine torque has to be controlled during the vehicle operation. The three powertrain elements have to be at different speeds only during the start-stop operation in "D" mode. Once the vehicle starts moving, potentially, the three elements could run at the same speed thus ensuring simpler control. In order to achieve this, an additional lockup clutch is provided between the sun and the planetary gear i.e., between the motor and engine. This lockup clutch closes beyond a certain speed to ensure that the engine and motor are running at the same speed. Once two elements of the PGS are running at the same speed, the third element also runs at the same speed. Thus, only torque from the engine and motor is to be controlled. The schematic is shown in figure 4.

High Voltage Battery Packaging - The high voltage battery is packaged in the passenger compartment below the third row seats. The challenge lies in ensuring proper cooling of the battery in the Indian conditions. The battery is air cooled externally using the cabin air. There are ducts provided to ensure the hot air coming out of the battery is thrown out. The BMS controls the internal liquid (water) cooling for the battery.

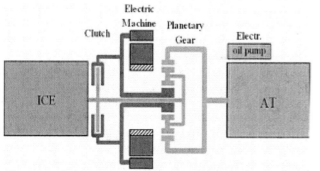

Figure 4: Schematic of the Interface between the Engine, Motor and Transmission.

SIMULATION STUDIES - In the current scenario where the development times for vehicles are shrinking, it is imperative to perform simulation studies to reduce both time and costs. The simulation studies also ensure better selection of the components. The vehicle level simulations were conducted using GT-Drive and AVL-Advisor software. The GT-drive software can be run in either dynamic ("forward") mode or in the kinematic ("backward") mode. In the dynamic mode the vehicle response is calculated based on the driver inputs. In case of kinematic mode, the engine state, motor state, battery state etc... are calculated based on the known vehicle state. AVL-Advisor runs only in the kinematic mode. The simulations were run to estimate the fuel economy of the vehicle. NEDC is considered as the reference drive cycle for simulation.

The various components that are modelled in simulation are

Engine: - 2.2 litre, common-rail diesel engine with a peak power of 87 kW and a peak torque of 270 Nm. The fuel map, emission maps, throttle map, peak torque-speed data, etc were tabulated.

E- Motor: - Electric motor considered has a peak power of 39kW and a peak torque of 320Nm. The efficiency data, peak torque-speed data, etc were tabulated.

Transmission: - The 6 speed automatic transmission gear ratios and the shift strategies were tabulated. The shift strategies were optimized for fuel economy and also for maximum performance.

Battery: - A 288V, 8.6 Ah battery is modelled. The chemistry considered was Ni-MH.

The degree of hybridization was incrementally increased and simulated. In the first stage, the conventional torque converter was replaced by the PGS. In reality, a double pinion PGS was used. However for the simulation purpose an equivalent single pinion PGS was assumed. In the next stage, start-stop was implemented. Finally, the recuperation functionality was added along with start-stop to simulate all the mild hybrid functions.

Scorpio VLX model with mHawk engine and a 6 speed automatic transmission was considered as a benchmark conventional vehicle.

The figure 5 shows the comparison of the engine speed between the conventional vehicle and the HEV. During the idling condition in NEDC, the hybrid engine is shut-off while in the conventional vehicle, the engine is running at idle speed. Also, at certain points the hybrid engine is running at a more optimum point (better BSFC) as compared to the conventional engine. The hybrid engine is running at lower speeds at certain points as it is possible to upshift to a higher gear as compared to the conventional engine (due to the additional motor torque).

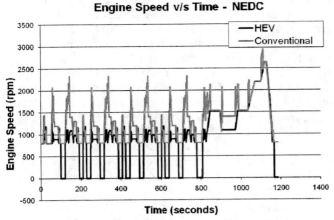
Figure 5. Engine Speed Variation in NEDC

The figure 6 shows the fueling rates of the conventional and the hybrid vehicle. As discussed for the above figure, the fueling rate is zero for the hybrid vehicle during idling in NEDC. Also, the fueling rate is less compared to the conventional vehicle at other operating points too as the engine is made to run in a better BSFC zone.

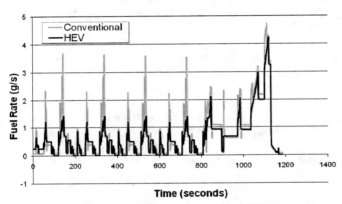
Figure 6. Fuel Rate Variation in NEDC.

The figure 7 shows the speed profiles of the hybrid and the conventional vehicle while simulating the acceleration test (0-100 kmph). The HEV was faster by about 16% as compared to the conventional vehicle.

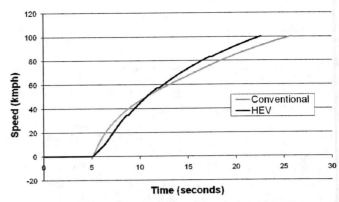
Figure 7. Acceleration Test Simulation Results

The fuel economy improvement in NEDC over the conventional vehicle was around 15%. This is shown in table 1.

FE improvement in HEV over conventional vehicle (NEDC)	15%

Table 1. FE Improvement in Simulation

Fuel economy tests were also simulated for other drive cycles like Indian Driving Cycle (IDC), UDDS, Indian Urban, Indian Highway, etc.... The results are shown in the table 2 below.

This shows that the chosen hybrid configuration, components and strategy yield a good fuel economy improvement in various drive cycles.

Drive Cycle	FE Improvement in %
NEDC	14.74
IDC	15.56
UDDS	20.59
Indian Urban	18.00
Indian Highway	22.73
Manhattan	29.94
FTP	21.57
J-1015	22.86

Table 2. FE Improvement in Various Cycles

HYBRID COMPONENTS – Hybridisation leads to additional components added to the vehicle. The various components of the hybrid system are discussed in detail in this section.

Motor: - The motor used in the vehicle is a PMSM machine with a peak torque of 320Nm and a peak power of 39kW. The motor is an internal rotor machine. The motor is water cooled by extending the cooling circuit of the engine.

Integrated Power Unit (IPU): - The IPU consists of the DC/DC converter, DC/AC inverter and the Motor Control Unit (MCU).

The DC/DC converter is an Insulated Gate Bipolar Transistor (IGBT) based system. The DC/DC converter converts the DC voltage from 288V to 14V. The 14V is used to power all the electric auxiliaries on the vehicle. The DC/AC inverter is used to convert the high voltage (288V) DC to AC for powering the motor. The MCU regulates the speed/ torque of the motor based on the request from HCU.

A separate liquid cooling circuit has been designed to cool the IPU. The IPU is packaged in the engine hood compartment. The water pump used for cooling the MCU is powered by a 12V battery.

Hybrid Control Unit (HCU): - HCU used in this vehicle is a rapid prototyping unit. HCU has an 8 MB main memory along with a 16 MB non-volatile flash memory. HCU is located under the driver's seat. The HCU interacts with the other controllers over CAN. A 12V battery powers the HCU, since this is the supervisory controller and has to be powered before any other system in the vehicle.

Auxiliary Electric Pump (for Transmission): - The auxiliary electric pump is used for transmission oil circulation at standstill and low speed conditions. The motor is a BLDC machine with its own controller. The pump is called as "Power On Demand" (POD) device.

Battery: - The high voltage battery is a 288V, 8.6 Ah Ni-MH battery. The operating temperature of the battery is -30 C to 65 C. Battery Management System (BMS) has a centralized battery control module whose main functions are

- Voltage sensing (at pack and module level)

- Current sensing

- Temperature sensing

- CAN communication with the other controllers

- Fault detection

- SOC monitoring

The battery enclosure provides electrical isolation, electrolyte resistance and allows for electromagnetic shielding. A liquid thermal system is incorporated in the battery. The battery is packaged in the passenger compartment below the third row seats.

CONTROL STRATEGY - HCU is the master controller which controls the flow of information to the other controllers in the system. The other controllers are

- Engine Control Unit (ECU): ECU controls all the functions of the engine. The main function is to meet the speed/ torque request coming from the HCU. ECU software has been modified from the conventional vehicle to accommodate stop-start functionality.

- Motor Control Unit (MCU): MCU regulates the speed/torque of the motor based on the request from HCU. Motor is used to crank the vehicle under normal conditions and also during the stop-start conditions.

- Transmission Control Unit (TCU): TCU performs the gear changes based on the gear request from HCU. It also locks the transmission during the start in "D".

- Battery Control Unit (BCU): BCU delivers the required power/ current to the motor or charges the battery based on the available power from the motor.

- Display Control Unit (DCU): The hybrid vehicle is also equipped with a central display depicting the mode of operation of the vehicle along with the energy flow. It also displays the fuel economy of the vehicle and also the improvement over the conventional vehicle. Also the torque of the motor/engine, SOC of the battery, and other important data are displayed. The display gets all the inputs from the HCU.

The main function of the HCU is to arbitrate the torque requests between the engine and the motor based on the pedal input and the battery SOC. Also, the gear change requests are controlled from the HCU. HCU controls the opening/closing of the lockup clutch between the engine and the motor. When the lockup clutch is open, the engine, motor and the transmission are rotating at different speeds, the mode is called as the 'PGS' mode. When the lockup clutch is closed, the engine, motor and the transmission are rotating at the same speed; it is called as the "Parallel" mode. HCU also controls the charging/discharging of the battery based on the SOC and torque request from the pedal. The stop-start requests are controlled from the HCU.

HCU also has a diagnostic subsystem which ensures the safe operation of all the various powertrain components and the battery. The POD speed request is sent from the HCU. HCU also controls the IPU water pump based on the temperature of the IPU. The various modes of operation of the hybrid vehicle are

Stop: - This is the vehicle stop condition. The conventional stop (and also "stop-start" stop) happens in "P" or "N". The "stop-start" stop happens in "D".

In case of the stopping in "P" or "N", the system would be in parallel mode i.e., lockup clutch closed before stopping. So while stopping, all the powertrain elements are slowed simultaneously.

In case of stop in "D", the system is in the PGS mode i.e., the lockup clutch is opened. The system also has to be ready for a start. So once the engine is stopped, the transmission is locked to facilitate a quick restart.

Start: - The conventional start (key start) and "start-stop" start happens in "P" or "N". This start is classified as called parallel / impulse start. The "stop-start" start happens in "D" and is called lever start.

During parallel start, the lockup clutch is closed before cranking the motor. Once the crank signal is given, the motor accelerates the engine to the idle speed. During cranking, the engine fuelling will not start until the engine is close to the idle speed. This results in fuel saving during cranking. This start happens when the engine is warm. Parallel start is depicted in figure 8.

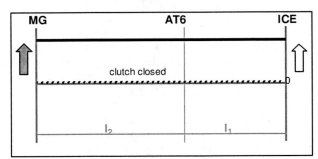

Figure 8: Parallel Start

During cold conditions, the oil flow is not sufficient in the transmission for the lock-up clutch to close. The motor runs in the positive direction and drives the main oil pump. Once sufficient pressure is built in the transmission, the lock up clutch will be closed and the engine cranked. This start is called as impulse start (cold start). Figure 9 shows the impulse start.

During lever start, the lockup clutch is open. The motor rotates in the negative direction, thus accelerating the engine in the positive direction towards the idle speed. The PGS equation is balanced to ensure the transmission speed is at zero (since the start occurs in "D"). During cranking, as described in parallel start, the fuelling will not occur until the engine speed is close to the idle speed. Lever start happens only in "D", during stop-start situation. This is shown in figure 10.

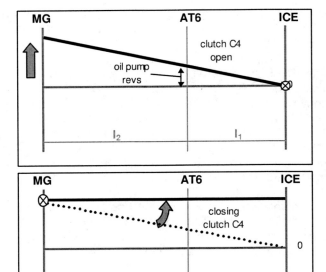

Figure 9: Impulse Start

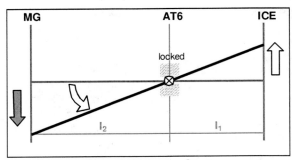

Figure 10: Lever Start

Idle: - Once the engine reaches the idle speed, this condition is reached. If the gear lever is in "P" or "N" during idling, the lockup clutch is closed. As the gear lever is in either "P" or "N", the transmission will not be propelled. In case of gear lever being in "D", the lockup clutch is open, so that as the vehicle starts to move, as the powertrain elements rotate.

PGS: - In this mode (in "D"), the engine, motor and the transmission are rotating at different speeds. This occurs at very low speeds. The speed/ torque of the engine and the motor have to be counterbalanced. This leads to some restriction on the total torque output.

In the PGS mode, when the vehicle starts to move the motor is rotating in the negative direction while engine and the transmission are rotating in the positive direction. As the speed of the vehicle increases, the motor moves from the negative region to positive region. The transition from PGS to Parallel mode takes place when the speed difference between the motor and the engine is below a threshold.

Parallel: - In this mode (in "D"), the engine, motor and the transmission are rotating at the same speed. There is no restriction on the torque supplied from the two aggregates (engine and motor). Based on the SOC, driveability, etc... the motor can either boost the engine torque or regenerate.

During coast-down, a light regeneration occurs. During braking, the regeneration torque varies based on the brake pressure and the vehicle speed. If the vehicle speed becomes low during braking, the mode will shift from parallel to PGS.

Hardware-in-Loop (HiL) testing was conducted to validate the control strategy. The HCU, ECU and the TCU were tested using the HiL setup. The network test was conducted to validate the CAN communication between the various controllers.

SIZING IMPACT ON FE AND PERFORMANCE- The size of the hybrid components along with the control strategy influences the FE and driveability of the vehicle. For example, if better acceleration performance is needed, a bigger motor and battery is required. Also, the control strategy should be designed to demand more torque from the motor. In this section we discuss the impact of the hybrid components size on fuel economy and performance.

Motor size is dependant on the amount of torque required to augment the engine torque. The motor mainly assists during cranking, hard acceleration and at points where engine torque is lowered to achieve a better BSFC.

The motor power was varied in simulation to study the impact of the motor size on fuel economy and performance. Figure 11 shows the effect of variation of motor power on the fuel economy. All the comparisons are against the base results with a 39 kW motor (offset to zero). As the motor size decreases, the fuel economy drops. This is due to the lesser torque available from the motor. The interesting aspect is that as the motor size increases, the fuel economy improves initially (until 45 kW) increases and then drops rapidly. This is due to the fact that as the motor power increases, it requires more energy from the battery. Since the battery was retained without changes, beyond a certain point, the battery is unable to meet the energy demand from the motor. This leads to a major drop in SOC, thus affecting the FE.

Figure 12 shows the effect of variation of motor power on the performance of the vehicle (0 – 100 kmph). As the motor power decreases, the performance decrease is not very significant. As engine contribution is the major factor during the acceleration test, the decrease in performance is marginal. But, as the motor power increases, the battery is not able to provide sufficient energy during hard acceleration. The increase in weight and drop in SOC are also factors for the drop in performance.

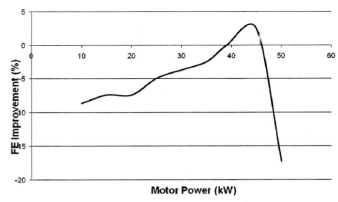

Figure 11. FE Variation due to Motor Power

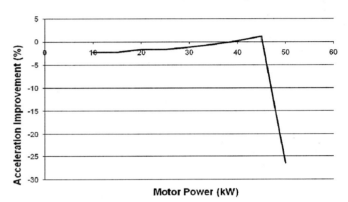

Figure 12. Performance Variation due to Motor Power

The increase in motor power will not always lead to a better fuel economy and performance as shown in the above figures. The battery size is very critical to get the best out of the motor. There is a limit on the size of the motor, for a given battery size, beyond which, the motor will load the battery (in turn loading the engine) and benefits of hybridisation is lost.

The high voltage battery is sized based on the energy requirements of the motor. This is dependant on the amount of boost and also the duration of boost required. Higher the boost required, bigger should be the battery capacity. The battery should also be able to absorb energy during generation and recuperation. This will be limited based on the maximum braking torque.

The battery voltage was varied to study the impact of it on the fuel economy and performance. Figure 13 shows the effect of battery voltage variation on the fuel economy. All the comparisons are against the results with a 288V battery (offset to zero). As the battery voltage decreases, the fuel economy drops substantially as lower power (voltage) battery is not able to provide sufficient energy to the 39 kW motor. This shows that there is a certain minimum voltage of the battery, below which battery requires considerable amount of charging leading to a decreased FE and thus benefits of hybridisation are lost.

As the battery voltage is increased, the fuel economy does not change very much. Since the motor power is kept constant (39 kW), increasing the battery power will not make much difference since the motor power constraints the amount of boost or regeneration.

Battery Voltage v/s FE Improvement

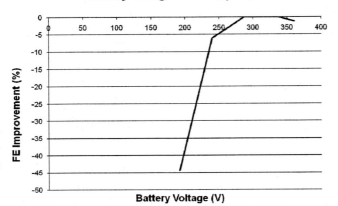

Figure 13. FE Variation due to Battery Voltage

Figure 14 shows the effect of variation of the battery voltage on the performance of the vehicle. As the battery voltage drops, the performance drops drastically, as the battery is not able to provide the necessary energy required for accelerating the vehicle. As mentioned earlier, increasing the battery voltage has very little impact on the performance as the amount of boost is restricted by the motor size.

Battery Voltage v/s Acceleration Improvement

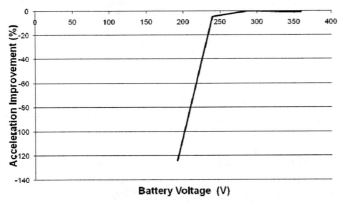

Figure 14. Performance Variation due to Battery Voltage

Fuel economy variation due to motor and battery power reveals that the battery and the motor specification go hand-in-hand. The battery should be able to provide enough energy to run the motor and at the same time, the motor size should be able to use the battery energy efficiently. If the motor is too big for the battery, the FE and performance decrease as the battery does not have sufficient energy. If the battery is too big for the motor,

the battery energy goes unused since the motor cannot utilize the energy.

SAFETY REQUIREMENTS - A hybrid electric vehicle along with all the advantages of better fuel economy and lower emissions also brings along additional safety requirements due to the high voltage system present in the vehicle. The safety algorithms have been incorporated in the software and also in the hardware.

Diagnostics - In the software, diagnostics have been implemented for the battery, motor, transmission and the engine. There are various levels of protection ranging from no boost from the motor, limp home with the high voltage system disabled to emergency shutdown in extreme cases. The emergency shutdown in most cases is related to the temperature being too low since the high voltage battery becomes very ineffective at low temperatures. When the battery temperature gets too high, there would be no electric boost or regeneration i.e., vehicle would operate like a conventional vehicle with the rotor of the electric motor freewheeling.

Control Strategy - When vehicle stops in "D", the vehicle starts automatically when the brake pedal is released. In case the driver tries to get out of the vehicle when stopped in "D", as a safety precaution the engine is started. This prevents the driver from getting out, which could be dangerous as the vehicle would move in "D". Also, when someone tries to open the bonnet while stopped in "D", the engine starts so as pre-empt the person from working under the hood when the high voltage system is active.

Hardware: - Three emergency buttons have been provided at three different locations to de-activate the high voltage system in case of emergencies. One is located near the high voltage battery, the second near the driver and the last one in the engine hood (in case when working under the hood). Also, the high-voltage cables, which run along the under body of the vehicle, have a special metallic conduit as a protection.

Torque Limiting Clutch - As mentioned earlier, the engine is retained from the conventional vehicle as well as the transmission. With motor as a secondary torque source, the torque at the input transmission shaft is higher as compared to the conventional vehicle. There could be instances where the composite torque of the engine and motor could be high enough to be detrimental to the transmission. In order to prevent such a scenario from occurring, a Torque Limiting Clutch (TLC) is installed before the transmission. This clutch will start to slip when the composite torque exceeds the upper limit. This prevents the high torque from being transferred to the transmission.

VEHICLE TRIALS- The hybrid vehicle was tested under real world conditions during summer in India. This was to evaluate the overall vehicle performance and also test the durability of the hybrid components in Indian conditions. The testing was conducted in a hilly terrain. The ambient temperature was around 38 C. In the graphs denoting temperature variation from the threshold temperature, the threshold temperature is offset to 0C and the temperature variation is denoted w.r.t. to the offset temperature.

The variation of the MCU temperature against the distance travelled is shown in figure 15. The MCU pump switches on at an upper threshold and then switches off at a lower threshold. These are indicated by the top and bottom lines in the figure. The MCU temperature reaches the upper threshold within 15 kms of travel. As shown in the figure the MCU temperature varies between thresholds all through the trip.

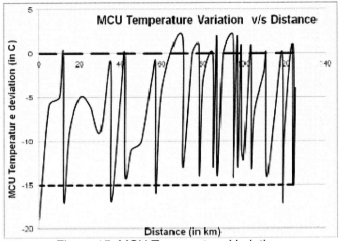
Figure 15. MCU Temperature Variation

The battery temperature variation against the distance travelled is shown in figure 16. The battery temperature steadily raises as the distance travelled increases. The battery temperature reaches about 8 C of the threshold temperature. This inidicates that the battery temperature would be with in the threshold during short distance driving, whereas during the long distance driving the battery temperature might increase to a level where safety precautions have to be taken.

The variation of the motor stator temperature against the distance travelled is shown in figure 17. Although the stator temperature increases rapidly initially (about 30 C in 10 km), the temperature steadies later. The cooling of the motor is coupled to the engine cooling system and thus the temperature is maintained in a steady range by the cooling sysetm. The motor is able to handle the higher temperatures better than the MCU and battery.

With temperatures reaching mid 40s in India, during summers, it is imperative that the components have to withstand such high temperatures. With most hybrids catering to North America and Europe, where the summers are mild compared to India, the suppliers have to design components taking into consideration, the Indian conditions.

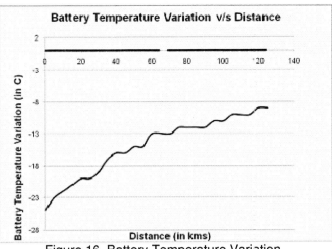
Figure 16. Battery Temperature Variation

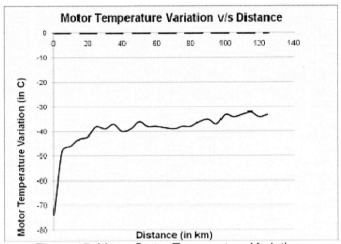
Figure 17. Motor Stator Temperature Variation

RESULTS - The hybrid vehicle was tested for emissions and fuel economy on a chassis dynamometer. The reference cycle considered was NEDC. The fuel economy is tabulated for the NEDC cycle, in table 3, after conducting repeated trials.

FE improvement in HEV over conventional vehicle (NEDC)	17%

Table 3: NEDC Fuel Economy Results

The actual fuel economy result deviates from the simulation result by about 2.5%. The vehicle emissions met the Euro III standards. The performance tests were conducted to evaluate the time taken to accelerate from 0-100 kmph. The results showed that the hybrid vehicle was about 22% faster than the conventional vehicle as shown in table 4. This contributes to the "fun-to-drive" factor.

Performance improvement in HEV over conventional vehicle	22%

Table 4: Performance Test Results

CONCLUSION

The diesel electric hybrid developed by Mahindra achieved a fuel economy benefit of 17% over the base Scorpio vehicle. This was achieved despite retaining the base engine and adding a secondary source i.e., increase in the input torque to transmission. This increase in the input torque also makes the vehicle fun to drive.

The fuel economy improvements in a diesel hybrid electric vehicle might seem lesser compared to the fuel economy improvements achieved on a gasoline hybrid electric vehicles. But, the gasoline engine is inherently less efficient as compared to the diesel engine. Gasoline engines fuel economy is about 30% lesser as compared to the diesel engines [4]. The gasoline with hybridisation achieves the same targets as a diesel engine, whereas hyrbdising the diesel engine takes it to a higher level. Thus, improving the fuel economy and thus reducing CO_2 emissions by 17% over an already efficient engine makes it a better candidate than a gasoline hybrid when looked at in absolute terms.

Diesel hybrids offer the ideal platform to bridge the gap between the conventional internal combustion engines (ICE) powered vehicles and fully electric vehicles. The topics of the overall cost of diesel system and comparison with gasoline hybrid would always exist. However if the quest is to bridge the gap with a sustainable solution which takes you away from horizon created by conventional ICEs, Diesel hybrid is the solution.

ACKNOWLEDGMENTS

Authors thank Nabal Pandey (Mahindra & Mahindra) for helping with the simulation runs.

REFERENCES

[1] A. Emadi, M. Ehsani, and J.M. Miller, Vehicular Electric Power Systems: Land, Sea, Air and Space Vehicles, New York: Marcel Dekker, Dec. 2003

[2] A. Emadi, K. Rajashekara, S.S. Williamson and S.M. Lukic, "Topological overview of hybrid electric and fuel cell vehicular power system architectures and configurations," IEEE Tran. On Vehicular Technology, vol. 54, no. 3, pp 763 – 770, May 2005

[3] M. Salman, N.J. Schouten, N.A. Kheir, "Control Strategies for Parallel Hybrid Vehicles", Proceedings of the American Control Conference, pp 524-528, June 2000

[4] H.J. Ecker, M. Schwaderlapp, "Downsizing of Diesel Engines: 3 Cylinder / 4Cylinder", SAE 2000 World Congress, March 2000

DEFINITIONS, ACRONYMS, ABBREVIATIONS

AT6: 6 speed automatic transmission

BLDC: Brushless DC

BMS: Battery Management System

BSFC: Brake Specific Fuel Consumption

CAN: Controller Area Network

C4: Lock-up Clutch

DCU: Display Control Unit

ECU: Engine Control Unit

FTP: Federal Test Procedure

HCU: Hybrid Control Unit

HEV: Hybrid Electric Vehicle

IGBT: Insulated Gate Bipolar Transistor

ICE: Internal Combustion Engine

IDC: Indian Drive Cycle

IPU: Integrated Power Unit

MCU: Motor Control Unit

MG: Motor/ Generator

NEDC: New European Driving Cycle

Ni-MH: Nickel Metal Hydride

OEM: Original Equipment Manufacturer

PGS: Planetary Gear Set

PMSM: Permanent Magnet Synchronous Motor

POD: Power-on-Demand

SOC: State of Charge

UDDS: Urban Dynamometer Driving Schedule

FUELS AND EMISSIONS

Demonstration of Power Improvements on a Diesel Engine Operating on Multiple Fuels	2010-01-1318 Published 04/12/2010

Evelyn Vance, Daniel Giordano, Jeffrey Rogers and Jeffrey Stewart
Sturman Industries

ABSTRACT

Continually increasing the maximum specific power output of engines used in military vehicles is vital to maintaining a battlefield advantage. An enabling technology for power optimization on existing engine architectures is advanced engine control based on real-time feedback control of the combustion. An engine equipped with intelligent controls and multi-fuel capable components has been used to demonstrate power improvements based on feedback control of the fueling by means of in-cylinder pressure measurements. In addition to optimized power output for the engine, the technology suite provides the capability to utilize both standard diesel fuel and alternatives such as jet fuel, biodiesel, or any mixture. A cylinder balancing algorithm adjusts the fueling to achieve even power distribution between cylinders for improved performance and durability, or to operate all cylinders at the cylinder pressure limit when maximum power is required. Tests performed at lug-line conditions show that closed-loop control reduced the cylinder-to-cylinder Indicated Mean Effective Pressure (IMEP) variation from ±8% to < ±2%. A multi-fuel algorithm adjusts and balances the cylinders to achieve optimal power output even when the fuel chemistry differs from standard diesel fuel, such as when operating on JP-8. This study quantitatively demonstrates that the normalized engine torque can be held within 2% for closed-loop operation between DF-2, JET-A and B100 fuels. The results of this work show improved power density of a diesel engine, with a wide range of fuels, through the combination of flexible, high pressure fuel injection technology and advanced engine controls with cylinder pressure feedback.

INTRODUCTION

The military has long sought improvements in specific power output for its vehicles. An engine may operate below its maximum achievable power output for various reasons, and its power output may degrade over time. Two potential sources of reduced power yield are uneven power contributions from each cylinder and a mismatch between the fuel chemistry and the factory engine calibration. Aside from development of new, more efficient engines, there are potential gains to be made on existing engines using innovative control methods that address these sources of power reduction and these solutions can be deployed rapidly. Using existing engine hardware, a controller has been developed and demonstrated that improves the power output of the engine using combustion feedback and algorithms for both cylinder balancing and multi-fuel use.

Due to part-to-part variability, engine asymmetries, and system degradation, each engine cylinder produces somewhat different power output. As a result, some of the cylinders will run below the cylinder pressure limit when the engine is operating at peak load. A closed loop controller using measured cylinder pressure traces enables all cylinders to operate at the cylinder pressure limit thereby increasing the maximum engine power. Furthermore, the closed loop controller can adjust to engine conditions that change over time, thus maintaining optimal power on each cylinder as the vehicle ages. The goal of the cylinder balancing algorithm is to use cylinder pressure feedback to continually adjust the fuel timing, pressure, and quantity for optimum power output.

The ability to compensate for chemical differences in the fuel has an additional benefit to military applications because deployed military diesel engines may encounter fuels of unknown properties. Closed-loop cylinder pressure control enables the use of locally available fuels, such as DF-2, JP-8, biodiesel, or any mixtures thereof. While certain engine hardware is robust to multiple fuel types, calibrations are not generally available for alternate fuels and thus a power decrease often results when non-standard fuel is used. Even if the properties of each fuel are known, a mixture of fuels may be present in the fuel tank. Onboard testing of the fuel is expensive and impractical, and may not even be feasible. Therefore, adjusting feed-forward calibration tables for the unknown fuel may not be possible. A closed loop (feed-back) control based on cylinder pressure measurements can be used to adjust injection timing, pressure, and quantity, to regain, at least partially, the peak power achieved with the standard fuel. This is the goal of the multi-fuel usage algorithm.

In this study, through the use of these new engine control strategies, power improvements are quantified through dynamometer testing showing that existing platforms with engine control modifications can provide power improvements for currently deployed military vehicles.

TECHNICAL OBJECTIVES AND TEST PLATFORM

The primary objective of this study is to demonstrate an increase in power output in a diesel engine using advanced engine control and flexible fuel injection technologies. The algorithms developed balance the power output from each cylinder, compensate for engine performance loss due to differing fuels, and achieve the target of 2% or less Indicated Mean Effective Pressure (IMEP) variation from the lowest to the highest cylinder IMEP. This section provides details about the flexible platform used to demonstrate the improved engine performance.

The test engine for this project was a modified International Model Year 2004 6.0L V-8 diesel engine. The stock engine was adapted to include both flexible high pressure fuel injection (HPFI) and hydraulic valve actuation (HVA). These systems, along with the electronics and algorithms, comprise the necessary enabling technologies for this project. The demonstration engine is shown installed in the dynamometer in Figure 1. A brief description of the component engine systems is given below.

Figure 1. Test Engine Installed in Sturman Industries' Dynamometer Laboratory

Sturman fuel injectors utilize a hydraulically intensified design in order to achieve high injection pressures and to enable use of alternative fuels. The Digital Injector incorporates multiple intensification ratios while maintaining multiple injection-per-cycle capability. The use of multiple intensifier stages enables injection rate-shaping to achieve more preferable heat release profiles if desired. For example, the digital injector is capable of a pilot injection followed by a rate-shaped main injection as shown in Figure 2. Many other injection profiles are also possible with this configuration.

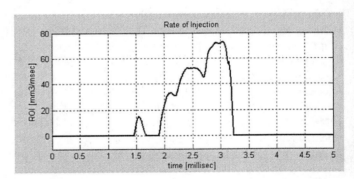

Figure 2. Pilot plus rate-shaped-main injection profile (multiple intensifications)

The engine platform used in this study is also equipped with HVA, and while this study did not exploit this added flexibility, future studies could incorporate the use of the HVA system for improving fuel economy[1,2,7,8], reducing emissions[1,4,5,9], improved power[1,6,8], Homogeneous Charge Compression Ignition (HCCI) operation[3,5,10,11] or further optimization of engine performance for multi-fuel operation.

The central controller for monitoring and controlling the various component (HVA, Injector, etc.) controllers is the Sturman Total Engine Controller (STEC). The electronics of the STEC are shown in Figure 3.

Figure 3. The STEC Electronics, Two Views

The architecture for the STEC consists of two Motorola MPC5554 microcontrollers coupled with an Altera Field Programmable Gate Array (FPGA), Controller Area Network (CAN), and Ethernet. The STEC controller communicates with each component controller via a CAN bus. In addition to the measurements received via CAN, the STEC also performs some direct measurements. The primary measurements the STEC uses for engine control are cylinder pressure measurements from each cylinder. The cylinder pressure measurements are used to calculate real-time data such as instantaneous torque. With this data, several algorithms can be used to tune the engine operation.

STEC ALGORITHMS

The basic cylinder balancing algorithm was developed in 2005 on a Sturman controller known as the CLCC, a "next-generation closed-loop combustion controller." That system controls the start of combustion (SOC) as well as individual cylinder power through the integrated heat release (IHR). The CLCC algorithm has the ability to adjust the timing and fuel quantity of the pilot, main, and post injector events, balancing all cylinders for level IHR values with an adaptive algorithm. The IHR output target for each cylinder is the average IHR output of all cylinders. Thus, some cylinders increase their IHR output while others decrease their output.

The CLCC cylinder balancing algorithm was ported to the STEC for use in this study. The immediate advantage of having the cylinder balancing algorithm reside on the STEC is that the STEC can directly monitor and control not only the injector drive module (IDM) but also the HVA system and other engine peripherals. Improvements made to the STEC-based cylinder balancing algorithm for this project include torque based calculations and balancing to the maximum cylinder output. The multi-fuel algorithm implemented in the STEC builds on the cylinder balancing algorithm described above. The objective of the algorithm is to produce optimum power in the engine even when a fuel of unknown chemistry is used.

Because production engines are calibrated for a particular fuel, the power output is expected to decrease when a fuel other than that assumed in the design is encountered. The cylinder balancing algorithm in maximum power mode can partially compensate for a loss in power by balancing all cylinders to the target of the highest performing cylinder. However, the multi-fuel algorithm further optimizes the performance of the engine. It uses as its target the known torque output of the engine with its expected fuel rather than the highest cylinder output resulting from standard fueling commands.

The mechanism for improving the performance of the engine to the original specification, in spite of downgraded fuel properties, is implemented via an additional controller table populated with empirical data from the engine running its standard fuel. It provides a map between the engine-out torque and the indicated torque from each cylinder for various engine speed and load conditions. The cylinder indicated torque is the target torque for the balancing algorithm.

STEC's Engine Metrics plug-in, illustrated in Figure 4, displays engine performance in terms of either torque or integrated heat release for each of the cylinders. A horizontal blue line depicting the target torque is displayed when closed-loop control is enabled. This provides a visual measure of the algorithm's effectiveness in real time.

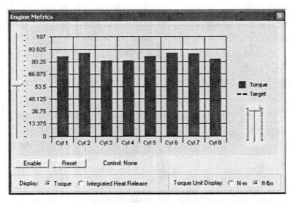

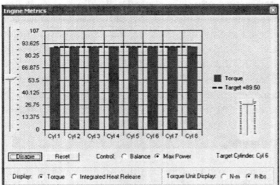

Figure 4. The Engine Metrics Plug-in

It should be noted that fuels with low energy content (low lower heating value) may not be capable of producing the same power as the original fuel. In this case, the algorithm should target the highest torque achievable rather than the baseline torque. The STEC currently implements a rudimentary solution to this issue. The table look-up for desired indicated torque is modified to keep the torque target within achievable bounds for the alternate fuel. In the long term, improvements should be made to the multi-fuel algorithm to more intelligently and dynamically develop maximum torque targets for alternative fuels.

PROGRAM TEST RESULTS

Three distinct data sets: baseline, cylinder balancing, and multi-fuel control, were collected during this project. The baseline results provide the basis for measurement, comparison, and evaluation of the engine performance for both the cylinder balancing results and multi-fuel control results. The dynamometer test cell used in this research is a 500 HP capable eddy current dynamometer combined with full data acquisition capability. The data acquisition consists of cylinder pressures, engine temperatures, fuel consumption and engine emissions measurements. Emission measurements were performed using Rosemount analyzers for NOx, THC (Total Hydrocarbons) and CO. Smoke measurements were performed by an AVL 415 smokemeter. Due to the altitude of the test facility, 8700 ft (2675 m), an air handling system

capable of pressurizing the engine intake to sea level conditions was used. The exhaust backpressure was also adjusted to simulate sea-level conditions for all the tests performed.

BASELINE ENGINE TESTING

Two sets of baseline tests were conducted to compare the open loop engine performance using both the stock Engine Control Unit (ECU) and the STEC to ensure that the STEC performed in the same manner as the stock ECU. Excellent agreement was found between the baseline tests, thus validating the STEC performance. The STEC baseline data was subsequently used for comparison in the evaluation of the engine under closed-loop control.

Figure 5 shows the baseline engine performance and evaluates the base engine with the HVA system and digital fuel injectors under open loop control. Since the system utilizes external pump carts to supply the HVA and injection systems, the raw torque values from the engine were adjusted by the pump cart power to provide a corrected horsepower (HP). This corrected HP was adjusted to match the overall torque curve on a stock engine that would utilize an on-engine hydraulic pump for the fuel injectors and a cam-based valve train. During the course of the engine testing, the HVA system was adjusted to provide valve open and close events that matched the camshaft timing. Figure 6 shows the intake and exhaust valve events compared with the camshaft baseline. A simulated piston position is also shown for reference to show the representative overlap at TDC.

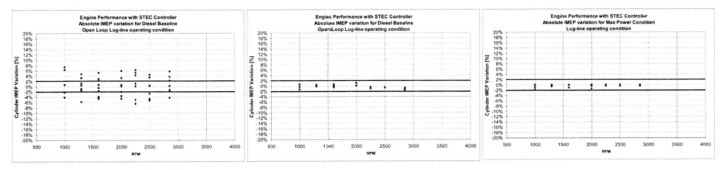

Figure 7. Comparison of the IMEP variation in the cylinders in open loop operation (left), Balance to Average operation (middle), and Balance to Maximum operation (right).

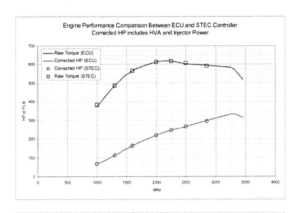

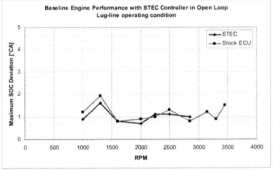

Figure 5. Power and SOC variation Comparison between the Stock ECU and STEC.

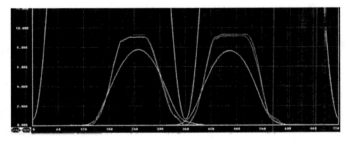

Figure 6. Valve Position vs. Crank Angle for HVA

In addition, the SOC variation was compared between the stock ECU and the STEC. Both curves match well indicating

that open loop performance repeats that of the stock controller.

CYLINDER BALANCING ALGORITHM TESTING (STANDARD FUEL)

The closed loop operation allows cylinder-to-cylinder balancing for both SOC timing and overall cylinder IMEP. Two separate closed-loop cylinder balancing algorithms underwent testing. The algorithms were:

• Balance all cylinder pressures to the average cylinder pressure (Balance to Average).

• Balance all cylinder pressures to the maximum cylinder pressure (Balance to Maximum).

Both algorithms were tested. Figure 7 shows the IMEP cylinder variations for the baseline open-loop test and each closed loop test condition. The closed-loop routine does an excellent job of balancing the individual cylinders at each operating condition, and as a result, all of the measured cylinder pressures fall within the desired boundary of ± 2% of the average IMEP for all the test points.

<figure 7 here>

The power output of the engine in the two closed loop cases is shown in Figure 8. As expected, the closed-loop "Balanced to Average" algorithm maintains consistent power output of the engine for each test point as compared to the open loop operation. The "Balance to Maximum" algorithm on the other hand is able to increase the engine torque by up to 7% and shows overall increases at all engine speeds. In general, the magnitude of the torque increase will depend on the severity of the cylinder to cylinder variation within the engine.

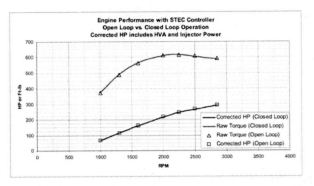

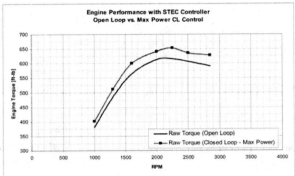

Figure 8. Comparison of open loop and closed loop torque and corrected horsepower (HP) output for the Balance to Average algorithm (top) and the torque output for the Balance to Max algorithm (bottom), using DF-2 Fuel.

MULTI-FUEL ALGORITHM TESTING

It is the goal of the closed loop controller's multi-fuel algorithm to optimize the engine performance for any fuel of unknown chemistry or quality. The demonstration of this algorithm was accomplished using both JET-A and three different biodiesel blends as the alternative fuels. JET-A fuel is essentially JP-8 fuel without three additives (an icing inhibitor, a lubricant/corrosion inhibitor and an anti static agent). Biodiesel was also added to the testing matrix for more rigorous testing of the multi-fuel algorithm.

The specifications for the fuels in the test are listed in Table 1.

Table 1. Fuel Properties

Fuel Name		DF2	JET-A	BioDiesel B100
	unit			SME
Cetane Index	-	44	46	50
Lower Heating Value	MJ/kg	43	43	38
Density	kg/m3	850	805	880
Kinematic Viscosity @ 40° C	mm^2/s	2.5	1.4	4.1
MW	(kg/km)	190	150	290
C/H ratio	-	0.583	0.524	0.576
Ignition Delay, Full load @ 2000 RPM	msec	0.44	0.41	0.36

RED for Biodiesel Cetane Index indicates it is the measured Cetane number, not the calculated Cetane Index.

Multi-fuel Testing Results

Each of the alternative fuels (JET-A, B20, B50, and B100) was tested for both open loop performance and closed loop performance, and compared to the baseline diesel performance. As with the diesel test sequence, the inlet fuel temperature for all data points was maintained at 70°F ± 5°F. The resulting engine torque curves were compared to the diesel baseline for open loop and closed-loop mode as shown in Figure 9. As can be seen, the engine torque was degraded as the percentage of biodiesel increased. This is consistent with the expectations since the lower heating value of the biodiesel is ~ 11% lower than diesel. An overall estimation of the normalized engine torque for these test conditions is shown in the bar charts on the lower half of Figure 9. Based on the measured results, a normalized torque loss of 9.5% was noted between operation on diesel and B100 in open loop mode.

<figure 9 here>

Each blend of biodiesel and Jet-A was tested in closed loop mode using the Multi-fuel algorithm with an average diesel torque target. The closed loop operation minimized the overall differences when compared to the diesel baseline. The torque loss using B100 in closed loop mode using the multi-fuel algorithm was 1.6% as compared to the 9.5% seen operating the engine in open loop mode.

The measured cylinder to cylinder IMEP variation for the operation on Jet-A, B20, B50 and B100 as compared to the average cylinder IMEP for diesel operation in closed loop mode using the Multi-fuel algorithm was calculated. The maximum variation for each of the operating conditions and each fuel was within the ±2% specification.

The SOC deviation from the diesel baseline is shown in Figure 10 for closed loop mode using the multi-fuel algorithm with average diesel torque target for operation with the different biodiesel blends. The results show minimal differences between the fuels. This is an indication that the controller is able to adjust the start of combustion location correctly independent of fuel composition.

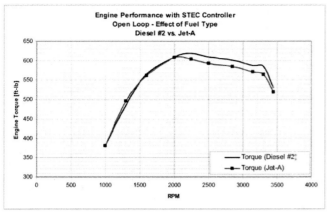
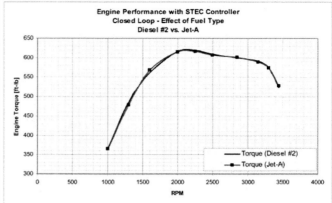
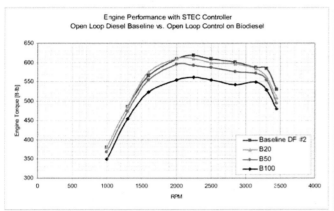
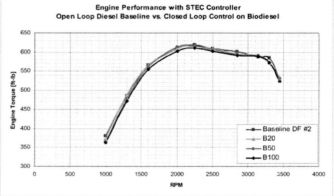
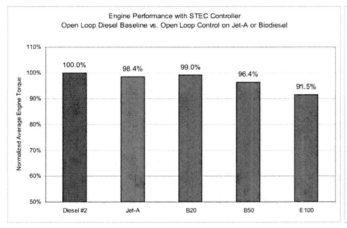
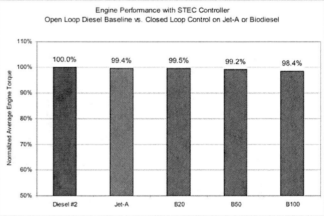

Figure 9. Comparison of open-loop (left) and closed loop (right) torque output for various test fuels using the multi-fuel algorithm with target values based on DF2 performance.

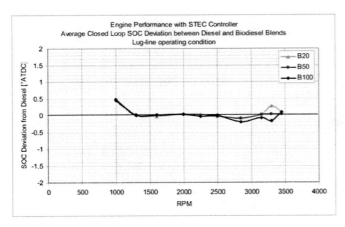

Figure 10. Average SOC Deviation from Diesel Operation in Closed Loop Mode using the multi-fuel algorithm with the average diesel torque as the target

In order to properly evaluate fuel consumption differences between the different fuels tested, the brake thermal efficiency was calculated based on the Lower Heating Value (LHV) for each of the fuels as shown in Table 1. The LHV for B20 and B50 was calculated based on the ratio of DF #2 to biodiesel. The closed-loop operation shows minor differences with respect to the normalized brake thermal efficiency as shown in Figure 11. This shows that the increase in the engine net power with closed-loop control occurred without a degradation in thermal efficiency.

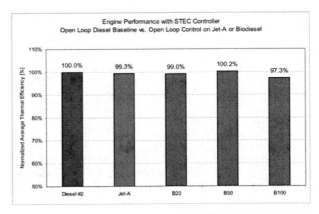

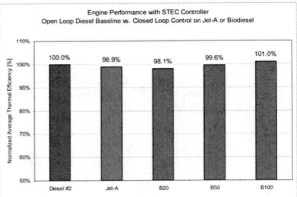

Figure 11. Comparison of open-loop (top) and closed loop (bottom) brake thermal efficiency for various test fuels using the multi-fuel algorithm with target values based on DF2 performance.

SUMMARY/CONCLUSIONS

This study effectively demonstrates that it is possible to improve the power density of a diesel engine using the combination of flexible, high pressure fuel injection and advanced engine controls with cylinder pressure feedback.

For the first time, the STEC controller and its algorithms were integrated with Sturman Industries' HPFI and HVA systems to demonstrate fully flexible fuel and air management with combustion feedback for the purpose of power improvement on a diesel engine. Through engineering development and engine test of these algorithms, data has been collected that documents the potential of the engine platform to enable power improvement.

An engine control algorithm was developed that provides real-time cylinder-to-cylinder balancing in two modes of operation. The first mode controls the individual cylinder IMEP to an average across the engine. This allows the individual cylinder balancing to maintain a constant engine power set-point and was demonstrated to meet the program specification of less than ± 2% deviation for all cylinders.

The second mode of operation is called the max power mode and results in an overall power increase in the engine. In this mode of operation, the cylinder with the highest IMEP is selected as the target, and the power from the other cylinders is increased to meet this target. Operation in this mode was also demonstrated to be within the ± 2% deviation for all cylinders using the highest cylinder as the target. This mode of operation also resulted in engine torque increases of up to 7%.

A multi-fuel algorithm was developed to compensate for engine power loss associated with different fuel composition. This algorithm uses a map of the performance of the engine running standard diesel fuel as the target set-point. The algorithm was tested using JET-A, B20, B50 and B100 and compared to the diesel baseline. The tests show that closed-loop operation on any of the fuels resulted in engine power within 1.6% of the diesel baseline with cylinder to cylinder IMEP variation within the ± 2% deviation.

REFERENCES

1. "Diesel Combustion by means of Variable Valve Timing in a Single Cylinder Engine", Osada Hideaki, Aoyagi Yuzo, Yamaguchi Takuya and Kobayashi Masayuki et.al., The Seventh International Conference on Modeling and Diagnostics for Advanced Engine Systems (COMODIA 2008), July 28-31, 2008, Sapporo, Japan

2. "A Study of a Variable Compression Ratio and Variable Valve Actuation System for Improving Fuel Economy", Tsuchida H, Tomita M, Aoyama S., Takemura S, Hiraya K., Tanaka D., Shigemoto S, Sugiyama T., Hiyoshi R., The Seventh International Conference on Modeling and Diagnostics for Advanced Engine Systems (COMODIA 2008), July 28-31, 2008, Sapporo, Japan

3. Strandh, P., Bengtsson, J., Johansson, R., Tunestal, P. et al., "Variable Valve Actuation for Timing Control of a Homogeneous Charge Compression Ignition Engine," SAE Technical Paper 2005-01-0147, 2005.

4. "Advanced combustion and engine integration of a Hydraulic Valve Actuation system (camless)", Lombard Benoît, Le Forestier Romain, SIA Conference on Variable Valve Actuation - November 30, 2006 - IFP Rueil.

5. Bression, G., Soleri, D., Savy, S., Dehoux, S. et al., "A Study of Methods to Lower HC and CO Emissions in Diesel HCCI," SAE Int. J. Fuels Lubr. 1(1):37-49, 2008.

6. "Transient torque rise of a modern light duty diesel engine with variable valve actuation", Tim Lancefield, 1998.

7. Lancefield, T., "The Influence of Variable Valve Actuation on the Part Load Fuel Economy of a Modern Light-Duty Diesel Engine," SAE Technical Paper 2003-01-0028, 2003.

8. Lancefield, T., Methley, I., Räse, U., and Kuhn, T., "The Application of Variable Event Valve Timing to a Modern Diesel Engine," SAE Technical Paper 2000-01-1229, 2000.

9. "Improving Pollutant Emissions in Diesel Engines for Heavy-Duty Transportation Using Retarded Intake Valve Closing Strategies", Benajes J., Molina S., Novella R. and Riesco M., International Journal of Automotive Technology, Vol. 9, No. 3, pp. 257-265, 2008.

10. "Cycle-To-Cycle Control of HCCI Engines", Shaver G., Gerdes J., IMECE 2003-41966, Proceedings of IMECE '03 2003 ASME International Mechanical Engineering Congress and Exposition, November 15-21, 2003, Washington, D.C. USA.

11. "Requirements for the Valve Train and Technologies for Enabling HCCI Over the Entire Operating Range", Milovanovic N., Turner J., 11th Diesel Engine Emissions Reduction Conference, August 21-25, 2005, Chicago, Illinois.

12. "On The Ignition and Combustion Variances of JP-8 AND Diesel Fuel in Military Diesel Engines", Schihl P. and Hoogterp L., September 22, 2008, Presented at the 26th Army Science Conference, Orlando, Florida, December 2008.

CONTACT INFORMATION

Dr. Evelyn Vance
Sturman Industries
1 Innovation Way
Woodland Park, CO 80863
evance@sturmanindustries.com

ACKNOWLEDGMENTS

This material is based upon work supported by the U.S. Army Tank-automotive & Armaments Command under Contract No. W56HZV-07-C-0528.

DEFINITIONS/ABBREVIATIONS

CAN
> Controller Area Network

CLCC
> Closed-Loop Combustion Controller

ECU
> Engine Control Unit

HCCI
> Homogeneous Charge Compression Ignition

HPFI
> High Pressure Fuel Injection

HVA

Hydraulic Valve Actuation

IDM

Injector Drive Module

IHR

Integrated Heat Release

IMEP

Indicated Mean Effective Pressure

LHV

Lower Heating Value

STEC

Sturman Total Engine Controller

Automotive Materials Engineering Challenges and Solutions for the Use of Ethanol and Methanol Blended Fuels	2010-01-0729 Published 04/12/2010

Pui Kei (P.K.) Yuen, John Beckett and William Villaire
General Motors

ABSTRACT

Economic market forces and increasing environmental awareness of gasoline have led to interest in developing alternatives to gasoline, and extending the current global supply for transportation fuels. One viable strategy is the use of alternative alcohol fuels for combustion engines, with ethanol and methanol in various concentration ranges proposed and in-use. Utilizing and citing data from this review, a comprehensive overview of the materials selection and engineering challenges facing metals, plastics and elastomers are presented. The engineering approach and solution-sets discussed will focus on production feasibility and implementation. The effects from the fuel chemistry and quality of fuel ethanol produced on the related vehicle components are discussed. Through the discussions and analyses outlined in this review, the industry can benefit from a better understanding of the materials challenges and solution sets, subsequently improving the products and the customer experience, hence furthering the acceptance and adaptation of usage of alternative fuels and biofuels such as ethanol and methanol.

INTRODUCTION

In the current automotive landscape, the needs for sustainable energy sources and alternative fuels have led to wide-scale development and implementation of flexible, alternative fuel spark-ignited engines that can operate on blends of alcohol and gasoline fuels. The legislative push in the US mandating widespread implementation of biofuels is a prime example and driver for this significant development. [1, 2]

A flexible-fuel vehicle (FFV) is an alternative fuel vehicle with an internal combustion engine designed to run on more than one fuel, usually gasoline blended with either ethanol or methanol fuel, where both fuels are stored in the same common tank. Flex-fuel engines are capable of burning any proportion of the resulting blend in the combustion chamber as the engine control system, along with fuel injection and spark timing, are adjusted automatically according to the actual fuel blend ratio detected by the vehicle's electronic sensors. The most common commercially available FFV in the market is the ethanol flexible-fuel vehicle, with more than 17 million automobiles and light duty trucks on the roads around the world by early 2009. FFV sales are concentrated in four primary markets, Brazil (8.2 million), the United States (almost 8 million), Canada (600,000), and Europe, led by Sweden (311,122). As ethanol FFVs became commercially available during the late 1990s, the common use of the term "flexible-fuel vehicle" became synonymous with ethanol FFVs. In the United States flex-fuel vehicles are also known as "E85 vehicles."[3]

For the purpose of this paper, we will use the term "alternative alcohol fuels" to cover all fuel blends of ethanol and/or methanol with gasoline. Also within the context of this paper, the term FFV will be used to cover what is defined as an "E85 vehicle," as it is the most popular type of FFV available.

Because of the presence of oxygen in their chemical structure, alcohol fuels have some unique fuel properties that can cause increased corrosion, wear, or degradation in conventional gasoline fuel handling systems[4] and engine components. Many FFV component materials upgrades are needed, hence the role of diligent materials engineering in product design and engineering is extremely critical to the function, performance, and successful implementation of FFV.

As the price of conventional gasoline increases and customers strive to reduce the cost of vehicle operation, misfueling E85 fuels into non-FFV's is known to occur. Vehicle operators may have a misconception that as long as they can put the fuel in the vehicle fuel tank, and the engine continues to feel normal and a Check Engine Light is not immediately illuminated that there are no concerns for their vehicles. From an OEM's perspective most non-FFV's hardware materials are not properly designed, nor validated to operate on fuels that are not recommended fuels in their owners' manual. As the materials engineering concerns are reviewed in this paper, hopefully one can understand that operating a vehicle on a fuel not defined in the owner's manual is very detrimental to the vehicle hardware. This behavior, although illegal in terms of emissions control and also non-compliant in terms of vehicle warranty coverage, has been widespread and often-discussed that even www.wikipedia.org has a page defining and referencing it.[5] Additionally with the growth and availability of ethanol fuels there has an emergence in the marketplace of aftermarket conversion kits that claim to provide their customers the ability to convert a conventional gasoline vehicle to be operate on high ethanol blend fuels such as E85.[6] A quick search on the Internet with the key words "E85 conversion kit" will result in numerous "hits" of conversion kit supplier.[7] Without full consideration of the materials compatibility impact and the engineering that is required, it is highly probable that a non-comprehensive conversion would compromise the expected durability, robustness, and capability of the vehicle, its engine, and its fuel handling components. For example, few or any of these conversion kits would comprehend the necessary upgraded fuel pump, engine valves, engine valve seats, fuel and oil sealing materials that are usually implemented by the automotive OEM's when engineering their vehicles to be a capable FFV. Many different stakeholders, including the US EPA, US DOE, GM and Ford, have stated the need to be cognizant of these potential issues. Dr. Coleman Jones, GM Biofuels Implementation Manager, summarizes this well, "Conversions that do not include these elements will result in vehicles with significantly reduced life that will probably not be [emissions] compliant for their useful life."[8]

Because of the potential issues that can arise out of misfueling, casual blending, and substandard or non-comprehensive conversions, it is very important for the automotive industry to clearly communicate the risks involved with improper ethanol fuels use. With these potential risks, when significant vehicle issues arise, consumers will be dissatisfied with the product and, by extension, with ethanol and other biofuels in general.

It is of note that substantial and similar challenges also exist for compression-ignited engines using alternative fuels such as biodiesel;[9] however this paper will only focus on the alternative alcohol fuels used in current spark-ignited engines.

MATERIALS COMPATIBILITY DEFINITIONS - OEM PERSPECTIVE

Compatibility is a contextual term, meaning it is defined to the intent or purpose of the writer. To an automotive engineer, it can and should reflect the ability to properly function in the intended environment for the life of the application to a design specification. It is important to communicate the scope or limits and provide something of value.

What makes the compatibility elusive is that it is subjective to the author's knowledge, experience and supporting data. There is no one definition that fits all situations.

When evaluating any decision-making compatibility definition, ask these questions:

1. What level or scope is defined; compound, part, component or system?

2. Does the definition relate to a desired target or actual use?

3. What properties (if any) were selected and why?

4. Is there a demonstrated link between the chosen properties to function?

5. What are the limitations or assumptions in the definition or scope?

6. Do the assumptions match the data?

7. To whom is the definition directed and with what intent? Does it fit your needs?

8. Is it serving to advance an agenda beyond selection?

The final compatibility definition should be specific enough to permit collection of data in the decision making or limit setting process. Not every criteria or choice is explained and some explanations remain restrictive for competitive advantage.

There is no perfect definition, some are better than others, but all have limitations. Relying on industry lab tests alone to predict compatibility of a component in a vehicle can be too much of a stretch. The best any lab bench test can do is give sufficient information to make an informed decision.

ALCOHOL FUELS OF INTEREST

In order to maintain the right focus and control the scope of the materials engineering required, it is very important to identify the key alternative alcohol fuel compositions of concern. For spark-ignited engines, the current dominant alternative fuel is ethanol, or ethanol-plus-gasoline blended fuels. And recently, there has been a re-emergence of interest in methanol usage in the global landscape. Although the use of methanol as a widely used alternative automotive fuel has been unsuccessful in the US and other major auto markets due to its toxicity concerns, China has been very active in promoting and implementing the use of methanol.[10][11] As China's vehicle sales and fuel usage are growing at a tremendous rate and is now one of the top three new car sales markets, its policies and vehicle needs are definitely becoming increasingly relevant globally.

E100 fuel, which is hydrous alcohol blend, nominally 95% ethanol and 5% water and is in warmer climates, such as Brazil.[12][13][14] In the US, Ed100 anhydrous denatured fuel ethanol is specified per ASTM D4806[15][16] and is most commonly used to blend with gasoline as E10 and E85.

E85 fuel, which is specified in the North America and parts of Europe like Sweden as an ethanol-gasoline blend varying in range from 60-85% ethanol depending on the seasonal and volatility requirements, is the predominant fuel of concern for most FFV materials development. For the US, the E85 fuel is specified per ASTM D5798.[17]

Gasoline (E0) and gasoline-ethanol blends containing up to 10% ethanol (E10) are considered conventional gasoline fuel. In the US, conventional gasoline for spark-ignited engines is specified per ASTM D4814.[18]

Mid-level ethanol blends of E11 to E60, which cannot be currently considered conventional gasoline fuels and are not legal for usage for non-FFV's in the US, have been a significant topic for discussion and debate as there has been a recent wave of activities concerning a Growth Energy's US EPA E15 fuel waiver request.[19] Significant similar materials compatibility and vehicle durability questions exist for the long term use of E20.[20][21][22].

M100 fuel in China, as specified via GB 23510-2009,[23][24] has been used in limited applications as it requires the use of dedicated M100 capable vehicles that are bi-fuel applications to enable cold engine start-up using conventional gasoline. This M100 standard has a national implementation date in China of Nov 1, 2009.

M85 fuel in China, as specified via GB 23799-2009,[25] has been used in limited applications as it requires the use of M85 capable flex fuel vehicles that are not widely available yet. This M85 standard has a national implementation date in China of Dec 1, 2009.

Efficient materials engineering for high volume global auto market production, needs to comprehend the value of selecting the right test and validation fuels.[26][27] As outlined by this recently published paper by Geng, P. et al., "Testing & Validation Fuels in Vehicle Development," "Because of the wide array of commercial or market fuels in use today, all components, engines, and vehicles must be vigorously tested to confirm emissions performance, drivability and material compatibility. To support the requirements of the modern global vehicle engineer, various testing and validation fuels are used to represent quality that ranges from good to poor for fuels that are currently in the market or that will be introduced into the market in the future."

The importance of selecting the correct fuels to develop robust designs that ensure vehicle performance, compliance and reliability in the global market place must be comprehended by the automotive engineers.

SYSTEMS AND COMPONENTS AFFECTED

There are many materials in vehicle systems and components that can be affected by use of alcohol fuel blends. Further detailed discussions surrounding some of these critical components are reviewed in subsequent sections.

For the vehicle fuel handling and storage systems (FH&S), the filler neck materials of FFV's may require change due to potential compatibility issues. Materials compatibility would have to be examined for seals, metal and plastic fuel tanks. Generally, the fuel pump must also be changed in terms of material compatibility, but also to accommodate the increased fuel flow rate necessary because of the lower heating value of the alcohol blended fuels. The fuel level sender is often changed and upgraded because of the increased electrical conductivity of alcohol fuel blends and its constant operational exposure to the fuel. Fuel line materials are often scrutinized for materials compatibility as both polymer lines and metal lines can be affected by the presence of alcohol in the fuel.

As is apparent with any vehicle FH&S components, detrimental corrosion, or degradation effects and subsequent leaks can result from improper materials selection. It is of note that any significant leaks in these subsystems can cause the On-Board-Diagnostics (OBD) system to set codes and trigger the check engine light to illuminate. Furthermore, a deterioration in the leak performance will either render it more difficult for the integrated engineering teams to reach critical evaporative emissions targets, or cause the vehicle to fail mandated evaporative emissions requirements.

Older conventional vehicles that do not have OBD systems may react differently due to their reduced ability to self-troubleshoot potential leaks and failure modes. Thus, the use of alternative alcohol fuels can be of higher risks to these older vehicles as their systems' capability and hardware materials robustness may be suspect.

For the engine components, the materials used in fuel rails, fuel injectors and their seals are directly affected by the use of alternative alcohol fuels. But within the overall engine structure, other hardware such as the engine valve, valve seat, piston rings, cylinder head intake port, intake manifold and the various oil seals can also be significantly affected. Some of these components may experience increased wear due to the different fuels and combustion characteristics. For example, although it may not be outright apparent that oil seal materials would be impacted by fuel, it is important to note that significant fuel dilution, up to 25% volatiles with alternative alcohol fuels usage, can occur to the engine oil in the crankcase during certain cold climate operating conditions when the engine oil does not have to chance to fully heat up to evaporate the fuel mixture.[28] To engineer for such extreme scenarios, it is prudent for materials engineer to consider using a certain level of diluted fuel mixture with the base engine oil to test the oil seal materials.

FUEL CHEMISTRY AND QUALITY

The effects from the fuel chemistry and quality of fuel are very important considerations when comprehending the engineering and robustness of the fuel handling and delivery systems. In many cases, it is very critical that testing of the materials must be done with the correct fuel sets.

Sulfates can affect the fuel injector operation and deposits control, as sulfates are indicated in filter plugging deposits and fuel injector deposits. Ethanol acceptability for motor fuel use includes sulfate content limits.[29][30][31] Chlorides can cause corrosion in the fuel delivery system and the base engine hardware Likewise copper is a concern as it can cause peroxides to develop in the fuel and can also affect the fuel delivery system and the elastomeric seals. Generally, these issues can directly affect the overall vehicle performance, fuel economy, and emissions.[32][33]

Philosophically, there must be a continued focus to develop improved fuel specifications for methanol and ethanol fuels, as exemplified by the often cited study, "The Development of Improved Fuel Specifications for Methanol (M85) and Ethanol (Ed85),"[34] by Brinkman, N. et al.

CORROSION

It has been well-documented that for ferrous metals, when exposed to ethanolic or alcoholic solutions, they exhibit more corrosion than from conventional gasoline.[35][36][37][38][39] It has been documented that general corrosion is often caused by ionic impurities, mainly chloride ions and acetic acid, with chloride ions being the most aggressive single contaminant in ethanol that contains water. And that a combination of three contaminants (chloride ion, acetic acid, and ethyl acetate) produces a synergistic effect in hydrous ethanol, and corrosion is many times greater than that by any single contaminant.[40] Formation of acetic and formic acids in the alternative alcohol fuel pathways are also common and can further increase the aggressive nature of the fuels. Wet corrosion can be caused by azeotropic water with the alcohol, which oxidizes most metals.[41]

Ethanol or methanol, by themselves and also in combination with absorbed water, can render the alternative alcohol fuel to be much more electrically conductive than that of gasoline, which can be considered to be a non-electrically conductive fluid by itself. And that any increases in contaminants, mainly ions that can further increase the alternative alcohol fuel's electrical conductivity, will render the fuel to be much more aggressive in corrosion behaviors that are dependent on conductive fluid behavior, such as electrochemical and galvanic corrosion.

Dry corrosion can be attributed to the ethanol molecule and its polarity.[42] De la Harpe reviewed reports of dry corrosion of metals by ethanol and found that magnesium, lead and aluminum were susceptible to chemical attack by dry ethanol.[43][44] For dry corrosion behavior, low water content is more critical, as higher water content increase the necessary activation energy to start this chemical process. As an example, similar problems can also occur in brake systems, as it would involve the reaction of the dry glycol based brake fluid with very clean aluminum surfaces.[45]

From studies conducted by K. Otto et al., they had found that, "as with the fuel contaminant effects in the fuel handling systems, contamination by organic chlorides enhances the corrosive nature of alcohol combustion products. Burning alcohols can also produce acids such as formic acid, which can cause corrosion at temperatures below the dew point of the exhaust gas of the engine. Peroxides, which can be produced during combustion, can also accelerate the corrosion process of burning alcohols. The specific corrosion product is hydrated ferrous chloride, which promotes rust formation further by its hydroscopic nature."[46][47]

STRESS CORROSION CRACKING

When metals are exposed to alternative alcohol fuels, potential failures and issues with stress corrosion cracking are very real and well-documented. "a recent survey of fuel grade ethanol storage and handling service experience in the USA conducted by the American Petroleum Institute (API), documented 24 failures attributed to stress corrosion cracking (SCC) in steel tanks, piping and related equipment in fuel grade ethanol service over the period 1990 through 2005."[43] [49] [50]

With the complexities and compounding factors that can influence the onset of stress corrosion cracking in the metals that are in contact with fuel, an interesting case-study, as exemplified by a recent recall that affected 34,343 vehicles,[51] offers some interesting insights as to how sensitive automotive fuel handling components are to the effects of stress corrosion cracking. As excerpted from the publicly available NHTSA Defect Notice,[52]

"Description of Problem: In a section of the fuel lines located in the engine compartment of the subject vehicles, due to the orientation of the fuel line's seam weld during a bending operation in fabrication, there is a possibility that stress at the seam weld location may become high, causing a small crack to develop as a result of the action of small amounts of corrosive ingredients in the fuel. If the vehicle is continuously operated in this condition, the crack on the fuel lines may expand as a result of fuel pressure fluctuation and, in the worst case, fuel could leak from the fuel lines.

Chronology of Principal Events

March 2005 - September 2005: A major global vehicle manufacturer received some field information from the Japan market which indicated the smell of fuel in a vehicle. They immediately began an investigation and recovered a fuel line in which a leak was found by a dealer, in order to identify the cause of the problem. As a result, it was found that there was a crack with small amounts of corrosive ingredients at the seam weld location. However, these ingredients were not found in the fuel, leading Them to assume that this stress corrosion crack might have developed due to the use of an inferior fuel or a fuel additive. In order to prevent any possibility of the stress corrosion cracking, the fuel line was changed to a seamless pipe in September 2005.

October 2005 - November 2007: This major global vehicle manufacturer continued its investigation, mainly on the fabrication process, since the vehicles which had experienced this problem were manufactured within a certain period. The result of the investigation on the manufacturing process showed that there was no process change or a possibility for the adhesion of corrosive ingredients during the manufacturing process. However, as a result, further investigation continued on the source of the corrosive ingredients, and it was found that these ingredients could be contained within the regular fuel. They conducted a repetition test on the stress corrosion crack. As a result, it was discovered that there is a possibility that stress at the seam weld location may become high, causing a small crack to develop as a result of the action of small amounts of corrosive ingredients in the fuel. In addition, it was found that

there is not a possibility for this kind of crack to occur on the new seamless pipe.

End of November 2007: As a result of the investigation above, The vehicle manufacturer decided to conduct a voluntary safety recall of all vehicles with the subject fuel lines within the affected range. This safety campaign will also be conducted in Japan, Canada, Australia, Europe, Taiwan and other countries."

Risks associated with stress corrosion cracking with fuel exposure must be comprehended by automotive materials engineers.

WEAR

With the use of alternative alcohol fuels, increased wear on metallic components can be a critical concern and various studies have been conducted to understand the implications and areas of sensitivity.

In the study performed by Browning, L.H. et al., it was found that, "corrosion and wear of fuel systems and engine parts is increased with methanol fuels over that for gasoline. This is due to the greatly increased diffusion rates through the oil film of the corrosive condensed combustion gases and water when this condensate contains methanol."[53] And that combustion intermediates, such as formic acid, can promote increased wear in the piston ring and cylinder liner assembly, as the oil film formation process can be disturbed by these constituents.[54]

Engine wear with alcohol-gasoline fuels has been reported in laboratory tests, with the wearing tendency of the engine with M15 found to be greater than with normal gasoline.[55] Corrosive attack on piston rings and cam followers with both E15 and M15 fuels gives an indication of the significant effects of the alternative alcohol fuels, even at relatively low blending levels. It is also of interest that in a study conducted by Yahagi, Y. et al., "the wear rate of steel in gasohol is closely related to the corrosion rate of the steel in gasohol, reaching a maximum at about 20 vol.% ethanol content."

VEHICLE FUEL HANDLING COMPONENTS - METALLICS

When specifying hardware to be used in the FH&S systems for alternative alcohol fuel applications, care must be taken both in the designs, and in the engineered materials selected to be used in those designs. This design process is especially critical for metallic components exposed to the conditions experienced in a vehicle system, i.e. varying fuel compositions (as mentioned previously), thermal, and dynamic (pressure) effects.

Many different metals have been used in FH&S designs for the fuel lines, fuel tanks, fuel filters, fuel pumps, level sensors, pressure control valves, regulators, liquid contacting sensors (composition and pressure) and fuel modules. These designs, and the types of metallic components used, have changed over time in response to lessons learned, and to various changing regulations. Many of these older designs for conventional gasoline usage have not been validated for use in alternative alcohol fuels, and therefore questions remain as to the compatibility of the metals, and to any protective coatings exposed to higher levels of alternative alcohol fuels. An optimal design for alternative alcohol fuels applications must take into account all the components in the FH&S system to ensure suitability to higher levels of ethanol, and methanol.

When these designs are designated for an alternative alcohol fuels application, it is important to conduct a complete review of the metals, and any protective coatings used. Particular attention should be given to the compatibility of the materials over the operating ranges expected for the vehicle system. This review should include a comprehensive Design-level Failure Mode Effects Analysis (DFMEA) and a validation plan. It should also be understood there are few short-term tests that can cover all the environmental conditions that may be experienced over the course of a vehicle life. Therefore end-of-life testing is necessary to assess the suitability of the materials used in the vehicle FH&S system.

FUEL TANK ASSEMBLY

When specifying a fuel tank assembly for an OEM vehicle to be exposed to alternative alcohol fuels, it is preferred to treat the program as entirely new. Initially, a review of the proposed designs should be conducted. This design review must include the tank itself, but also the attached fill spouts, filler pipe(s), internal baffles, or all other metallic component used in, or attached to the tank. Special care should be given to the attachment methods as frequently, these interfaces are overlooked. A comprehensive list of the metals and associated protective coatings should be collected and reviewed in detail with respective in-house and supplier materials engineering departments. Careful consideration of lessons learned should also be included. After the list is generated, a complete DFMEA and detailed validation plan should be conducted with all the affected parties.

In the case of the fuel tank and its components, there are many resources at hand to assist the engineer. Some of these include industry groups such as The Strategic Alliance for Steel Fuel Tanks[56], government publications such as those found in the DOE Handbook for Handling, Storing, and Dispensing E85[57], and reference publications such the ASM Metals Handbook, Ninth Edition: Volume 13 - Corrosion, or R.W. Revie, Uhlig's Corrosion Handbook 2nd Edition, as well as other engineering publications related to materials compatibility and corrosion resistance. If there is a lack of information on the material, it will be necessary to begin an immediate probe and/or validation testing to determine compatibility. This validation plan can be accomplished through long-term soaks in various fuels listed previously, and by using specific time periods and temperatures that are related and would best represent the potential operating conditions of the vehicle in customer's hands.

Most metals used in or attached to the fuel tank are protected carbon steels. In general, there are many available coatings for carbon steels used for the fuel tank, fill spouts and filler pipes. Apart from the compatibility of the coating in an alternative alcohol fuel system, there should also be a focus on the ability of the coating to withstand the assembly (attachment) processes, and the adequate coating of internal surfaces found in tubes such as the fill spout, or filler pipe. It is important to consider that these local areas can be attacked by the various corrosive components of the alternative alcohol fuels being used.

FUEL TUBE BUNDLE

Another area of the FH&S system that must be scrutinized is the metallic portion of the tubes used to transport fuel to and/or from the engine. The main focus here should be the feed line, and if present, the return line, as well as the other components that may be attached to the assembled fuel tube bundle. These components may include fuel filters, liquid pressure sensors, fuel composition sensors and metallic quick connects. In general, many of the metals used in alternative alcohol fuels applications use an austenitic stainless steel construction, but there may be some carbon steel components with a corrosion resistant coating. Similar to the process mentioned above for the metallic fuel tank, a detailed review of the tube bundle and its components must be completed.

When the material of the component is primarily stainless steel, particular attention must be paid to the interfaces of the attached components. If other metals are present, even in small quantities, they may be affected by alternative alcohol fuels. When coated carbon steels are used in alternative alcohol fuel applications, additional scrutiny must be paid to the interior of the tubes.[58] Many standard coating processes used in the past concentrated only on the ability to withstand corrosion from the outside or underbody of the vehicle, and did not adequately coat the inner wall surfaces of metal tubes. If this step is not accomplished properly, corrosion originating from the inside may become an issue. Standard coatings for metallic tube to be considered may be zinc-alloy based or nickel based.

FUEL PUMP MODULE

Typically the component in the FH&S system with the greatest variety of metals is the fuel pump module assembly

(FPM). The FPM metallic materials are generally found in the pump itself, as contained within structural members, springs, float arms, fuel filters, electrical terminals, pressure regulators, and even flanges, including any metallic filled polymer based flanges. The metallic materials used in the FPM may include coated steel, copper (exposed or coated), aluminum, stainless steel, silver, platinum, palladium, and gold. Even though the FPM is part of the fuel tank assembly, it should be treated separately when specifying and reviewing its usage in an alternative alcohol fuels program. Again, a detailed design review must be completed which includes a comprehensive materials review, a complete DFMEA and a validation plan for the alternative alcohol fuel being designed for. This step is especially important when considering an existing FPM for use in an alternative alcohol fuel application. Older FPM designs may contain metals or coatings that are incompatible in this type of fuel exposure, which can greatly shorten the life, or affect the performance of the system.

A common mistake made in the selection of a FPM for use in alternative alcohol fuels, is to overlook the internal components of the fuel pump. These components include the electrical contacts, pumping section, and armature or other motor features. In general, when specifying an alternative alcohol fuels FPM, great care must be taken to protect all exposed copper, eliminate any zinc plating, and to properly protect any aluminum components because those types of metals and coatings are very susceptible to attack from the alternative alcohol fuels. The electrical devices used to suppress the electronic noise generated by the fuel pump's electric motor can be a major source of concern in an alternative alcohol fuels application. Care must be taken to ensure these components, their electrical leads, and solder/ weld joints are protected, either by coating the exposed metals after sub-assembly with a compatible coating, or by encapsulating the entire sub-assembly. The armature wire and its attachment points must also be coated with a compatible material. The commutator face used in a brush type motor is often formed from copper, especially in older model fuel pumps. It must be changed to carbon to prevent degradation in an alternative alcohol fuel application. Another option may be to consider a brushless motor design.

Another area in the fuel pump that must be scrutinized is the pumping section. Many suppliers use aluminum for the plates in the turbine style pumping sections, and this must be properly protected to avoid a potential pump failure. In general aluminum that is properly anodized and sealed will have no issues in alternative alcohol fuels used in automotive applications. Aluminum pumping sections that are not protected will prematurely wear, and degrade from the effects of alcohol exposure. Older pump designs such as vane style, or gear style, used unprotected steels for the pumping sections. These older designs would need to be modified to insure durability with higher levels (>10%) of alcohol based fuels.[59]

The fuel filter and the pressure regulator in a FPM, if made from metallic materials, can pose a particular challenge. Moisture accumulated in the fuel can collect inside the filter when the FPM is inactive. This coupled with a component that has a tendency to create a static field, and the presence of an electrically conductive fuel, which can allow more ionic exchange, can accelerate any detrimental corrosion effects from prolonged immersion in an alternative alcohol fuel. The pressure regulator must maintain a good seal when the fuel pump is off, and also maintain good pressure control when the pump is on, so internal metallic corrosion must be avoided for key functional purposes. In general, metallic fuel filters included in the FPM, and the fuel pressure regulator should be fabricated from austenitic stainless steels to avoid any adverse corrosive effects. Many older designs used 400-series stainless steel for the fuel filter, and zinc coated carbon steel regulator construction. These older designs have been changed for various reasons over the years, and have minimal data associated with exposure to alternative alcohol fuels.

Other areas of the FPM that may use metals include the structural components (guide rods, and springs), float arms, wires, electrical terminals, and level sensor contacts. These components must also be reviewed to insure compatibility with alternative alcohol fuels. Structural components are generally made from austenitic stainless steel to maintain integrity. Float arms with their stresses from the various bends necessary for a design also require an austenitic stainless steel to avoid failure. Wires and terminals must be properly protected for use in alternative alcohol fuels. This usually means using compatible wire insulation, and at a minimum, tin plating the terminals. The level sensor terminals also must be considered. Even though there is a light electrical load on this circuit, the residual stresses created when forming these components can make them susceptible for corrosion attack in high alcohol fuels.[60]

Other non-obvious areas within the FPM that may create challenges are the high voltage (11-14V) electrical interfaces. Inside most fuel pumps in use today for automotive systems, there are carbon brushes that contact the carbon, or copper commutator face of the armature. With the higher conductivity commonly found in alternative alcohol fuels, proper care must be taken with the timing of the motor to avoid excessive wear at the carbon interfaces. An improperly timed motor can generate arcing at the commutator segment interface which will wear away the brush very quickly. Also, the pump electrical connector must be designed to separate the terminals to avoid electrical cross-talk which can lead to premature failure of the terminals from corrosion.

VEHICLE FUEL HANDLING COMPONENTS - POLYMERS

When polymers are used for the FH&S components, like for the metallic components a detailed design review of the materials used, a complete DFMEA, and a comprehensive validation plan will need to be completed. Most polymers used today in this environment are chosen to withstand the ever changing fuel environment they are exposed to including alternative alcohol fuels. However, in the past, there were polymers used that may not withstand the rigors of alternative alcohol fuels.

Some main properties to look for in a polymer for FH&S components are: compatibility with the wide range of fuels, and fuel by-products that can be present in the system; ability to withstand a wide range of temperatures and pressures observed in the system; and ability to stand up to the physical vibrations and movements that come with functioning in an automotive system.

When specifying the polymers to be used in a FH&S system, it is important to keep in mind the entire temperature, pressure, and fuel operating ranges that will occur. Many compatibility charts only relate capability at one temperature. For example, the Cole-Parmer Chemical Resistance Database[61] is only helpful at room temperature; however, many of the chemical resistance sources available on IDES "The Plastics Web®"[62] contain information over a wider temperature range, and often times at various fuel concentrations. Another source of information will come from the polymer suppliers themselves. The manufacturers may have already tested the physical characteristics of their product in various fuels to generic specifications (such as ASTM D543, D790, or D2990)[63], and will already know general compatibility of their materials. These are a good start, but there is always a need to validate the design to the system being specified. This can be accomplished with long-term soaks, powered and un-powered to add the pressure component, in various fuels listed previously, using specific time periods and temperatures that are related to, and would best represent the potential operating conditions of the vehicle in customers' hands.

FUEL TANK ASSEMBLY

Many of today's fuel tanks and its attached components are made from plastic. In general, these are blow molded multilayer units made primarily from high-density polyethylene with a barrier layer of ethylene vinyl alcohol. These polymers stand up well in alternative alcohol fuels, and can meet most if not all of the OEM specifications for fuel tanks. However, some older designs used monolayer tanks that were either treated with sulfonation, or fluorination. These designs have not been validated for use with alternative alcohol fuels, and questions exist as to the ability to withstand permeation when used with alternative alcohol fuels.[64]

FUEL TUBE BUNDLE

As mentioned in the previous section on metallics, fuel bundles can be made from various metals, but more commonly, there is a combination of metals, and polymers used for these tubes, and the components attached. Many current fuel systems use conductive multilayer plastic tubes to both provide an evaporative emission barrier, and also as a means to reduce the buildup of static charges associated with pumping fuel through a relatively small conduit.[65]

When designing the fuel tube bundle, attention also must be paid to the components attached such as the various quick connects, filters, or sensors. Many of these components contain polymers that come into direct contact with the fuel, and are subject to the rigors (temperatures and pressures) of a fuel system. A design review of these components must be included with the overall tube bundle to ensure compatibility, and proper charge management especially in alternative alcohol fuels.

FUEL PUMP MODULE

Much like the section relating to metallics, there can be many types of plastics used in an FPM. These can be found in structural components such as: flanges, reservoirs, or pump retainers; or can be functional components such as: filters, fuel tubes, floats, or internal pump components. Again, a detailed design review, complete with a DFMEA, and a validation plan must be performed when considering any un-tested plastic for an alternative alcohol fuel system.

The use of high alcohol fuels can pose a challenge to any existing FPM design. Many plastics currently used today can withstand the typical fuel system environment, but care must be taken when investigating a carryover design for usage towards a new alternative alcohol fuel design. Some polymers used in the past such as polybutylene terephalate, or polyurethane foams, used in floats, may pose issued when alcohol levels increase. These materials can degrade in a high alcohol content fuel causing a premature performance failure. In general, most polyoxymethylene (acetal) co-polymers, and phenolic foams can be used in the FPM, or as the level sensor float without issue in an alternative alcohol fuel system.

Fuel pumps contained in the FPM and the coatings used for wire insulation, also must be considered as these components may make use of polymers with an unknown performance in alternative alcohol fuels. Again, validation testing must be conducted to ensure performance and durability of these components over the expected usage life of the fuel system in alternative alcohol fuels.

Fuel filters contained in the FPM are also becoming more common, and are often made from polymers, or a combination of different polymers. Also, the FPM generally contains plastic tubes used to direct the fuel from the pump to the flange. Like the fuel tube bundles, these fuel flowing components must also take into account the charge management aspect of flowing fue. It is very important to construct these components from conductive versions of compatible polymers, and provide an adequate ground path to avoid degradation of the walls of these parts by migrating static charges.

In general, when designing a FPM to be used in alternative alcohol fuels, it is important to conduct the proper validation in the various fuels, and operating conditions that will be seen in a FH&S system. Many times, this has already been accomplished by the FPM supplier, but when investigating an existing design that has not been validated for alternative alcohol fuels, a detailed validation plan must be drawn up, and tested to completion. At the end of this validation plan, attention must be paid to the appearance, and dimensional qualities of the exposed plastics, and these must be listed in the final report. Again, it should be understood that short-term tests are inadequate at simulating all of the environmental conditions that the polymers may see in an automotive application. End-of-life testing will be necessary to establish the suitability of the plastic in the application being specified.

ENGINE COMPONENTS AND HARDWARE - METALLICS

FUEL DELIVERY SYSTEM: FUEL RAILS AND INJECTORS

Due to the sensitive and critical nature of the engine subcomponents that handle fuels, such as fuel rails and fuel injectors, General Motors has been selecting stainless steels as the metallic materials of choice for fuel-contacting components. The proper function of these parts is essential with regards to engine performance and operation.

Testing and studies have shown that there are significant effects on port fuel injector durability from E-85 fuel corrosivity.[66]

With the use of stainless steels, basic concerns with red-rust corrosion can be alleviated and the desired general robustness can be achieved. It is important to note that while stainless steels are generally very corrosion resistant, they are not corrosion-proof against alternative alcohol fuels.[67][68] Additionally, chloride contaminations, which can occur much more readily in alternative alcohol fuels due to its apparent solubility, can also cause significant corrosion issues,

especially when combined with stress, raising concerns with regards to stress corrosion cracking.

With the increased implementation of spark-ignited direct-injection (SIDI) technology, the stresses exerted on the fuel delivery components are even higher than conventional port fuel injection (PFI) systems, as the fuel pressures used are significantly increased. A typical PFI system would experience pressures in the 400kPa ranges, whereas a SIDI system would see fuel pressures up to 15,000 kPa or even higher with future systems.

Studies have shown that aluminum can corrode significantly in varying different conditions when exposed to different alternative alcohol fuels.[69][70] The risks associated with the improper usage or materials design and preparation of aluminum components in direct contact with alternative alcohol fuels, even with as low alcohol content as E10 fuels, can be showcased in this example, whereas actual field problems were exhibited and had led to a voluntary recall by the vehicle manufacturer. This affected their aluminum fuel delivery pipes used in 214,570 vehicles produced.[71][72]

In the public documentation for this recall, it was noted that "For the aluminum fuel rails (delivery pipes) located in the engine compartment of the subject vehicles, when E10 ethanol fuel is used, chloric ions are generated from the attached fuel injectors. For the combination of these ions and certain ethanol fuels, which have a lower moisture content among the fuels sold in the U.S., there is a possibility that the inside of the fuel rail(s) may corrode. Corrosion products may then plug the injector(s) and cause a rough engine idle and/or the illumination of an engine warning lamp. If the vehicle is continuously operated in this condition, the corrosion in the fuel rail(s) may expand and, in the worst case, a pinhole may develop on the fuel rail which could result in fuel leakage."

"The vehicle manufacturer received some field information from the U.S. and Thailand markets which indicated fuel leakage from the fuel rail. ... As a result, it was found that some portions of the inside surface of the fuel rail corroded, and that ethanol fuel was being used for those vehicles. In addition, They identified that the corrosion is a dry corrosion which can occur due to a reaction between alcohol and aluminum. Although They had not identified the actual cause and mechanism of the corrosion, in order to eliminate any possibility of the occurrence of the corrosion, the coating on the fuel rail was changed in September 2007.

The vehicle manufacturer focused on the moisture content of the ethanol fuel, which greatly contribute to the dry corrosion, and investigated the ethanol fuels sold in the U.S. and Thailand. In addition, They investigated other factors which could contribute to the corrosion. As a result, it was found that there are ethanol fuels which have very low moisture content among the fuels sold in Thailand, which could lead to the corrosion without any other factors. Therefore, The manufacturer initiated a campaign in Thailand to replace the fuel rails with ones with the new coating in June 2008.

June 2008 -December 2008: ... since the vehicle manufacturer had not found any ethanol fuel sold in the U.S. with low moisture content which could lead to the corrosion by itself, they continued the investigation on the ethanol fuel sold in the U.S., but also investigated other factors which could contribute to the corrosion. As a result, although they did not find any ethanol fuel with a low enough moisture content which by itself could lead to the corrosion, it was confirmed that, due to chloric ions generated from the fuel injectors, there is a possibility that the corrosion may occur inside of the fuel rail with ethanol fuels which have a lower moisture content among the fuels sold in the U.S."

With alternative alcohol fuels that can exhibit aggressive dry corrosion phenomenon and also aggressive wet corrosion from water and contaminants, it is imperative that automotive materials engineers are cognizant of the risks and optimize the materials robustness required. Due care must be taken in designing and selecting the metals to be used in direct fuel contact.

ENGINE VALVE AND VALVE SEAT

When compared to gasoline, the use of alternative alcohol fuels would often result in significant increases in valve seat insert and valve face wear. This phenomenon is widely recognized and the engine manufacturer is tasked to identify and incorporate appropriate valvetrain materials and design features that can meet the ever increasing life expectations of the end-user.[73] In most cases, higher levels of alcohol, such as E85, E100, M85, and M100 would promote the most engine valve to valve seat wear compared to gasoline usage in production intent engine designs.

Considering the function of a valve seat and how it would be manufactured and operated in a FFV engine, it would be apparent that the materials engineering would be a fine balancing act. When selecting and designing these materials for production FFV engines, the materials engineer must comprehend several highly conflicting parameters, such as cost, machinability, general abrasive wear resistance, corrosive or tribo-oxidative wear resistance of the sintered powder metal materials.

Depending on the powder metal technology and wear-resistant philosophy used, there are several technical paths in which valve seat makers have been following, and they often involve the use of increased tool steel content in the powder metal mixes, and/or the utilization of selective powder metal

components that can come from the inclusion of elements such as cobalt and molybdenum in the powder metal mixes.

Even at intermediate levels of ethanol, significant increased wear can occur in the combustion sealing interfaces of these valves and valve seats. In an internal probe testing of E20 fuels usage conducted by General Motors Powertrain Engineering, it was found that seven out of eleven production engines tested, with similar models that are already in service in the field, exhibited questionable numbers in leakdown and cold-cranking compression checks at the end of test. The engine dynamometer test plan that was run is based upon the standard engine durability cycling regimes. These results may indicate possible engine misfire and unacceptable operating conditions, and have led to many technical discussions on whether the use of E20 can lead to significant problems in the field in terms of valve face to valve seat wear.

ELASTOMERS

The elastomer characteristics described in this paper are limited to sealing applications in gas or alternative alcohol fuel operated internal combustion engines for passenger cars and light duty trucks. The purpose of this review is to provide essential material information to parts responsible engineers, assisting them in producing more functionally capable parts and components.

This paper will summarize what we and others have discovered in last 35 years of alcohol flex fuel testing to answer the following;

1. What tests and properties should one uses to provide a proper elastomer choice for seals and gaskets?

2. What test fluids are sufficiently aggressive to provide a safe margin over commercial customer fuel exposure?

3. Why should one use test fluids or liquids as opposed to certification test or commercial fuels?

4. What test conditions are appropriate or practical?

5. How important are these factors in choosing the best blend of economics and durability?

PERFORMANCE CHARACTERISTICS

Reliable sealing is a systems approach involving the proper material, design, joint surface finish, and load to seal for the desired time at temperature. Since the 1970's the auto industry tested methanol and ethanol in engines and vehicles as the potential for commercial use approached. During this time many investigated the risks of switching to dedicated or supplemental blended fuels with alcohol.

The driving forces for the use of blended fuel then were much the same as they are today;

- Reduce use of gasoline

- Increasing cost of gasoline

- Readily available supply of alcohol from a variety of domestic sources [e.g. crops, coal, biomass, natural gas].

What we are striving for is simply, can one use any commercially and legally available fuel for use in my vehicle and get the economy, performance and durability one has come to expect? Those three expectations are moving targets; driven by availability, technology and/or legislation.

Elastomeric materials selection is focused on finding the elastomer type that is low risk in the chosen media and conditions, reasonable cost, best processing (cycle time) and readily available.

Fluorocarbon elastomer (FKM/FPM) is the choice for a flex fuel environment where the polymer is in direct contact with the fuel and is expected to last for the design life[74]. What makes FKM/FPM the performance choice of flex fuel gaskets and seals?

- Low swell (≤30%) in a wide range of methanol concentrations with minimal extraction from prolonged fuel exposure reducing strain on the seal

- Low permeation of fuel vapor enables smaller cross section designs, less rubber, and supports emission regulation compliance

- Uniform degradation behavior so smaller test specimens can be used and larger volume of data generated

- Temperature resistance (CUTL per SAE J2236) 200 to 225°C, well beyond fuel system needs

- Variety of cure systems and hardness (50 to 90 Shore A) to aid in processing and good cross link density for optimum properties

- Retains moderate to high (30 to 90) sealing force (CSR) in air under compression for extended time (>4000h) and temperature (150°C)

- Cost is high to expensive especially to improve low temperature flexibility

Critical Elastomer Properties

To seal effectively in static applications, the elastomer must contribute a force sufficient to resist the passage of media through and around the compound. Environmental effects over time at temperature result in elastic memory loss, excessive swell, and set (permanent loss in height) causing leakage. A compound with good fluid permeation resistance is desirable to reduce internal chemical attack. Maintaining a sufficient contact pressure and width is both a design and material property function. In a static sealing joint the load

deflection response varies for each compound type, hardness and crosslink density. Design analysis focuses on keeping the elastomer within its acceptable range of physical stress and strain under load and temperature. In dynamic uses (against a moving surface) the load must be light enough to minimize wear, but with sufficient contact to function. During compression, press-in-place elastomeric seals may change shape and bend or twist (lateral stability is profile dependent) reducing sealing force depending on the joint operating conditions and the available free volume to expand.

Let's address the questions posed at the beginning of this section. For the static application compound;

What tests and properties should one uses to provide a proper elastomer choice for seals and gaskets?

Table 1 lists each property including two shaded elements for CSR and Fuel Resistance to emphasize the criteria needed for reliable and comparative conditions. Each property is meant to connect the material behavior with the functional elements of a successful seal for good decision making.

<table 1 here>

Volume Change

Over 25 years ago Dr. Nersasian, a respected research scientist at DuPont Elastomers Division (now DuPont Performance Elastomers) established volume change as a most significant material property. *"Although changes in tensile properties and hardness are important factors related to performance [of elastomers], the controlling physical property is probably that of volume increase or rubber/fuel compatibility[75] The volume increase of a rubber part can serve as a practical primary guide in predicting functional performance. Actual performance predictions should be based on vehicle or bench tests which simulate vehicle dynamic conditions. In the total immersion test used to measure volume increase (ASTM D471) the rubber is exposed to more fuel than it is in most fuel handling operations. The true value of these tests is the characterization of the swelling behavior of the rubber which can be related subsequently to vehicle testing and ultimately to establishing proper specifications."* Similar statements are made in SAE J1681, 7.1.2 and 7.1.4 and SAE J1748 by industry fuel systems experts.

What test fluids are sufficiently aggressive to provide a safe margin over commercial customer fuel exposure?

Data on the elastomers most commonly used for engine seals and gaskets, ethylene acrylic (AEM), polyacrylate (ACM), fluorocarbon (FKM), hydrogenated nitrile (HNBR), nitrile butadiene (NBR), and silicone (MVQ) indicates methanol blends have a more severe effect than ethanol blends. The

Table 1. Critical Properties for Developing Gasket & Seal Compounds

Critical Property	Test	Impact	Equipment	Comments
Compression Stress Relaxation (CSR)	ISO 3384 ASTM D6147 Continuous or Manual	Ability to retain sealing force under compression at time and temperature	Elastocon EB02 continuous 3M/Dyneon manual fixture	Test at higher compressive load if application is 35% or more. Test times 1500h or more in air.
	One common test fixture, rigidity of fixture, specimen size aspect ratio to parts, center & fix specimen during measurement. Assumes degradation is uniform through test specimen.			
Fluid Resistance	ISO 1817 ASTM D471	Elastomer growth due to total immersion fluid interaction. Extraction and chemical attack likely if high swell	Tube & Block Testers	Static oil. Labs often combine tensile and volume specimens. Watch area to fluid volume ratios. Immerse for 168h or more.
	Volume Change, Mass Loss			
	Compound must reach equilibrium when measuring immersion properties.			
Finite Element Analysis	Tension; uniaxial, planar, & equibiaxial. Compression; Volumetric	Keep elastomer within its strain limits while maintaining sufficient sealing force	Extensiometer, laser measurement of small deflections	Measure through the entire temperature range of use. Response and strength varies with temperature.
Low Temperature Flexibility	ISO 2921 ASTM D1053 ASTM D1329	Ability to remain flexible for sufficient sealing force under compression	As defined in standards	Often overlooked in specifications. T10 per D1053 gives lower (warmer) values.
Permeation	SAE J2695	Effectiveness in resisting fluid or vapor migration	As defined in standard	Modified Thwing–Albert Cup
State of Cure	GMW15117	Reaching maximum cross link density / network for strength, function	Thermal Analysis, DTMA	NMR is also effective, but limited in use due to high cost.

most aggressive concentration range for both ethanol and methanol blends in fuel are 15 to 35% by volume.

Historical Test Fluid Studies

According to Dr. Abu-Isa, GM Polymers Department, most elastomers [16 in this study] were more severely affected by mixtures of gasoline and methanol rather than straight gas or methanol[76]. The high aromatic content in the mixtures led to additional property deterioration. Indolene HO-III [30% aromatic], a toluene spiked Indolene to raise the aromatic level to 50% were used as they closely represented commercial fuels in the U.S. and Europe at that time. Tests were run for 72h at room temperature per ASTM D471 [except test specimens were weighed in closed bottles to prevent evaporative loss]. Methanol concentration levels were 2, 5, 10, 25, 50 and 75% with Indolene. Extractables were determined by weight at start of test and at end of test after drying specimens in a vacuum oven at 100°C for 24h. The report found the following;

Fuel Resistant Polymers

1. Viton FKM AHV from DuPont had 100% swell in 100% methanol with no extractables observed. The higher aromatic spiked Indolene was slightly more aggressive on property loss. A Type A or 1 grade and GLT would not be recommended.[77]

2. FMVQ had volume swells of 8 to 23% in both fuels throughout the range of methanol. Extractables were 2 to 4%.

3. NBR had volume swells of 26 to 53%, the highest volume swell occurring at 10 to 25% methanol. Extractables were in the 1 to 4% range.

Medium Fuel Resistant Polymers

4. Polyacrylate (not the AR-12 Zeon ACM) had 155% max swell in the 25 to 50% methanol with 0 to 1% extractables observed. The higher aromatic spiked Indolene was more aggressive on property loss. The maximum swell reached 193% at 25% methanol. This Hycar 4042 grade would not be recommended. ACM's are not recommended for current fuel

environments. ACM (AR12 type) is acceptable for 10% ethanol fuel dilution in engine oil.

Poor Fuel Resistant Polymers

5. Silicone (MVQ) had volume swells of 37 to 254% in both fuels throughout the range of methanol. The highest swell occurred at the 2 to 5% methanol range. The lowest volume swell occurred at 100% methanol. Extractables were 1 to 8%.

Although some of the compounds in these historical referenced studies no longer exist, the basic chemistry of current formulations remains the same. In the paper **Effects of Ethanol Fuels on Common Static Sealing Polymer Systems, F. J. Walker, Freudenberg-NOK, 2007,** Mr. Walker looked at current Freudenberg formulations. Polymer families of concentrations of ethanol are examined at one-week and six-week intervals at 23°C. Effects of ethanol content 10, 25, 50, 85 and 100% in a reference fuel on the physical and mechanical properties of five common engine sealing rubber constructions based on HNBR, AEM-D, MVQ, HT- ACM, and t-FKM (terpolymer). Mr. Walker found that;

• Due to volume swells in excess of 25%, HNBR should be cautiously used where direct exposure to ethanol-containing fuels with ethanol content less than 85%. Likewise, validation of seal designs should be conducted to address the substantial loss of both tensile and elongation in all but 100% ethanol.

• Excessive volume swell and mechanical property losses suggest that the ACM and AEM-D polymer system is not suitable where direct exposure to ethanol-containing fuels exist.

• MVQ represents the most robust material for retention of mechanical properties and resistance to volume swell for fuels containing 85% or greater ethanol. It should not be used where ethanol concentration is less than 85%.

• t-FKM exhibited consistent volume swell over the entire ethanol concentration range. However, its tensile properties are substantially affected at ethanol concentrations <85%. Elongation retention over the entire ethanol concentration range was moderate with values chiefly in the 70% range. Validation of function is recommended for applications where direct exposure to ethanol blended fuels at ethanol concentrations <85%.

To support a robust compound development, laboratory immersion testing must include test fluids (see GMW14914 or ISO 1817) in the most aggressive concentration range. General Motors runs testing using the appropriate GMW and ISO test fluids in the most aggressive concentration range for each elastomer.

The optimum fuel dilution level in engine oils is a product of engine field and dyno experience. Gaskets and seals used in engine oil must consider these levels for compound selection.

Dilutions in the 5 to 15% by volume fuel level are a good starting point.

Methodology Development for Testing
What test conditions are appropriate or practical?

Test methodology must consider the scope of your intended purpose. The limitations of each test should be clearly understood. Table 1 is an example of how to establish a robust elastomer compound using property selection that relates to function. Equally important is the precision and accuracy behind the procedure and its equipment. It takes time and dedication to improve a particular test method after it is published. An example is the progress made on CSR fixturing by Paul Tucker (formerly of 3M/dyneon).[78][79] The test fuels selected must be sufficiently aggressive to promoting a durable elastomer, and not contribute to test measurement variation.

Benefits of Earlier Work

Equilibrium in Immersion Testing. The majority of laboratory testing focused on the concentration range influences of ethanol and methanol over time and temperature. DuPont reported[80] on the importance of reaching equilibrium when testing. *"When comparing data at different temperatures it is essential that equilibrium swell is reached. Depending on the type of FKM you are testing, the time to reach equilibrium is different. Comparisons should only be made under identical test conditions and with sufficient time to reach equilibrium."*

Conventional immersion testing [ASTM D471] has its place however component bench tests are essential to functional success. ASTM D471, Subsection 4.3, *"This test method attempts to simulate service conditions through controlled accelerated testing, but may not give any direct correlation with actual part performance, since service conditions vary too widely. It yields comparative data on which to base judgment as to expected service quality."*

Predictive finite element modeling to predict sealing performance is the future direction.

The limitations of traditional immersion tests is illustrated in the report; ***NBR Diaphragms for Fuel Handling Systems***[81]. This report evaluated four DuPont FAIRPRENE™ NBR production compounds with three used in fuel pump diaphragms and tested for 70h at room temperature in ASTM Ref Fuel C (50% toluene) in various concentrations of ethanol and methanol. At the time of this study (1980), NBR and coated fabric diaphragms were used in fuel fill caps, carburetors, ECV's and fuel pumps. The report concluded;

• Data indicates methanol blends have a more severe effect than ethanol blends. This is true for all four formulations.

• The data is in basic agreement with data previously reported that fuel/alcohol blends do have an effect on elastomer systems. However, these effects differ on different formulations as well as the property measured. For the following reasons it is our opinion that static tests alone are not a sound basis for judging the suitability of diaphragm material for use in fuel systems:

• Diaphragms are not usually subjected to fuel under total immersion conditions.

• Diaphragms in dynamic applications are usually supported by fabric which reduces the total swell in the area of 50%.

• With the exception of volume swells Fuel C has the greatest influence on wet properties. While the addition of alcohol further decreases physical properties, these decreases do not affect the suitability of a product under dynamic conditions.

• If rubber parts are involved under total immersion conditions, then static data can be considered as the basis for judgment as to a compound's suitability for a given end use. High swells could cause a homogeneous rubber valve to malfunction.

Why should one use test fluids or liquids as opposed to certification test or commercial fuels?

The main reason is to minimize the variation in test results caused by the test media. The second is test fluids are specifically created for material development decisions. The distinctions are that commercial fuels in service stations are too variable for elastomer selection in lab testing. Due to the presence of large number of 'boutique' fuels that waste automotive resources trying to determine compatibility and which is most aggressive. What about certification test fuels that control the ingredients for component or system evaluation? General Motors uses these, however because they are standard formulations that must work in a functioning engine, their focus is not on material selection or evaluation.

SAE Test Fuels

In 1990 SAE published ***'Gas / Methanol Mixtures for Materials Testing'*** the result of a Cooperative Research Program Project Group 2. This provided test fluid standards using a mutually agreed upon set of mixtures of ASTM Ref Fuel C, ethanol and methanol (e.g. CM0 = 100% C, CM15 = 85%C/15%M, etc.). SAE J1681, SAE J1748, ISO 1817 and GMW14941 also provide consistent test fluids so results can be compared with minimal concern for fuel affecting the result.

Failure Modes and Their Implications

The material and design objective in seal or gasket development is to maintain the sealing function for the design life. To accomplish this, a seal must remain flexible and retain a sufficient force against the sealing surface. Typically oil resistant elastomers like nitrile (NBR, HNBR), polyacrylate (ACM), and fluorocarbon (FKM/FPM) harden with prolonged exposure to oil at high temperature. Silicone rubber (MVQ) experiences reversion or softening under the same conditions. Hardness and modulus increase with an increase in the base polymer crosslink network and elongation decreases. Reversion is a deconstruction of the base polymer crosslinks resulting in loss of hardness and modulus. In a crankshaft seal, early softening would reduce oil pumping efficiency then progress to loss of radial load to seal and/or excessive seal wear of the contact lip due to low modulus.

The major concerns for elastomers in oil and flex fuel systems are severe volume swell and ingredient extraction due to prolonged fluid exposure. A new elastomer crosslink network that is dense to resist absorption and/or permeation of fluid is desirable. Solvent swell tests at equilibrium (constant weight) are used to evaluate the crosslink network with lower swell an indication of good density. Increased swell indicates more of the fluid is entering the compound, increasing the risk of catalytic effects and increased extraction of sensitive compound ingredients. Extended time at higher temperatures only accelerates these effects. Extraction of phthalates (DOP) used to promote low temperature flexibility in NBR can result in higher hardness and reduced elongation. Early studies evaluated extraction usually at 168h (equilibrium) for a variety of temperatures. The better test approaches required drying the test specimens after immersion, then retesting essential as-received properties like low temperature retraction (ASTM D1329 TR10) for any change.

Engine oil chemistry alone or mixed with fuel contains additives that will reactively affect the polymer network. Oxygen in the oil from internal moving parts and flow, combined with heat and water vapor create acidic fluid conditions. An increase in Total Acid Number (TAN) of oil is an indication of aggressive change. Surface cracking of immersed test specimens is an obvious indicator as is the loss of physical properties. Some immersion tests bend the specimen to create a stress concentration as a site for crack propagation. Elongation or modulus changes after immersion are considered key indicator of chemical resistance.

Laboratory testing using ASTM D471 is limited in that no aeration of the oil occurs as in an operating engine. Most FKM/FPM polymers show improved property retention in aerated oil than in static[82]. The amount of blow-by gas into the engine crankcase oil is never adequately represented in standard industry immersion tests. Engine oil changes during the life of the engine are only mimicked if the tester specified oil change intervals. A classic example of the value of practical part bench testing is found in the low temperature performance of FKM o-rings[83]. The paper's authors state,

"The rubber chemist uses several tests to define low temperature characteristics of elastomers...however, these tests and their associated results are less clear to a design engineer. They can't make full use of the low temperature data because there is no direct correlation between laboratory tests and the actual service performance of the seal due to the varying nature of end use conditions."
Building a leak fixture that spaced the radial load on the o-ring, air pressure was applied at various low temperatures until the pressure and temperature leak boundary conditions of the o-ring were reached at 10 SCCM.

Compatibility Parameters

With historical data on elastomers, general flex fuel compatibility guidelines began to take shape from multiple sources. Material compatibility began to be defined around these specific property retention levels;

• Loss of tensile and elongation not exceeding 50% of original properties. A 70% retention level is good.

• Volume change, 168h ≤5% (dynamic seal) ≤ 45% (static seal) or ≤ 60% (if permeation is not an issue). FKM tests in CM15 at various times between 168h to 5000h at 60°C indicate no significant change so 168h is a representative test time for determining FKM maximum swell. The optimum test temperatures for fuels is 60°C to 80°C and 150°C for fuel contamination in engine oil in a pressure vessel.

• Retained Elongation ≥ 150%

• Limit soluble extractables per unit area to less than 2.5 grams/m^2 for fuel hoses

• Weight loss minimal (≤5%)

• 20 to 50% retained sealing force at 150°C

• 21 to 30 day at 23°C, Permeation, SAE J2665, ≤140 g/m^2/day with M15-M85

Blending effects - non-linear behavior (higher blend does not always equal worse)

Since fuel purchasing decisions will be cost based, a North American consumer's fuel tank is likely to see the entire range of ethanol concentration from E10 to E85. This will continue as market development occurs. In China, a concentration range of M5 to M100 is likely. This means two things, testing with the most aggressive concentrations is necessary for elastomer selection. If commercial formulations fall into the most aggressive range for elastomers (e.g. E20 or M20) it will drive use of the most resistant elastomer types.

Repeated lab studies[84] [85] [86] with AEM, ACM, NBR, HNBR, FKM type 1, and MVQ confirm that the most aggressive concentration range for ethanol and methanol was 15 to 35%. In all cases methanol was more aggressive.

In March 1990 for the SAE Congress, Dow Corning Corporation presented work on the effects of methanol/gas blends at elevated temperatures on fluorosilicone elastomers (FMVQ).[87] Testing was at 23°C and 60°C with blends of methanol and Fuel C or Howell EEE for 3 months. Ten Dow Corning Silastic™ FMVQ compounds were compared with (37% ACN) NBR, a blend of FMVQ (40%) and FKM (60%), DuPont Viton™ GF and GFLT. All FMVQ compounds were post cured. The 60°C results were;

• M25 was the most aggressive fuel at both test temperatures

• Tensile loss was roughly 70%, elongation 50% Volume change 35% and hardness change −22 pts at M25

• The lowest tensile loss was in straight Fuel C

• The lowest loss in elongation, volume and hardness was at M100

• FMVQ property losses maximized in 24 to 48h and did not vary for 3 months

• Increasing the test temperature increases volume swell

• In M85 the FMVQ/FKM blend tensile loss stabilized at 70% at 23°C, but at 60°C loss continued until no tensile strength remained at 3 months

• From the FMVQ/FKM data, room temperature testing does not fully indicate material compatibility, and the elevated temperature testing is required

• Fluorosilicones cannot be judged as a class in compatibility testing for methanol/gas blends. Differences occur in both chemical formulations and manufacturing process conditions. Some are better than others.

• FKM had lower volume swell with the methanol content below 50%. At levels greater than 50%, FMVQ had lower volume swell

• At M25 and 60°C the NBR volume change was 53%

• Fluorosilicones are good candidates for methanol/gas applications

DISCUSSION AND CONCLUSIONS

How important are these factors in choosing the best blend of economics and durability?

There are two systems for which elastomer compatibility to alcohol must be developed; fuel-only systems and oil systems where fuel dilution will occur. Test temperatures at 150°C are typical for fuel in oil while 60°C to 80°C is the range for fuel-only in a pressure vessel environment.

Fluoroelastomer formulations are available and can work in any methanol fuel blends. There is a significant piece cost increase associated with their use.

Since there are several grades of fluoroelastomers, Type 1 through 5, [A, AL, B, G, F, GFLT-S, GLT-S, etc.] the choices of which grade are appropriate for fuel versus oil/fuel dilution will be different. It would not be recommended to use a type 1, A or GLT (pre-2006) for any gasket immersed in a 15% methanol fuel diluted in oil system. As DuPont data suggests, any concentrations below 10% in methanol (similar to fuel/oil dilution situation) would not be a problem for any grade of fluoroelastomer.

Zeon Chemicals, supplier of AR-12 based polyacrylate (ACM), is currently working on an oil/fuel dilution resistant formulation. Results in ethanol look promising. ACM would not be used for a static gasket in methanol fuel-diluted engine oil.

From this analysis, it would be recommended that focusing on methanol blends would be the best choice if the most aggressive mixture for elastomer evaluation is desired.

SUMMARY AND CONCLUSIONS

The paper provided a high-level and comprehensive review, from an automotive OEM's perspective, of the following materials topics and components categories as it relates to the use and implementation of alternative alcohol fuels:

- *Alternative alcohol fuels' relevance in the global automotive landscape*

- *Automotive materials compatibility and its implications*

- *Corrosion and wear of metallics*

- *Effects on Polymers and Elastomers*

- *Fuel storage & handling components*

- *Engine hardware and subsystems*

As with any engineering challenge, one can't ignore the political, social and environmental impact on that body of work. Preparation for materials solutions to alternative alcohol fuels has spanned decades yet the universal implementation of those solutions still depends on the collective will of society to use it. The same applies to the supply and distribution of the intended fuels. Manufacturers are prepared to respond, but as complexity in the market place increases, compliance comes with increasing cost and time necessary to respond with a solution.

Until the uncertainty of which fuel based solution remains, OEM manufacturers have a difficult choice. They implement for past and current reality. The potential fuel reality is that the automotive engineers are also tasked to develop global engines three to four years in advance of their use.

Consumers want transparency. They anticipate to use any commercially and legally available fuel for use in their vehicle and to get the economy, performance and durability that they expect. Some will be willing to pay for easy

modification tools with the illusion that they are now flex fuel capable. For now manufacturers take a decisive and selective approach by properly labeling and certifying a vehicle as "flex fuel capable." Meanwhile the pursuit to explore the most aggressive fuels and the most robust materials to meet the challenge continues.

CONTACT INFORMATION

Pui Kei Yuen
puikei.yuen@gm.com

John Beckett
john.beckett@gm.com

William Villaire
william.villaire@gm.com

ACKNOWLEDGMENTS

The authors thank researchers past and present for their contributions to alternative alcohol fuels and materials knowledge. We would like to extend our special thanks to Joe Walker (Freudenberg-NOK) and the following colleagues from General Motors for their contribution and assistance for this paper, William Studzinski, David Moore, Coleman Jones, Pat Geng, Andrew Bucyznsky, Melissa Schulz, Phil Yaccarino, Kristin DeMare, Rich Novaco, Mitch Hart, Norm Brinkman, Steven Kemp, Dale Gerard, Heather Simmons, and the GM Library Staff.

REFERENCES

1. US EPA, "Renewable Fuel Standard Program," http://www.epa.gov/OMS/renewablefuels/

2. US Government, "Public Law 110-140 - Energy Independence and Security Act of 2007," http://www.gpo.gov/fdsys/pkg/PLAW-110publ140/pdf/PLAW-110publ140.pdf

3. Wikipedia, "Flexible Fuel Vehicle," http://en.wikipedia.org/wiki/Flexible-fuel_vehiclehttp://en.wikipedia.org/wiki/Flexible-fuel_vehicle, September 2009

4. Bologna, D.J. and Page, H.T., "Corrosion Considerations in Design of Automotive Fuel Systems," SAE Technical Paper 780920, 1978.

5. Wikipedia, "E85 in Standard Engines," http://en.wikipedia.org/wiki/E85_in_standard_engines

6. Wikipedia, "E85 in Standard Engines - After-market Conversions," http://en.wikipedia.org/wiki/E85_in_standard_engines#After-market_conversions

7. Google Search Results for "E85 conversion kit," www.change2e85.com, www.ffie85.com, www.whitelightning.net, www.driveflexfuel.com, www.e85fuel.com, etechmn.com, www.e85conversionkits.net, September 2009

8. American Lung Association of Minnesota, "Clearing the Air on Vehicle Conversions: Vehicle E85 Conversion Kit Meeting," http://www.cleanairchoice.com/news/E85ConversionKitSummaryBook.pdf, 2008

9. Geng, P.Y., Buczynsky, A.E., and Konzack, A., "US and EU Market Biodiesel Blends Quality Review - An OEM Perspective," *SAE Int. J. Fuels Lubr.* 2(1):860-869, 2009.

10. Dolan, G., "China Takes Gold in Methanol Fuel," http://www.ensec.org/index.php?option=com_content&view=article&id=148:chinatakesgoldinmethanolfuel&catid=82:asia&Itemid=324, Journal of Energy Security, October 2008.

11. Dolan, G., "China Takes Gold in Methanol Fuel," http://www.ensec.org/index.php?option=com_content&view=article&id=148:chinatakesgoldinmethanolfuel&catid=82:asia&Itemid=324, Journal of Energy Security, October 2008.

12. Wikipedia, "Ethanol in Brazil," http://en.wikipedia.org/wiki/Ethanol_fuel_in_Brazil, September 2009

13. Wikipedia, "Ethanol in Brazil," http://en.wikipedia.org/wiki/Ethanol_fuel_in_Brazil, September 2009

14. Agência Nacional do Petróleo, Gás Natural e Biocombustíveis (ANP) (2005-12-06). "Resolução ANP N° 36, DE 6.12.2005 - DOU 7.12.2005, Specifications for AEAC and AEHC" (in Portuguese). ANP., http://nxt.anp.gov.br/NXT/gateway.dll/leg/resolucoes_anp/2005/dezembro/ranp%2036%20-%202005.xml?f=templates$fn=default.htm&sync=1&vid=anp:10.1048/enu, Retrieved November 2008.

15. ASTM Specification, "D4806 - Standard Specification for Denatured Fuel Ethanol for Blending with Gasolines for Use as Automotive Spark-Ignition Engine Fuel," 2009.

16. ASTM Specification, "D4806 - Standard Specification for Denatured Fuel Ethanol for Blending with Gasolines for Use as Automotive Spark-Ignition Engine Fuel," 2009.

17. ASTM Specification, "D5798 - Standard Specification for Fuel Ethanol (Ed75-Ed85) for Automotive Spark-Ignition Engines," 2009.

18. ASTM Specification, "D4814 - Standard Specification for Automotive Spark-Ignition Engine Fuel," 2009.

19. US EPA, "Notice of Receipt of a Clean Air Act Waiver Application to Increase the Allowable Ethanol Content of Gasoline to 15 Percent; Request for Comment (published April 21, 2009)," Federal Register / Vol. 74, no. 75, 2009.

20. "E20: The Feasibility of 20 Percent Ethanol Blends by Volume as a Motor Fuel," Executive Summary, http://www.mda.state.mn.us/news/publications/renewable/ethanol/e20execsumm.pdf, 2009.

21. Environment Australia, "Setting the ethanol limit in petrol," 2001/2002.

22. Orbital Engine Company, "Environment Australia Project: 'Market Barriers to the Uptake of Biofuels - Testing Petrol Containing 20% Ethanol (E20),'" Environment Australia, Tender 34/2002, 2002.

23. Standardization Administration of P.R. China, "Announcement of Newly Approved National Standards of P.R. China 2009 No.:4," http://www.sac.gov.cn/templet/english/zmCountryBulletinByNoEnglish.do?countryBulletinNo=20094144

24. Standardization Administration of P.R. China, "Announcement of Newly Approved National Standards of P.R. China 2009 No.:4," http://www.sac.gov.cn/templet/english/zmCountryBulletinByNoEnglish.do?countryBulletinNo=20094144

25. Standardization Administration of P.R. China, "Announcement of Newly Approved National Standards of P.R. China 2009 No.:4," http://www.sac.gov.cn/templet/english/zmCountryBulletinByNoEnglish.do?countryBulletinNo=20094144

26. Geng, P.Y., Reichenbaecher, L., Spangenberg, A., Burnett, D.E. et al., "Testing & Validation Fuels in Vehicle Development," *SAE Int. J. Fuels Lubr.* 1(1):1397-1418, 2008.

27. Harrigan, M.J., Sr., Banda, A., Bonazza, B., Graham, P. et al., "A Rational Approach to Qualifying Materials for Use in Fuel Systems," SAE Technical Paper 2000-01-2013, 2000.

28. West, B.H. and McGill, R.N., "Oil Performance in a Methanol-Fueled Vehicle Used in Severe Short-Trip Service," SAE Technical Paper 922298, 1992.

29. ASTM Standard, "D7318 - 07 Standard Test Method for Total Inorganic Sulfate in Ethanol by Potentiometric Titration," 2009.

30. DuMont, R.J., Cunningham, L.J., Oliver, M.K., Studzinski, W.M. et al., "Controlling Induction System Deposits in Flexible Fuel Vehicles Operating on E85," SAE Technical Paper 2007-01-4071, 2007.

31. Devlin, M.T., Baren, R.E., Sheets, R.M., McIntosh, K. et al., "Characterization of Deposits Formed on Sequence IIIG Pistons," SAE Technical Paper 2005-01-3820, 2005.

32. ASTM Standard, "D7318 - 07 Standard Test Method for Total Inorganic Sulfate in Ethanol by Potentiometric Titration," 2009.

33. Dumont, R.J, Cunningham, L.J., Oliver, M.K., Studzinski, W.M. et al., "Controlling Induction System Deposits in Flexible Fuel Vehicles Operating on E85," SAE Technical Paper 2007-01-4071, 2007.

34. Brinkman, N.D., Halsall, R., Jorgensen, S.W., and Kirwan, J.E., "The Development of Improved Fuel

Specifications for Methanol (M85) and Ethanol (E_d85)," SAE Technical Paper 940764, 1994.

35. Rawat, J., Rao, P.V.C., and Choudary, N.V., "Effect of Ethanol-Gasoline Blends on Corrosion Rate in the Presence of Different Materials of Construction used for Transportation, Storage and Fuel Tanks," SAE Technical Paper 2008-28-0125, 2008.

36. Krings, N., Abel, J., Hebach, A., Ochs, H., Reitzle, A., Virtanen, S., "Corrosion in Ethanol Containing Gasoline," Electrochem.org Abstract, http://www.electrochem.org/meetings/scheduler/abstracts/214/1695.pdf, 2009.

37. Lou, X., Goodman, L.R., "Pitting Corrosion of Carbon Steel in Fuel Grade Ethanolic Environment," National Association of Corrosion Engineers International - International Corrosion Conference Series, ISSN: 03614409, 2009.

38. Rawat, J., Rao, P.V.C., and Choudary, N.V., "Effect of Ethanol-Gasoline Blends on Corrosion Rate in the Presence of Different Materials of Construction used for Transportation, Storage and Fuel Tanks," SAE Technical Paper 2008-28-0125, 2008.

39. Kane, R.D., Papavinasam, S., "Corrosion and SCC Issues in Fuel Ethanol and Biodiesel," National Association of Corrosion Engineers International - International Corrosion Conference Series, ISSN: 03614409, 2009.

39. Krings, N., Abel, J., Hebach, A., Ochs, H., Reitzle, A., Virtanen, S., "Corrosion in Ethanol Containing Gasoline," Electrochem.org Abstract, http://www.electrochem.org/meetings/scheduler/abstracts/214/1695.pdf, 2009.

40. Walker, M.S., Chance, R.L., "Corrosion of Metals and Effectiveness of Inhibitors in Ethanol Fuels," GM Research Report, 1983.

41. Brink, A., Jordaan, C.F.P., le Roux, J.H., Loubser, N.H., "Carburetor corrosion: the effect of alcohol-petrol blends." Proceedings of the VII International Symposium on Alcohol Fuels Technology, Paris, France, 1986.

42. Hansen, A., Zhang, Q., Lyne, P., "Ethanol-diesel fuel blends-a review," Elsevier Bioresource Technology 96 (2005) 277-285, 2005.

43. De la Harpe, E.R., "Ignition-improved ethanol as a diesel tractor fuel." Unpublished MSc. Eng. Thesis, Department of Agricultural Engineering, University of Natal, Pietermaritzburg, South Africa, 1988.

44. De la Harpe, E.R., "Ignition-improved ethanol as a diesel tractor fuel." Unpublished MSc. Eng. Thesis, Department of Agricultural Engineering, University of Natal, Pietermaritzburg, South Africa, 1988.

45. Leber, E., GM Powertrain Europe, Email Communication. January 22, 2009.

46. Otto, K., Carter, R.O., III, Gierczak, C.A., and Bartosiewicz, L., "Steel Corrosion by Methanol Combustion

Products: Enhancement and Inhibition," SAE Technical Paper 861590, 1986.

47. Otto, K., Bartosiewicz, L., Carter, R.O., III, and Gierczak, C.A., "A Simple Coupon Test for Analyzing Corrosion Caused by Combustion Products of Liquid Fuels," SAE Technical Paper 880039, 1988.

48. Kane, R.D., Papavinasam, S., "Corrosion and SCC Issues in Fuel Ethanol and Biodiesel," National Association of Corrosion Engineers International - International Corrosion Conference Series, ISSN: 03614409, 2009.

49. Kane R.D., Eden D.A., Sridhar N., Maldonado J.G., Brongers M.P.H., Agarwal A.K., Beavers J.A., "Stress Corrosion Cracking of Carbon Steel in Fuel Grade Ethanol," Publication 939D, American Petroleum Institute, Washington, D.C., 2007.

50. Beavers, J; Sridhar, N; Zamarin, C., "Effects of steel microstructure and ethanol-gasoline blend ratio on SCC of ethanol pipelines," National Association of Corrosion Engineers International - International Corrosion Conference Series, ISSN: 03614409, 2009.

51. National Highway Traffic Safety Administration, "Recall Acknowledgement for NHTSA CAMPAIGN ID Number: 07V545000," http://nhthqnwws112.odi.nhtsa.dot.gov/acms/docservlet/Artemis/Public/Recalls/2007/V/RCAK-07V545-5582.pdf.

52. National Highway Traffic Safety Administration, "Defect Notice (Part 573) for NHTSA CAMPAIGN ID Number: 07V545000," http://nhthqnwws112.odi.nhtsa.dot.gov/acms/docservlet/Artemis/Public/Recalls/2007/V/RCDNN-07V545-2142.pdf.

53. Browning, L.H., Pefley, R.K., "Engine Wear and Cold Startability of Methanol Fueled Engines," Proceedings of the Twenty-Second Automotive Technology Development Contractors’ Coordination Meeting - SAE, p 11-16, 1985 CODEN: PSOED4 ISBN-10: 0898837162, 1985

54. Mattsson, L., Olsson, B., Nilsson, P.H., Wirmark, G., "Wear and Film Formation in the Presence of Methanol and Formic acid," Wear, v 165, n 1, p 75-83, May 1 1993 ISSN: 00431648 CODEN: WEARAH, 1993.

55. Saarialho, A., Juhala, M., Leppamaki, E., Nylund, O., "Alcohol Gasoline Fuels and Engine Wear in Cold Climates," Proceedings: 19th International FISITA Congress: Energy Mobility - SAE Paper No. 82037, 1982

56. The Strategic Alliance for Steel Fuel Tanks Technical Publications, www.sasft.org.

57. U.S. Department of Energy's National Renewable Energy Laboratory (NREL), "Handbook for Handling, Storing, and Dispensing E85," www.eere.energy.gov/afdc/pdfs/41853.pdf, April 2008.

58. Marsala, V., General Motors Corporation, personal communication, August 2009.

59. Naegeli, D.W., Lacey, P.I., Alger, M.J., and Endicott, D.L., "Surface Corrosion in Ethanol Fuel Pumps," SAE Technical Paper 971648, 1997.

60. Kane, R.D., Papavinasam, S., "Corrosion and SCC Issues in Fuel Ethanol and Biodiesel", NACE Technical Paper 09528, 2009.

61. Cole-Parmer: Chemical Resistance Database, www.coleparmer.com/techinfo/ChemComp.asp.

62. IDES "The Plastics Web®", www.ides.com.

63. Galipeau, R. James, "Predicting the Effects of Contact Materials and Their Environments on Thermoplastics Through Chemical Compatibility Testing," Presented before the ANTEC Conference, May 1995.

64. Kathios, D.J., Ziff, R.M., Petrulis, A.A., and Bonczyk, J.C., "Permeation of Gasoline-Alcohol Fuel Blends Through High-Density Polyethylene Fuel Tanks with Different Barrier Technologies," SAE Technical Paper 920164, 1992.

65. SAE International Surface Vehicle Recommended Practice, "Fuel Systems and Components-Electrostatic Charge Mitigation," SAE Standard J1645, Rev. Aug. 2006.

66. Galante-Fox, J., Von Bacho, P., Notaro, C., and Zizelman, J., "E-85 Fuel Corrosivity: Effects on Port Fuel Injector Durability Performance," SAE Technical Paper 2007-01-4072, 2007.

67. Matthias, S., Thomas, L., "Investigation of the Pitting Corrosion Behaviour of Stainless Steels in Ethanol Containing Fuels," National Association of Corrosion Engineers International - International Corrosion Conference Series, ISSN: 03614409, 2009.

68. Shifler, D. A., "Possible corrosion aspects for the use of alternative fuels," National Association of Corrosion Engineers International - International Corrosion Conference Series, ISSN: 03614409, 2009.

69. Scholz, M., Ellermeier, J. "Corrosion behaviour of different aluminium alloys in fuels containing ethanol under increased temperatures," (In German) Mat.-wiss. u. Werkstofftech. 2006, 37, No. 10, DOI: 10.1002/mawe. 200600074, 2006.

70. SGS Industrial, "Report: Testing the Corrosive Effect of Ethanol Blended Petrol on Aluminum Alloys," Report to New Zealand Ministry of Transport, SGS File Ref No. INZ26942-02, 2009.

71. National Highway Traffic Safety Administration, "Recall Acknowledgement for NHTSA CAMPAIGN ID Number: 09V020000," http://nhthqnwws112.odi.nhtsa.dot.gov/acms/docservlet/Artemis/Public/Recalls/2009/V/RCAK-09V020-4420.pdf.

72. National Highway Traffic Safety Administration, "Defect Notice(Part 573) for NHTSA CAMPAIGN ID Number: 09V020000," http://nhthqnwws112.odi.nhtsa.dot.gov/acms/docservlet/Artemis/Public/Recalls/2009/V/RCDNN-09V020-2280.pdf.

73. Mantey, C.A., Messarano, A., and Kolkemo, A., "Exhaust Valve & Valve Seat Insert - Development for an Industrial LPG Application," SAE Int. J. Commer. Veh. 2(2): 1-11, 2009.

74. Balzer, J.R. and Sohlo, A.M., "Effects of Long-Term Flex-Fuel Exposure on Fluorocarbon Elastomers," SAE Technical Paper 900118, 1990.

75. Nersasian A., "Compatibility of Fuel-Handling Rubbers with Gasoline/Alcohol Blends," Elastomer Chemical Dept., DuPont, Elastomerics, October 1980, pp. 26-30.

76. Abu-Isa, I.A., "Effects of Mixtures of Gasoline with Methanol and with Ethanol on Automotive Elastomers," SAE Technical Paper 800786, 1980.

77. DuPont Performance Elastomers Viton Selection Guide, p. 5, January 1999.

78. Tuckner Paul, "Compression Stress Relaxation Testing - Comparisons, Methods, and Correlations," SAE Paper 2001-01-0742, March 2001.

79. Tuckner, P., "The Effects of Configuration on Sealing Force Measurement and Compression Stress Relaxation Response," SAE Technical Paper 2003-01-0946, 2003.

80. Nersasian A., "Compatibility of Fuel-Handling Rubbers with Gasoline/Alcohol Blends," Elastomer Chemicals Dept., DuPont, Elastomerics, October 1980, p. 26-30.

81. Gatcomb G.L., "Performance of FAIRPRENE Fuel Pump Diaphragm Materials in ASTM Fuel C and ASTM Fuel C/Alcohol Blends," DuPont, Technical Bulletin No. 1, June 1980.

82. Bauerle, J.G. and Bruhnke, D.W., "The Effects of Aeration of Test Fluids on the Rentention of Physical Properties of Fluoroelastomer Vulcanizates," SAE Technical Paper 890362, 1989.

83. Stevens, R.D., Thomas, E.W., Brown, J.H., and Revolta, W.N.K., "Low Temperature Sealing Capabilities of Fluoroelastomers," SAE Technical Paper 900194, 1990.

84. Gatcomb G.L., "Performance of FAIRPRENE Fuel Pump Diaphragm Materials in ASTM Fuel C and ASTM Fuel C/Alcohol Blends," DuPont, Technical Bulletin No. 1, June 1980.

85. Nersasian A., "Compatibility of Fuel-Handling Rubbers with Gasoline/Alcohol Blends," Elastomer Chemicals Dept., DuPont, Elastomerics, October 1980, p. 26-30

86. Fiedler L.D., Knapp T.L., Norris A.W., and Virant M.S., "Effect of Methanol/Gasoline Blends at Elevated Temperature on Fluorosilicone Elastomers," Dow Corning Corporation, March 1990.

87. Fiedler L.D., Knapp T.L., Norris A.W., and Virant M.S., "Effect of Methanol/Gasoline Blends at Elevated Temperature on Fluorosilicone Elastomers," Dow Corning Corporation, March 1990.

A Study on Refrigerant Irregular Emission from China Mobil Air Conditioning Vehicles Based on JD Power Result	2010-01-0479 Published 04/12/2010

Bing Li
Shanghai Jiaotong Univ. & PATAC

William Hill
General Motors & SAE

Copyright © 2010 SAE International

Key Words

Refrigerant emission, China vehicles, JD Power

ABSTRACT

The purpose of this article is to study current refrigerant emission levels in China with reasonable accuracy of the first year vehicles. This is an initial survey on refrigerant irregular emissions based on JD Power investigation and warranty data from OEMs in 2008. Totally 49 brands and 8881 vehicles were included for the study, covering almost all the kinds of passenger vehicles in China market.

Irregular emissions represent the refrigerant losses due to accidents and other environmental related failures of the mobile AC system. This paper also wants to draw people's attention on irregular refrigerant emissions related to system design and reliability which is not focused yet.

According to the calculation of irregular emissions from China vehicles by J.D Power result, the irregular emission is 5.8g/yr, which can be a reliable number used in the **GREEN-MAC-LCCP©** model for China vehicles refrigerant emissions.

1. INTRODUCTION

With more and more attention on global warming, the whole world is studying the various ways to reduce CO2 emissions. Using alternative refrigerant is one of the solutions, reducing current refrigerant -R134a leakage is another consideration to control global warming gas emission. So it is quite necessary to have an overall study on the leakage level of MAC (Mobil Air Conditioning) in China before figuring out the right way to reduce refrigerant emissions. Due to the cost of future alternate refrigerants, refrigerant leakage rate will be an important criterion for all future refrigerants in MAC design.

China, as a member of the Kyoto Protocol, has experienced a dramatic increase of passenger vehicles in the past ten years. Most of R134a industry output in China which is around 20,000 tons goes to mobile air conditioning systems in 2007 and equivalent amount in 2008.[1]

It is widely accepted that there are five kinds of emissions: emissions before the registration of the vehicle, **regular** (gradual or steady loss), **irregular** (sporadic loss due to traffic accidents or environmental system failure), emissions during service and **end-of-life** (EOL) emissions (occurring when a vehicle is dismantled)[2]. Each of these will be discussed in more detail in section 3.

Laboratory measurements have been used to evaluate the regular emission, especially for the exporting vehicles required by EU regulation. While for the irregular emissions, there is no detail calculation or measurement yet, but mostly rely on experts' estimation.

The present study has employed a statistical method to determine the annual rate of R-134a irregular emission from MAC of all new vehicles in China in 2008 by using the JD Power result and customer complaints in dealers.

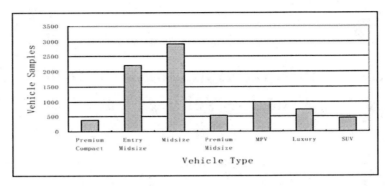

Chart 1. Vehicle distribution under survey

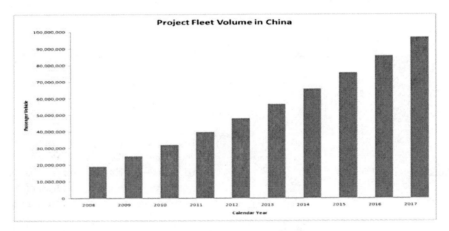

Chart 2. Vehicle Fleet Volume in China

The study concentrates mainly on passenger-car air-conditioning systems because they account for more than 95% of refrigerant emission from vehicles. The irregular result can be an input to the **GREEN-MAC-LCCP**© model as well, which was developed to assess the lifecycle energy and GHG emissions associated with the production, usage and disposal of refrigerants and MAC components.[3]

2. NUMBER OF VEHICLES FOR STUDY

In 2008, there were 6,725,634 passenger cars sold in China. Among these vehicles, 98.5% were equipped with AC units, thus 6,619,395 new cars will have potential contribution to refrigerant irregular emissions because of customer usage.

The distribution of new vehicles is shown below in Table 1[4]:

Table 1. Vehicle sales volume statistics in 2008

Vehicle Type	Sales Volume in 2008	Equipped with A/C
Cars	4976736	100%
MPV	196428	100%
SUV	490075	100%
Mini-Van	1062395	90%
Total	**6725634**	**6619395**

Among all these new vehicles, 8881 vehicle owners accepted the investigation on whole vehicle performance including A/C performance with the questionnaires of 'Is A/C cold enough?' and 'Is A/C not cold at all?'

Vehicle owners under survey covered all kinds of vehicles in the market, from compact to luxury vehicles. Below is the distribution chart:

The vehicle fleet in China is estimated to be over 25,000,000 vehicles. Emissions of this total fleet needs to be considered as well which will be covered in the future studies. Below chart shows the vehicle fleet over time as shown in the **GREEN-MAC-LCCP**© model.

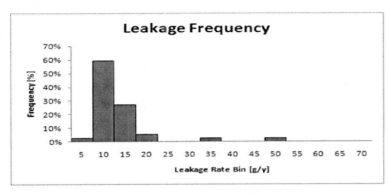

Chart 3. Leakage Frequency of Fleet

3. REFRIGERANT LOSS CONSIDERATIONS

3.1. EMISSIONS BEFORE THE REGISTRATION OF THE VEHICLE

These emissions include losses of refrigerant during the refrigerant manufacturing process, the shipment of the refrigerant to the vehicle assembly plant, and losses during the vehicle assembly process.

3.2. REGULAR EMISSION

Regular leak of MAC happens on all the joints of the system, hoses and shaft seal of the compressor in refrigerant circuit under all conditions whether the system is operating or not due to the higher pressure in system than atmosphere.

The weak spots of the system can lead to a gradual steady loss of R134a from 10 to 40 grams/yr. Based on studies that have been conducted on new vehicles, most vehicles range between 10 and 20 grams, which is a range tolerated by car manufacturers.

Below is an example of fleet vehicles showing leakage rate percentage:

The regular annual leakage rate increases over the lifetime of the vehicle, this is not the subject of this paper and limited data is available on how it changes. The limit as outlined in **COMMISSION REGULATION (EC) No 706/2007**[5] is 40 g/y. A reserve charge of 200 grams is added to the vehicle in consideration of this leakage rate. This reserve charge should provide 7-10 years of service without a requirement to charge the vehicle due to this leakage amount. For the dual systems in MPV, the regular leakage requirement from the EU regulation is 60g/yr and additional reserve charge might be expected in this type of vehicles due to increased regular leakage amounts.

To estimate regular emissions, a special test set-up is needed. According to **COMMISSION REGULATION (EC) No 706/2007**, the test has to be undertaken in a sealed enclosure including an equipment to ensure a homogeneous concentration of gas is detected and the use of a gas analysis method is adopted. The accurate numbers for China vehicles based on EU test procedure will be carried out on benches in the near future in Shanghai Jiao Tong University in 2009.

SAE has also developed a standard that estimates leakage based on system design characteristics. [SAE J2727] It has been shown to result in slightly higher results than those obtained using the EU test procedure. This means that this is a conservative value of leakage and to some extent accounts for some variation in the assembly process for vehicles. This result is required to be reported for all vehicles sold in the state of Minnesota in the USA.

3.3. IRREGULAR EMISSION

Irregular leak occurs during the car usage when system has failures caused by internal or external reasons. External reasons such as car accident or condenser damage by stones during driving or emissions by service are well understood to be part of irregular emissions. However, few considerations were taken on bad part quality, or bad vehicle assembly which also leads to system irregular leaks. Now this number is estimated to be 3.5g/yr in **GREEN-MAC-LCCP**© model. This number should be a function of the vehicle lifetime and the expected exposure to severe climatic and road conditions which is not yet comprehended in the model.

What is the number in China? Study will be done on the irregular leaks related to part quality and vehicle assembly in this paper.

According to the data from more than 50 dealers and the survey result from 8881 car owners in China, this study will take into account of this kind of irregular emissions to calculate the average refrigerant irregular emissions happened

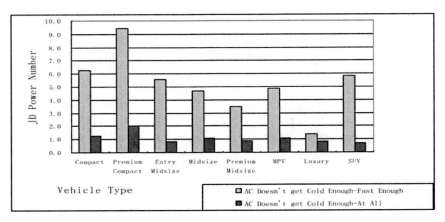

Chart 4. JD Power Performance on Vehicles

on one-year new vehicles due to part quality and vehicle assembly.

3.4. EMISSIONS DURING SERVICE

This leakage element is one to comprehend the loss of refrigerant that occurs during servicing of the vehicles. The **GREEN-MAC-LCCP©** estimates the number of services in a vehicle lifetime and considers this leakage for each service over the life of the vehicle. This model also considers a different loss in the case of professional repair and Do-It Yourself service. The losses are attributed to loss in connection of the service hoses to the system and any losses from the service cylinder that may occur.

3.5. END-OF-LIFE (EOL) EMISSIONS

At end of life, not all countries require recovery and reclamation of the refrigerant and even where it is required by law it is not always enforced. The losses in this case include losses similar to those in service but also consider the reclamation cylinder losses that may occur.

4. IRREGULAR REFRIGERANT LEAKAGE RATE CALCULATION

4.1. JD POWER DATA SUMMARY

JD Power data has been published for 2008. According to the questionnaire that was conducted by JD Power Company, it contained information on vehicle owner's age, vehicle mileage, vehicle type and complaints on air conditioning system. Two items were related in the questionnaire sheet which is 'A/C is not cold at all' and 'A/C not cold enough'. Totally 49 brands were investigated and 8881 vehicles were included in 2008's JDP report. This survey is taken at the end of one year of ownership so it represents only the first year of customer assessment and the leakage rate may change over time and the actual leakage is greater in reality.

System charging of refrigerant varies with vehicle size, which leads to different emission values. The calculation will be based on the following vehicle types: compact, premium compact, entry midsize, midsize, premium midsize, luxury, MPV and SUV. Chart 4[6] shows the details:

4.2. IRREGULAR REFRIGERANT EMISSION

Based on the warranty data from OEM and dealers, 'A/C not cold enough' has several reasons including system component is malfunction, air flow from the A/C outlet is not sufficient and refrigerant in system is not enough. Refrigerant leaks from the joints due to assembly failure or bad O-Ring sealing has been estimated to be 30% among all the complained vehicles; while there is around 20% of the complaints on 'A/C not cold at all' are due to the refrigerant leakage has reached 50% of the total charge.

Below are the formulas used for this calculation:

$$E_{ir1} = JDP_1 / 100 * S_{ample} * L_{eak1} * C_{harge} * K$$

$$E_{ir2} = JDP_2 / 100 * S_{ample} * L_{eak2} * C_{harge}$$

Table 2. Irregular Refrigerant Emission from JD Power

Total PPH	Compact	Premium Compact	Entry Midsize	Midsize	Premium Midsize	MPV	Luxury	SUV
Sample Size	621	392	2212	2924	541	989	733	469
AC Doesn't get Cold Enough-(JDP1)	6.3	9.4	5.6	4.7	3.5	4.9	1.4	5.8
AC Doesn't get Cold -At All-(JDP2)	1.3	2.0	0.8	1.1	0.9	1.1	0.8	0.7
Sales Volume within investigation	386031	110159	1158438	1563577	829373	949293	217815	324085
Charging volume(gram)	400.0	400.0	550.0	600.0	650.0	850.0	650.0	650.0
Refrigerant emission of JDP1 (g)	2343.7	2210.9	10219.4	12368.5	1846.2	6178.8	1000.5	2652.2
Refrigerant emission of JDP2 (g)	626.0	627.2	1946.6	3859.7	633.0	1849.4	762.3	426.8
Total refrigerant emission of the volume (Kg)	1846.0	797.6	6371.4	8677.9	3800.6	7705.9	523.8	2127.6

From where:

S_{ample}: Sample size of the vehicle in China sold in 2008

L_{eak1} : Leakage percentage among vehicles that A/C is not cold enough, around 30% of the total vehicles with this complaint

L_{eak2} : Leakage rate among vehicles that A/C is not cold at all, around 20% of whole vehicles in this complaint had the leakage of whole charge

C_{harge} : Vehicle charge volume, varying from different vehicle size and type.

K :   Consider that the customer does not complain until at least half of the total charged volume is leaked from his vehicle, thus K is estimated to be 0.5.

So the total leakage amount of all the sample vehicles will be calculated by this formula:

$$E_{irS} = (E_{ir1} + E_{ir2}) * V_{olume} / 1000 \quad (kg)$$

E_{irS}: Emissions from the sample vehicles under investigation

V_{olume}: Total vehicle volume that represented by JD Power survey

Take the Compact vehicle as an example to explain the calculation:

$$E_{ir1} = JDP_1 / 100 * S_{ample} * L_{eak1} * C_{harge} * K$$

$$= 6.3/100 * 621*0.3*400*0.5$$

$$=2343.7 \text{ gram}$$

$$E_{ir2} = JDP_2 / 100 * S_{ample} * L_{eak2} * C_{harge}$$

$$=1.3/100*621*0.2*400$$

$$=626 \text{ gram}$$

$$E_{irS} = (E_{ir1} + E_{ir2}) * V_{olume} / 1000$$

$$= (2343.7+626.0)*386031/1000$$

$$= 1846.0 \text{ Kg}$$

By using the same calculation as above, all the irregular refrigerant emissions are carried out as shown in Table 2.

All the irregular leakage will be calculated by the accumulative of all the vehicles under investigation with JD Power numbers, thus the total emissions would be:

$$E_{irTotal} = \sum_{i=1}^{8} E_{irT} / 1000 \quad (Ton)$$

i: vehicles types listed in above table.

After applying this calculation to all the sales vehicles in China in 2008, the total irregular emission would be 38.1 Ton from the JD Power result, then the average irregular refrigerant emission due to part quality and vehicle assembly would be 5.8g/yr, which can be a reliable number used in the **Green-MAC-LCCP** model for China vehicles refrigerant emissions.

<table 2 here>

5. CONCLUSION

This study used the JD Power data in 2008 and the warranty estimation from different car manufacturers to calculate the irregular refrigerant emissions by part quality and vehicle assembly. The calculation result can be an input to the **GREEN-MAC-LCCP©** model of the China market in the Leakage Input sheet.

Since JDP is the complaints on the vehicles only in the first year so we should consider extrapolating this failure rate to later years in the vehicle life.

Although the irregular refrigerant emission accounts for a smaller percentage of the total emission than the regular one, when considering the huge amount of the vehicle volume, it can not be ignored. High attentions on system robust design and reliability are necessary and actions need to be taken now.

6. REFERENCES

1. Chen Jiangping. Prof. Shanghai Jiaotong University, **Introduction of China MAC Industry**, MAC Workshop in Shanghai, Nov.2008

2. Schwarz Winfried (Öko-Recherche) & Harnisch Jochen (Ecofys), Final report **Establishing the Leakage Rates of Mobile Air Conditioners**, Apr. 17, 2003

3. Papasavva, S., Hill, W.R., and Brown, R.O., "GREEN-MAC-LCCP®: A Tool for Assessing Life Cycle Greenhouse Emissions of Alternative Refrigerants," *SAE Int. J. Passeng. Cars - Mech. Syst.* 1(1):746-756, 2008.

4. Statistics on China vehicle sales volume in 2008 published on Jan. 8, 2009

5. **COMMISSION REGULATION (EC) No 706/2007 of 21 June 2007**, Official Journal of the European Union

6. JD Power result for China vehicles published in Oct., 2008

7. CONTACT INFORMATION

Jiangping Chen
Shanghai Jiaotong University, Institute of Refrigeration & Cryogenics Engineering
Tel: 0086-21-3420-6775
jpchen70@yahoo.com.cn
jpchen@sjtu.edu.cn
Add: No.800, Dongchuan Road, Minhang District, Shanghai, China
Post code: 200240

8. ACKNOWLEDGEMENTS

The authors acknowledge of the support from China Vehicle Secretary for the vehicle sales data in 2008 of China. Thanks to Stella Papasvva's explanation on her model to evaluate the refrigerant leakage. Appreciation will also go to Tam Nguyen's help on the paper composition.

When the paper was written in the beginning of 2009, Bill Hill was working in GM while he got retirement in Oct. 2009, special thanks goes to his help with the vehicle fleet data and guidance on the paper.

Review of CO$_2$ Emissions and Technologies in the Road Transportation Sector	2010-01-1276 Published 04/12/2010

Timothy V. Johnson
Corning Incorporated

ABSTRACT

The topic of CO$_2$ and fuel consumption reductions from vehicles is a very broad and complex issue, encompassing vehicle regulations, biofuel mandates, and a vast assortment of engine and vehicle technologies. This paper attempts to provide a high-level review of all these issues.

Reducing fuel consumption appears not to be driven by the amount of hydrocarbon reserves, but by energy security and climate change issues. Regarding the latter, a plan was proposed by the United Nations for upwards of 80% CO$_2$ reductions from 1990 levels by 2050. Regulators are beginning to respond by requiring ~25% reductions in CO$_2$ emissions from light-duty vehicles by 2016 in major world markets, with more to come. The heavy-duty sector is poised to follow. Similarly, fuel policy is aimed at energy diversity (security) and climate change impacts. Emerging biofuel mandates require nominally 5-10% CO$_2$ life cycle emissions reductions by 2020. Processes that utilize plant cellulose and waste products show the best intermediate term potential for meeting these goals, but long term trends are towards biofeedstocks for refineries.

Vehicle technologies are emerging to meet the regulatory mandates. Light-duty engine efficiency gains will result in about 30% fuel and CO$_2$ reductions by 2015. Many of the reductions will come from the use of direct injection technology in gasoline engines, and downsizing diesel and gasoline engines for more specific power. CO$_2$ savings shows a general linear relationship with cost. Diesel hybrids offer the greatest CO$_2$ reduction potential. Plug in hybrids can lead to heavy electrification of the fleet for energy diversity and greenhouse gas reductions, but their CO$_2$ reductions are moderate and expensive. Battery performance is generally acceptable, but cost will be a recurring issue. Most light-duty efficiency technologies return money to the consumer over the life of the vehicle, so the CO$_2$ reductions also come with an economic gain to the owner.

In the heavy-duty sector vehicle and operational improvements offer the best gains at 16 to 28% fuel reductions. Engine technology trends are indicating nominally 15% reductions using advancements in currently utilized technologies. Research is shifting to gasoline engines, wherein upwards of 20-25% CO$_2$ reductions might be realized. Heavy duty hybridization is emerging for vocational and urban vehicles, and can offer a 2 to 4 year payback period.

Black carbon reductions from vehicles can have a profound effect on GHG impact, accounting for upwards of ~20% of CO$_2$ reductions proposed by the Intergovernmental Panel on Climate Change (IPCC) by 2050.

INTRODUCTION

For more than 35 years vehicle emissions regulations have focused primarily on criteria pollutants, such as hydrocarbon (HC), NOx, CO, and particulate matter (PM). These regulations had and still have a profound impact on powertrain technologies, ranging from fuels and lubrication oils, to engine technologies and emission control systems, across all vehicle and equipment categories. Many argue that no other trend has influenced powertrain technology more than this 99% tightening (nominal) of pollutant emissions. However, in some sectors, such as for gasoline multi-port injection engines, the rate of emissions technology progress has slowed as a result of maturity. For example, in 1999 there were roughly 30-40 SAE papers related to advanced catalysts for gasoline engines. In 2009, the number was about a quarter of this.

On the other hand, tightening CO_2 regulations are just starting and are poised to have a similar impact on powertrain technologies as the historic tightening of criteria pollutants. California regulators finalized the first CO_2 regulations for passenger cars in 2005, followed by Europe in 2009. (In both cases, proposals were made much earlier.) The US Environmental Protection Agency (EPA) is proposing similar regulations through 2016. The United Nations Intergovernmental Panel on Climate Change (IPCC) is proposing CO_2 targets for 2050 and beyond, which will likely drive CO_2 regulations into other vehicle and equipment segments. As such, it is reasonable that it could be more difficult to meet emerging CO_2 regulations than emerging criteria pollutant regulations.

There are numerous technologies being considered to reduce CO_2 emissions. These included low carbon fuels, (e.g. biofuels); advances in engine technologies, like direct injection gasoline engines, cooled exhaust gas recirculation (EGR) for gasoline engines, and downsizing; and electrification of the drivetrain, such as with hybrid electric vehicles (HEV), plug-in hybrid electric vehicles (PHEV), and battery electric vehicles (BEV).

The objective of this paper is to provide an introductory, high-level review of the current status of mobile CO_2 regulations and technologies to address them. The paper is intended to provide a broad perspective on the topic rather than a deep analysis on any given topic. It should be noted that one can look at reducing CO_2 emissions more broadly - reducing energy consumption, increasing fuel diversity and energy security, and improved efficiency. The points made in this paper are generally applicable to all these issues.

Because fuel diversity and CO_2 emissions directions are being driven by fossil fuel availability and climate change issues, the review begins with global fuel production and consumption trends, along with CO_2 projections proposed by the IPCC, followed by the regional regulatory response. Then comes a technology overview on fuels, and light-duty and heavy-duty engine technologies, including powertrain hybridization.

REGULATORY OVERVIEW

This section will summarize the environmental and resource drivers for reducing CO_2 and fuel consumption, and the regulatory framework for moving forward.

The first subsection outlines issues regarding petroleum reserves and consumption, which, via energy diversity and energy security arguments, drive fuel economy regulations. The second subsection covers climate change drivers behind specific CO_2 emissions regulation, or "equivalent" CO_2

regulation, which encompasses most climate change agents. The third subsection will summarize fuel economy or CO_2 emission regulations.

PETROLEUM

Dwindling oil reserves and expected increases in global consumption are part of the argument for the need to regulate fuel consumption (and thus CO_2) in the transportation sector. The availability of oil is a complex dynamic of economics, technology, and distribution. This section will look at the gross impacts of oil reserves and consumption.

Global oil consumption is about 85 million barrels per day or 31 billion barrels per year. As shown in Figure 1, non-OECD (Organization of Economic Cooperation and Development) oil consumption is increasing at more than 3X the rate of OECD countries (1). Given that half of oil production goes to transportation, and that vehicle penetration rates in non-OECD countries is very low compared to OECD countries, significant future oil demand will come from the developing countries. Globally, oil consumption is predicted by EIA (US Energy Information Administration) to grow from 0.5 to 1.5% per year, and reach about 33 to 44 billion barrels per year by 2030, depending on price (2).

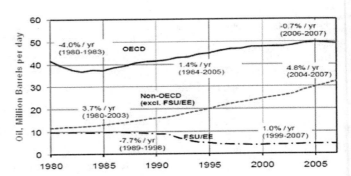

Figure 1. Non-OECD countries' oil consumption has been growing ~3X faster than OECD countries.

On the other hand, largely due to technology advancements in discovery and recovery, proven oil reserves have increased from 998 billion barrels in 1988, to 1068 billion barrels in 1998, to 1261 billion (excluding 150 billion barrels of Canadian oil sands) in 2008 (3), for an average growth rate of 1% per year. This is similar to the consumption rates. As such, the proven oil reserves to production ratio has held relatively constant at 40-43 years since the mid-1980s, excluding the oil sands. (Pre-1980 the ratio was at the mid-30 year level.)

Looking forward, insights can be made by stepping away from the conservative "proven reserves", which have a >90% probability of being exploited with current technology at current prices, to "probable reserves", which have a >50%

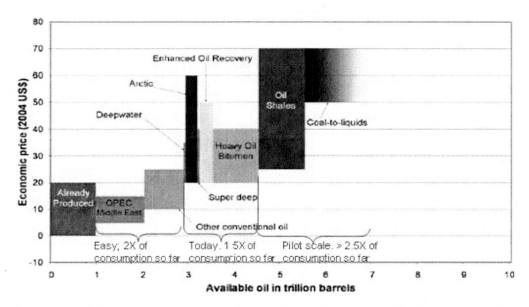

Figure 2. Estimates of probable oil reserves as a function of price (4). Notes were added by author. At EIA consumption estimates for 2030 of 44 billion barrels per year, these reserves would last >100 years.

probability of being extracted. Figure 2 shows one such estimate (4). At today's price of about $85 per barrel, reserves might be on the order of 6 trillion barrels, or 6X what we've consumed thus far. At oil consumption rates of 44 billion barrels per year (year 2030, EIA high consumption), the probable reserves to production ratio is >130 years.

From this perspective, it appears that there are enough hydrocarbon reserves to last for the foreseeable future. However, political, environmental, resource (like water) and other constraints might limit production, reducing these reserves considerably. Also, energy diversity and security issues are significant motivations for tightening fuel consumption standards.

<figure 2 here>

CLIMATE CHANGE AND REGULATORY RESPONSE

Given the UN IPCC consensus statement that there is a >90% chance that anthropogenic greenhouse gases are warming the globe (5), CO_2 emissions from the burning of fossil fuels would appear to drive vehicle fuel consumption and CO_2 regulations more than depletion of hydrocarbon reserves. This subsection will look at the climate change issues and the regulatory response.

Climate Change

Figure 3 summarizes the IPCC's analyses on anthropogenic CO_2 and the impact on CO_2 concentrations in the air (5). The Panel's analyses show that the 450 ppm goal for CO_2 in the

atmosphere is the best balance of warming potential (40-60% probability of stabilizing at ~2C° above pre-industrial levels) and reasonable reduction measures (−3% per year to 2050). To attain this point of stability, nominally 80% CO_2 reductions are needed by about 2050. This brings total global emissions roughly equivalent to 1910 levels. From a vehicle industry perspective, to do its share the 80% reduction is for the whole in-use fleet, not just new vehicles.

At the 2009 UN Climate Change Conference in Copenhagen, the US pledged (nonbinding) to drop CO_2 emissions by 1.3% by 2020, 3.1% by 2030, and 80% by 2050 versus 1990 levels; the EU pledged unconditional 20% reductions by 2020, and 30% if other developed countries follow; Japan: 25% by 2020; China: 40-50% reductions in carbon intensity (CO_2 normalized to GDP) versus 2005 levels, by 2020. All these proposals are nonbinding and will be negotiated in subsequent meetings. However, the pledges suggest CO_2 reductions are in serious discussions and now a matter of negotiation.

The IPCC also made estimates of greenhouse gas (GHG) sources and proportional impact. About 13% of GHGs come from the transportation sector. About 77% of all GHG contributions are from CO_2, 14% from methane; and 8% from N_2O. About 57% of the CO come from fossil fuels, while most of the CO_2 balance is from deforestation.

<figure 3 here>

Black carbon is a short-lived climate-forcing agent, staying in the atmosphere for weeks. However, the IPCC estimates it is

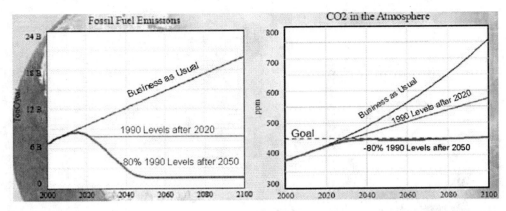

Figure 3. The UN IPCC models show that 80% reductions in anthropogenic CO_2 are needed to stabilize CO_2 levels to 450 ppm in the atmosphere (5).

roughly equivalent to methane in its warming potential at present levels. Because of its impact on snow melting and in the atmosphere, in the arctic black carbon might represent about 20% of total GHG impact (6). Globally, about 17% of black carbon emissions come from transportation (7). As black carbon has about 2000X more atmospheric warming potential than CO_2 on a mass basis (8), about 20-25% of an unfiltered diesel vehicles carbon footprint is in black carbon. Remediation of diesel soot today is primarily done to minimize the adverse health effects, but the climate forcing impact could further increase interest.

Regulatory Response

In Europe, road transportation accounts for about 20% of CO_2 emissions, and passenger cars are about 60% of this or 12% of the total inventory (9). In the US, road transportation is about 28% of GHG emissions with about 15% coming from light duty trucks and passenger cars (10). From 1990 to 2004, transportation GHG emissions increased 24% (11). As such, and provided that emissions from passenger cars and trucks are closely regulated with an established framework, the industry has emerged as the first industry to have binding CO_2 regulations. California emerged first with automobile regulations in 2004, dropping emissions from about 240 g CO_2 (equivalent - includes all GHG normalized to CO_2) per km in 2009 to 170 g/km in 2016. However, these required a waiver from the US EPA before they could go into effect. The waiver was granted in 2008 and the regulations went into effect. However, the US government put forth a Notice of Proposed Rulemaking (NPRM) in September 2009 that tightens average new vehicle fleet requirements 25% to 170 g CO_2 (equivalent) per km in 2016 for all of the US. The goal is to have this regulation finalized in March 2010. If implemented as planned, California agreed to forego their requirements as the EPA requirements require very similar reductions. As such, a 50-state CO_2 regulation would be in effect. (Note: To provide the reader with an equal basis of

comparison, the CO2 values were normalized to the New European Drive Cycle using the methods developed by F. An, et al. (12).)

Europe finalized their first CO_2 mandates in March 2009 at 130 g CO_2/km for an average sized car of 1372 kg in 2015 (13). In the next round of tightening the European Union set a target of 95 g CO_2/km for 2020 (12), to be reviewed in 2013. Similarly, California (and 13 other states following its lead; about 40% of the US market), is targeting 40-50% reductions from 2009 baseline levels by 2025 (14).

F. An, et al. (13), compared the fuel economy and CO_2 standards around the world on a normalized basis - CO_2 emissions on the New European Drive Cycle (NEDC). Results are shown in Figure 4. On this basis, Japan's fuel consumption regulations are similar to Europe's CO_2 regulations. The US is about 30% higher than Europe or Japan. The US CO_2 regulation is based on a vehicle's footprint (area between the tires), while that in Europe are based on a vehicle's mass. (Every automaker will have a different fleet average CO_2 emission dependent by mix.) However, when this author looked at about 20 automobiles for footprint and weight, and adjusted for test cycle differences using An's method, the European regulation is about 15% tighter car-for-car than the US regulation. So, roughly half of the difference between the US levels and those in Europe is due to larger vehicles in the US fleet.

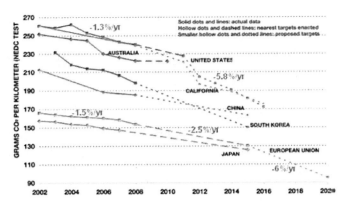

Figure 4. Historic CO_2 emissions (solid lines), and enacted (dashed lines) or CO_2 regulations (dotted line) for various countries with fuel consumption or CO_2 regulations (13). Data were normalized to CO_2 emissions on the NEDC test. Percent improvements per year were added.

Also shown in Figure 4, without mandated regulations the average rate of improvement shown here is about 1.5% per year. This might be regarded as a market rate of improvement that is largely consumer driven. Note that moving forward, the regulations are forcing much greater rates of improvement, up to 6% per year reductions. In other words, the governments are requiring greater improvements than the automotive companies perceived were needed from market forces. In this way, CO_2 reductions are being mandated similarly to criteria pollutants, a paradigm shift that should result in faster technology evolution.

Heavy Duty

In 2006 Japan introduced the first heavy-duty vehicle fuel economy standards in the world (15). It calls for nominally 12% increases in fuel economy (km/liter) from a 2002 baseline by 2015. To estimate improvements, computer simulations of various vehicles are used to conduct and analyze engine dynamometer tests. Most of the reductions are expected to come from engine improvements.

In the US, in July 2008 the EPA published an Advanced Notice of Proposed Rulemaking (ANPRM) concerning vehicle CO_2 regulations, and requested comments on regulating the heavy duty truck and non-road sectors (16). Based on this information, the EPA is targeting a proposal in the first half of 2010 for regulating CO_2 from heavy-duty trucks. Converse to light-duty applications, wherein vehicle weight and engine technologies can have the biggest impacts, for trucks, chassis and vehicle improvements, like aerodynamic cowls and low rolling resistant tires can have the biggest impacts. In this regard, the EPA has proposed fuel consumption chassis test cycles for a variety of applications as part of their Smartway program. Also, in December 2009

California finalized HD tractor-trailer truck greenhouse gas regulations, with phase-in beginning in 2010 and proceeding through 2017 (17). The rule requires EPA Smartway cowling and tire technology on all such trucks operating in the state. Smartway is the US EPA program that encourages adoption of fuel efficient technologies.

REGULATORY CONCLUSIONS

Regulations governing vehicle emissions of CO_2 (eq.) are just now emerging, but are poised to be a long term trend. They are primarily being driven by increasing concern about anthropogenic impacts on climate change, but are also synergistic with energy diversity policy (e.g., EISA). The long term regulations will directionally follow the IPCC's recommendation to drop CO_2 (eq.) emissions by 80% from 1990 levels by 2050 for the entire in-use fleet. In the US and Europe, light duty regulations are in place or being proposed to drop new vehicle emissions about 20-25% from 2008 levels by 2015-16, with increased annual percentage reductions targeted through 2020 in Europe. Given that technologies for reducing criteria pollutants, like NOx, hydrocarbon, particulates, and CO, have been commercialized for more than 30 years, it is reasonable that meeting the future CO_2 emissions standards will be more challenging than future criteria pollutant standards.

BIOFUELS

One of the cornerstones to decreasing transportation CO_2 emissions and reducing dependency on petroleum is to move towards low-carbon fuels. This can mean such diverse fuels and sources as low-carbon intensity electricity, hydrogen, and natural gas. Given the interest, legislation, and attractive short term potential of biofuels, this section will deal exclusively with this low-carbon fuel and petroleum replacement. The more significant mandates that will drive the field are in the US and Europe.

GOVERNMENT MANDATES

In November 2007, the US passed the Energy Independence and Security Act (EISA). It calls for 36 billion gallons of biofuels per year (~20% penetration) for the transportation sector by 2022. Figure 5 shows the general requirements for the fuels that can be used (18). In May 2009, as required by the act, the EPA proposed annual ramp up rates for each fuel type, and life cycle GHG calculation methodologies (19). Significant additional GHG is emitted for land use changes, but with time, the CO_2 reductions make up for this. As such, the time horizon and a way to compensate for early reductions versus later ones (discount rate) becomes important. Figure 6 shows the results of the full net CO_2 life cycle analyses for a 100 year time horizon at a 2% discount rate for a variety of biofuels. In these analyses, corn ethanol using combined heat and power (CHP) and sugar cane

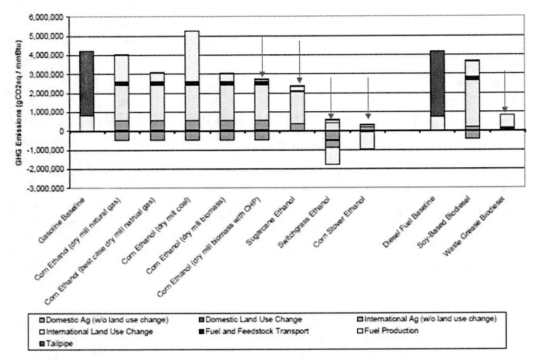

Figure 6. CO₂ life cycle analyses for various biofuels over a 100 year time horizon at a 2% discount rate ([19]). Using these analyses, the identified fuels meet the cellulosic (>50% reduction) or advanced biofuel (>60%) requirements of EISA.

ethanol meet the cellulosic fuel requirement (>50% GHG reduction) of EISA, and cellulosic ethanol from switchgrass and corn stover, or biodiesel from waste grease meet the advanced biofuel requirement (>60% reduction). Economic impacts are estimated from $4 to $18 billion per year or < $0.11 per gallon for oil priced at >$53 per barrel.

California adopted the Low Carbon Fuel Standard (LCFS) in April 2009 ([20]). Its aim is to reduce greenhouse gas emissions from California's transportation fuels by 10% by 2020. The standard sets carbon intensity (CI) targets for each year and is phased-in gradually, with the bulk of the reductions required in the last five years (3.5% penetration in 2016 going to 10% in 2020). This allows for the development of advanced, lower CI fuels and more efficient advanced-technology vehicles. The baseline gasoline fuel consists of reformulated gasoline containing 10% corn ethanol and ULSD is the baseline diesel fuel. The LCFS takes a cradle-to-grave model analysis to calculate full life-cycle GHG emissions associated with producing, transporting, and burning the fuels, including both direct and indirect land-use effects. In December 2009, governors of 11 Northeast states signed a memorandum of understanding to develop a regional LCFS, with the goals to be determined.

Type of Fuel	BGY
Total Renewable Fuels by 2022	**36 BGY**
Corn Ethanol	15 BGY cap
Advanced Biofuels – Includes imported biofuels and biodiesel. Includes 1 billion gpy biodiesel starting in 2009. All must achieve ≥ 50% reduction of GHG emissions from baseline*	21
Cellulosic Fuels – Includes cellulosic ethanol, biobutanol, green diesel, green gasoline. All must achieve ≥60% reduction of GHG emissions from baseline*	16

*Baseline = average lifecycle GHG emissions as determined by EPA Administrator for gasoline or diesel (whichever is being replaced by the renewable fuel) sold or distributed as transportation fuel in 2005

Figure 5. Biofuel mandates (BGY, billion gallons per year) by the US Energy Independence and Security Act (EISA, November 2007). The mandate represents about 20% of transportation fuels ([18]).

<figure 6 here>

In December 2008 the European Council approved the Renewable Energy Directive, which will be implemented by November 2011 ([21]). It calls for >10% biofuel mandate in the transportation sector, and each member state needs to implement it into law (ramp up rates etc) by November 2010. The biofuel must be sustainable and decrease GHG emissions by 35% in 2010 and 50% by 2017 (60% for new installations). Second generation biofuels (cellulosic and wastes) get a 2X credit, and renewable electricity gets a 2.5X credit. The GHG reductions include cultivation, processing, and land use changes, but do not encompass indirect land use changes, which are being analyzed. Examples of default GHG reductions range are 19% for palm oil biodiesel, 52% from sugar beet ethanol, and 83% from bio-waste biodiesel. Given that the EU imports diesel fuel and has an excess of gasoline for export, current biofuel production splits are about 75% biodiesel and 25% ethanol.

BIOFUEL TECHNOLOGIES AND PROPERTIES

Biofuel technology is rapidly evolving to try to meet the regulatory mandates outlined in the previous section. The technology is quite diverse and dynamic. This is illustrated in Figure 7 (22). Feedstocks range from sugar sources (cane, beets, corn) to organic waste to "designer" crops like algae. Processing routes evolve from the current ethanol processes (fermentation and purification) and biodiesel route (vegetable oil esterification and purification), to second generation processes using enzymes or thermal processing to utilize plant cellulosic materials. Third generation systems use energy crops (like algae) and biofeedstocks that are integrated into the refinery process.

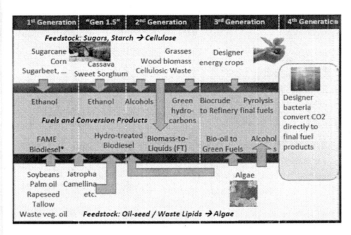

Figure 7. Illustration of biofuel process diversity and evolution (22). Gasoline based fuels are on the top, and diesel fuels are on the bottom. Eventual evolution of biofuels could be to biocrudes for refinery feedstocks.

Virtually all of the engine manufacturers and oil companies desire a refinery-integrated biofuel approach. Engine makers favor this approach because most gasoline engines are designed to operate on <10% ethanol, but higher blends will be needed to meet EISA targets. This potentially can cause backward compatibility issues in a wide variety of applications. Examples might be difficulty in air:fuel management for open-loop control automotive engines (23); too lean operation of fixed carbureted small non-road engines; fiberglass fuel tank failures in marine applications;

and a variety of other material compatibility issues (24). Glycerine in biodiesel blends can precipitate out in cold conditions, plugging fuel filters. Ash can cause fuel injector corrosion or fouling and prematurely plug diesel particulate filters. Also, biodiesel can cause higher levels of lube oil dilution because it has a higher distillation temperature (more heavy hydrocarbons) than diesel fuel. On the refinery side, biofuel blends need to meet standards to allow transport by pipeline, requiring terminal blending; and in general, oil companies desire efficiencies of scale offered by refinery operations.

These potential problems are addressed if biofeed stocks are catalytically hydroprocessed to produce biodistillates, generally known as renewable diesel. Several processes for renewable diesel production are now in commercial use. These include stand-alone processes by Neste Oil (to produce NExBTL™) and UOP (Ecofining™), as well as ConocoPhillips' co-processing of triglycerides with petroleum diesel feedstocks. All these processes require hydrogen and are conducted under high pressure. The products are hydrocarbons (not oxygenates), that are very similar to those found in petroleum diesel (25).

One issue with this approach is that most raw biofeedstocks are not efficiently transported more than about 100 miles before CO_2 benefits are consumed by transportation emissions. To expand feedstock access to processing facilities, they are concentrated close to the source. Such a scheme is depicted in Figure 8 (26), wherein Archer Daniels Midland (ADM) is teaming with ConocoPhillips (COP).

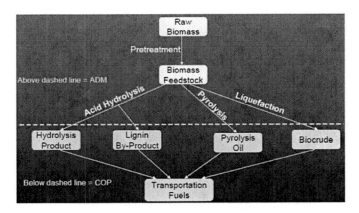

Figure 8. Schematic of a joint effort by Archer Daniels Midland (ADM) and ConocoPhillips (COP) to synthesize and process biocrude (26).

Regarding properties, biofuels have a range of energy densities. Figure 9 shows comparisons of the energy density of a variety of transportation fuels (27). Because it has more oxygen and lower carbon and hydrogen contents, biodiesel is variable but has up to 10% lower volumetric energy density, as shown here, but is more typically 5 to 6% lower than

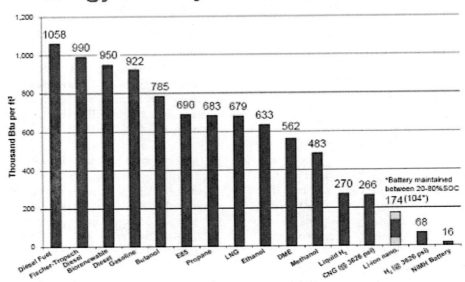

Energy Density of Various Fuels

Figure 9. Comparison of volumetric energy densities of a variety of current and potential transportation fuels (27). Biodiesel has 5 to 10% lower volumetric energy content than diesel, while ethanol has 30% lower energy content than gasoline.

diesel, according to a literature survey by the CRC (25). When blended with petroleum fuels, B20 (20% biodiesel) will have very similar fuel economy (e.g., miles/gallon) as diesel fuel. Conversely, ethanol has 30% lower volumetric energy than gasoline so E85 will have nominally 25% lower fuel economy than gasoline.

Figure 10 shows a comparison of key properties of fatty acid methyl ester based biodiesel (FAME) and renewable biodiesel with No.2 ultra-low sulfur diesel fuel (25). Noteworthy is the increased cetane level of both biodiesel and renewable diesel relative to diesel. Emissions of CO, particulate matter (PM), and hydrocarbons are generally reduced about 10-20% for B20. NOx emissions are roughly the same as with diesel.

Ethanol has an octane number, (R+M)/2, of about 100, so it is an octane enhancer to gasoline. When used with legacy vehicles, there is a trend towards lower non-methane hydrocarbon, PM and CO emissions, but acetaldehyde and formaldehyde emissions increased (28). Catalyst temperatures for engines with closed-loop control decrease. The vapor pressure of E3 is about 13% higher than that of gasoline, but then is flat up to E20 levels (29). Hot soak and diurnal evaporative emissions were tied to the vapor pressure but were ~6X higher for E3 than for the base fuel. However, they were still within the 2 gram/test regulatory requirement for the vehicles.

<figure 9 here>

Property	No. 2 Petroleum ULSD	Biodiesel (FAME)	Renewable Diesel
Carbon, wt%	86.8	76.2	84.9
Hydrogen, wt%	13.2	12.6	15.1
Oxygen, wt%	0.0	11.2	0.0
Specific Gravity	0.85	0.88	0.78
Cetane No.	40-45	45-55	70-90
T_{90}, °C	300-330	330-360	290-300
Viscosity, mm²/sec. @ 40°C	2-3	4-5	3-4
Energy Content (LHV)			
Mass basis, MJ/kg	43	39	44
Mass basis, BTU/lb.	18,500	16,600	18,900
Vol. basis, 1000 BTU/gal	130	121	122

Figure 10. Some general properties comparisons between diesel fuel, FAME, and renewable diesel. Cetane levels are increased for the biofuels, but renewable diesel has more-similar properties to diesel (25).

BIOFUEL CONCLUSIONS

Governments are beginning to address energy diversity, energy security, and climate change issues with biofuel mandates. The US will be requiring that about 20% of the transportation fuels have >50% of the GHG emissions reductions of conventional fuels by 2022. Europe will require 10% penetration of similarly effective fuels by 2020. Methods for determining the carbon intensity of the fuels are in the proposal stage. Candidates to meeting these requirements use the plant cellulose or are based on waste products. It appears that the long term trend is towards integrating biofeedstocks into refinery operations. Biodiesel has about 5-10% lower energy content than diesel but a higher cetane value. Fuel economy impacts will be negligible for common blends. HC, CO, and PM emissions are about

10-20% less for B20 than for diesel fuel, but NOx emissions are roughly similar. Ethanol has about 30% lower energy content than gasoline, but higher octane levels. Similar to biodiesel, HC, PM, and CO emissions trend lower as ethanol levels increase, but evaporative emissions go up.

LIGHT DUTY POWERTRAIN TECHNOLOGIES

This section will summarize the developments on engine performance and hybridization. The regulatory mandates governing fuel economy and CO_2 are forcing the automotive companies to aggressively move to increase efficiency. Leading approaches involve improving the powertrain with improved engine performance, hybridization, and new transmissions; and improving vehicle performance by reducing weight, reducing drag and reducing rolling resistance.

IMPROVED ENGINE PERFORMANCE

The opportunities for improved efficiency are illustrated in Figure 11 for a diesel engine (30), but the general breakout is similar for gasoline engines. Most of the reductions, upwards of 65 to 75%, come from improved combustion and thermal management. As such, this section will focus on this opportunity.

Vehicle Energy Flow

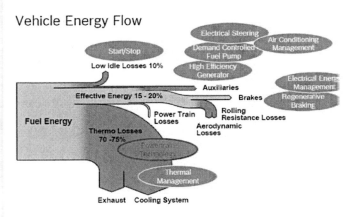

Figure 11. Most of the light-duty vehicle efficiency improvement opportunities reside with engine combustion and thermal management (30).

Measured fuel efficiency depends on the drive cycle being used. Figure 12 shows simulated fuel economy values for an E-class Mercedes equipped with a gasoline or diesel engine and a full hybrid (110 kW engine, 31 kW electric motor), driven on city and highway certification cycles, and on the high load US06 cycle (31). A full hybrid approach uses the engine and battery to the full synergistic extent, but has low or no all-electric range. The diesel performs best in the high-load operation, and the hybrid performs best in the stop-go urban cycle. The US Corporate Average Fuel Economy

(CAFE) and CO_2 values for certification are based on a weighting of 55% city and 45% highway driving. On the other hand, surveys from electronically monitored real-world driving show most of the driving falls between the highway and US06 tests (32). The connection between test cycle CO_2 measurements and real-world emissions will depend on vehicle choice and driver patterns.

In a similar context, it should be noted that certified CO_2 emissions could vary from actual real world emissions due to fuel differences. Certification fuel is often much different from fuels on the market.

The leading gasoline and diesel technology choices for meeting the tighter European and US CO_2 standards include direct injection gasoline, gasoline turbocharging, dual clutch transmissions, and stop-start systems (33, 34). Enablers to these technologies are cooled-EGR (exhaust gas recirculation) for gasoline engines, cylinder de-activation, and variable valve technology. Overall, specific power will increase, enabling significant engine downsizing.

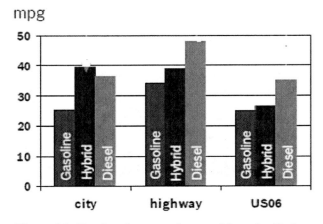

Figure 12. Simulated test results on a Mercedes E-class with the same powertrain power, showing the importance of drive cycle in measuring fuel economy (30). MGP is miles per gallon.

In that regard, engine downsizing is emerging in Europe as a significant technology package for meeting the 2015 regulations. The principle is to run the engine at higher Brake Mean Effective Pressure (BMEP; or load) for lower fuel consumption per unit of energy. As shown in Figure 13, this allows reduced cylinder size to deliver the same net power to the crankshaft (35). In this example at 2000 RPM on a diesel engine, the BMEP is increased from 2 bar to 2.5 bar (+20%) when dropping engine displacement 20%. The smaller engine is consuming 10% less fuel as a result of the more efficient load point. It is estimated this smaller engine will save 20% fuel on the NEDC. (More typical levels are on the order of 10%.) To maintain the same performance as the larger engine

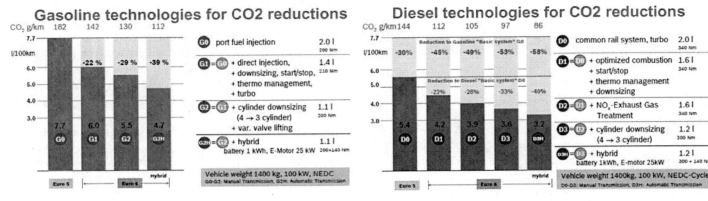

Figure 14. Comparison of incremental gasoline and diesel engine technologies for CO₂ reductions. The diesel 30% CO₂ advantage will be maintained or slightly increase in the near future (31).

at acceptable NOx levels, new technologies for the diesel engine need to be utilized such as increased turbocharging, higher peak cylinder pressure, higher injection pressures, variable valve technology, and more charge air cooling (36). Even so, there are limitations and trade-offs, namely NOx increases may overwhelm the incremental fuel savings, and added cost may limit the technology to higher-priced cars. Also note that the test cycle or drive conditions become quite important. The fundamental fuel consumption savings is limited to low-load operation, as the specific fuel consumption flattens at the higher loads in Figure 13. However, the enabling technologies may offset this, somewhat.

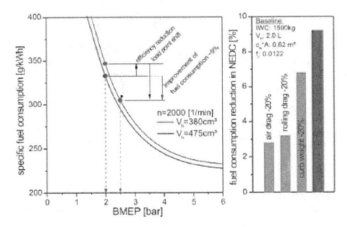

Figure 13. The principle of downsizing a diesel engine. The smaller engine is run at higher BMEP to give a lower specific fuel consumption (35).

Rueger (31) compared incremental technologies for gasoline and diesel engines, and concluded that the diesel advantage on CO₂ (~30%) will be maintained in the near future, albeit with a 9% smaller torque advantage (+100 Nm vs. +120 Nm today). Figure 14 shows the analyses. Note that for an average sized European car (1400 kg), the 2020 CO₂ target

value of 95 g/km is nearly attained for the diesel without hybridization.

One of the more attractive gasoline engine technologies is emerging from the SwRI (Southwest Research Institute) research consortium called HEDGE™ (High-Efficiency Dilute Gasoline Engines). Turbocharging is used to improve efficiencies, and a large amount of cooled-EGR, in the range of 25-45% depending on design, is used to reduce auto-ignition under the higher compression ratios (~14:1 vs. 9-11:1 for other gasoline technologies). It is a stoichiometric engine using standard multi-port injection. If the hallmark of the technology is cooled-EGR, as implied by the name, the technology is already being partially implemented in the 2010 Toyota Prius. Figure 15 shows the BSFC (Brake Specific Fuel Consumption) map for a 2.4 liter spark-ignition engine that was retrofit with the technology (37). Given that high-load fuel consumption is 10-30% lower than for the base engine, it has good low-end torque, and peak BMEP is quite high (+25% vs. direct injection engines), the concept is quite amenable to downsizing and downspeeding. Issues to resolve include ignition stability and slow flame propagation caused by high levels of EGR; boosting issues raised by lower exhaust temperatures (EGR), high mass flow, and high pressure ratios; and EGR control.

<figure 14 here>

Looking into the long term, several research efforts are aimed at recovering waste energy from the exhaust. BMW reported on using a Rankine turbo-steamer on a 2 liter stoichiometric gasoline engine (38). Steam is generated in the exhaust heat exchanger, which is then used to power a turbine. The water is cooled and returned back to the heat exchanger. Total engine power was increased ~10% in the mid-speed moderate load regime. Exhaust temperatures dropped 300C°. A more advanced system might increase power by 15%.

Thermoelectrics are also being evaluated. The US Department of Energy sees thermoelectrics replacing generators and air conditioner units in light duty vehicles in 7 to 15 years (39). In the first vehicle tests, 1 to 5% fuel consumption gains were reported, with the higher values coming at high load (40).

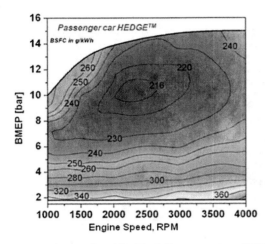

Figure 15. Brake Specific Fuel Consumption (BSFC) map for a 2.4 liter spark-ignition stoichiometric gasoline HEDGE™ engine with multi-port injectors. High load BSFC is 10-30% lower than for the base engine (37).

HYBRIDIZATION

Hybrid electric vehicles (HEVs) are now in their third generation, and have been in the market for 13 years. They are available in virtually every vehicle class. Full hybrids have the lowest CO_2 emissions, with the best delivering >50% reductions from conventional vehicles. However, despite incentives, their market penetration was flat in 2009 at only about 2.5%. Europe is favoring the diesel car over hybrids, and fuel prices are too low in the US to make HEVs appear attractive. As shown in the previous section, they have their best performance in light load, stop-and-go traffic primarily due to their ability to recover braking energy.

Figure 16 shows the well-to-wheel CO_2 emissions estimated using the GREET model (Argonne National Laboratory) for a variety of mid-size HEVs operated on the US city cycle. A number of interesting observations can be made. First, full hybridization drops emissions by about 48% for both gasoline and diesel models. Second, diesel HEVs have the lowest emissions, except for battery electric vehicles (BEVs) on the California grid. For this reason, many European diesel automobile manufacturers are developing diesel HEVs. Under a tight CO_2 regulatory mandate, the diesel HEV looks attractive. Third, the plug-in HEV (PHEV), wherein the electrical grid helps charge the battery, has ~14% higher CO_2 emissions on the US grid relative to HEVs, 5% lower in

California, and 5% higher in Europe (not shown). The PHEV modeled here has a moderate battery size in the power split configuration (vs. series) and powers about 1/3 of the travel in EV mode. More recently, Argonne scientists showed PHEVs might get 47 to 62% of its energy from the grid in typical drive patterns if equipped with 4 to 8 kW-hr batteries (41), and close to 90% of energy from the grid for the largest batteries being considered (16 kW-hr). As such, the PHEV has stimulated significant interest in increasing energy diversity. However, despite their CO_2 emissions they are seen as an important step in moving cars toward the grid to help attain the ultimate 2050 goal of 80% reductions (13).

There are several different system architectures and strategies for PHEVs. Figure 17 shows three major types (32), ranging from low or no all electric range (AER) to significant AER, perhaps up to 65 km (40 miles) as in the upcoming Chevy Volt. It is important to note that in all the examples the vehicle reverts to a typical full HEV when the battery reaches a low state of charge (SOC). In the first case with zero or very low AER, the vehicle is closest to an HEV in operation with all blended operation (engine and battery), except the battery is drained to the low SOC and then recharged on the grid. At the other extreme, the last example has the longest AER (biggest battery), and all the needed power in this range comes from the battery/motor, because it is sized for the series configuration with the engine. The Chevy Volt is an example, wherein it has a 136 kW, 16 kW-hr battery (42). The middle example is a compromise, and the most common configuration emerging for PHEVs with moderate AER (8 to 25 km, 5 to 15 miles), requiring a 2 to 6 kW-hr battery. More transportation fuel is shifted to the grid when going from top to bottom in Figure 17. (Some useful parameters in estimating batteries relative to AER are that about 60 to 70% of the battery capacity is used, and a mid-size car will consume about 0.25 to 0.30 kW-hr per mile in normal driving.)

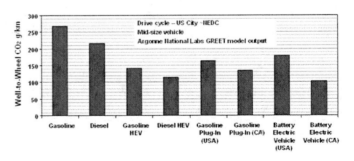

Figure 16. Well-to-wheel CO_2 emissions for various electric powertrains compared to gasoline and diesel vehicles as estimated using the GREET model.

<figure 17 here>

117

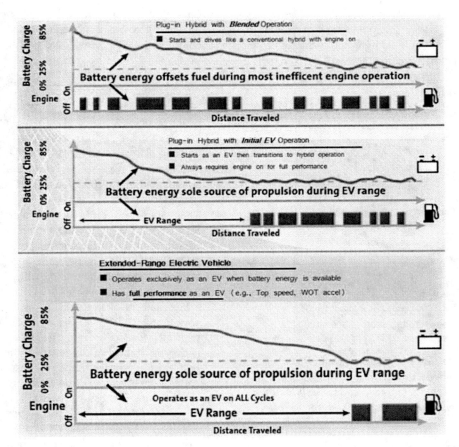

Figure 17. Illustration of the general types of PHEVs being proposed. All concepts revert to HEV mode after the battery is drained to 25% state-of-charge. (13)

Batteries are the single most expensive component and performance determinant in electric drives, thus they are the focus of much work. PHEV lithium ion (Li-ion) batteries are meeting most goals put forth by the US Advanced Battery Consortium (USABC), as shown in Figure 18 for one manufacturer (43), but full electric vehicle batteries fall short on several of these key parameters. The consensus from a 2009 battery workshop concurred with this evaluation, finding that cost was the greatest gap for PHEV batteries; calendar life trends looked good, but there simply was not enough data to confirm attaining the goal (44). Gaps are much more significant on full battery electric vehicles. Weight needs to be taken out, more energy needs to be stored, and costs need to be significantly reduced (~45-50% less than PHEV).

	PHEV 10	PHEV 40	BEV
Battery Size (kWh)	3.4	11.6	40
Battery Weight (kg)	60	120	265
Specific Power - Acc./Regen. (W/kg)	750/500	320/210	300/150
Specific Energy (Wh/kg)	57	97	150
Energy Throughput- X times Size	5000	5000	1000
Cost (US$/kWh)	500	290	150

Figure 18. Current battery status relative to USABC goals for PHEVs and BEVs. Batteries fall short on cost and life (43). Green: goal met; Yellow: goal not yet met but progressing; Red: much more work needed.

Looking further at cost, Nelson (45) evaluated Li-ion battery manufacturing costs for several compositions and capacities. Battery pack costs increased linearly with AER, as shown in Figure 19. Specific usable battery costs for the lowest cost formulation (lithium manganese oxide) decreased from $400/kW-hr for 20 km AER batteries to $300/kW-hr for batteries delivering 65 km AER. Looking at the prognosis for further cost reductions, Nelson estimates abut 47% of the cost is in materials and 17% is in purchased items. Direct labor was only 6% of the total manufacturing cost. Base raw material costs show potential for little change, either way (46, 47). Mock (47) estimated a 90% learning curve for batteries, wherein there is a 10% cost reduction for each doubling of cumulative volume. His survey of costs plateaus at about $250/kW-hr for the battery pack, ($360/kW-hr usable battery), agreeing with Nelson.

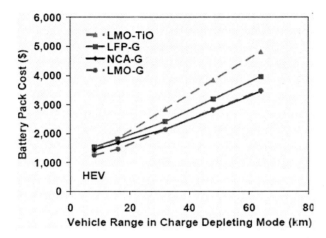

Figure 19. Li-ion battery pack costs increase nearly linearly with AER. 60 kW batteries at 25% SOC. LiMnO, LiFeP, LiNiCoAl; 100,000/yr manufacturing capacity (45)

COST OF FUEL EFFICIENCY FOR LIGHT DUTY APPLICATIONS

There have been numerous studies looking at the cost of fuel savings in the light duty sector. The results are difficult to normalize and assess, as different baselines are used, cost estimates vary, and usually cost values are expressed as percentage increases from the base. Instead of choosing one or two representative studies, the author attempted to normalize baseline comparisons to a mid-sized Tier 2 Bin 5 car and use studies that looked at the different kinds of powertrains, at least to provide internal consistency within each study. The author's estimate of a PHEV with 20 mile AER is also shown. The GREET model provides the CO_2 reductions, and incremental costs to the HEV are for an 8 kW-hr battery at $360/kW-hr (low end), in addition to an $800 on-board grid charging system. The results are shown in Figure 20. The first observation is that there is a wide range of expert estimates for both cost and CO_2 reductions within each powertrain group. However, the group averages provide a relatively tight relationship between incremental cost and incremental CO_2 reductions. After an initially high value for advanced gasoline engines, emissions savings increase linearly with cost increase.

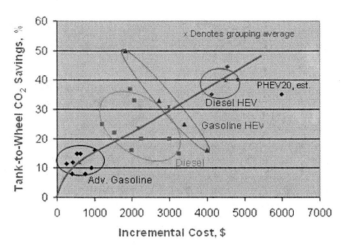

Figure 20. Based on the average results from the literature, CO_2 savings increase linearly with cost. Tier 2 Bin 5 midsize cars.

The above analysis did not take into account fuel savings in the cost estimates, nor did it look at all CO_2 emissions (no well-to-tank or life cycle analyses). In Figure 21 the net cost of well-to-wheel CO_2 (eq) reductions, on a $/tonne basis, is compared for a variety of powertrain and fueling options (48). As others have found (49, 50), incremental improvements to the powertrains on the road today using standard gasoline or diesel fuels save money, so incremental CO_2 reductions have a negative cost. However, as alluded to earlier in the discussion of Figure 4, these fuel savings might not directly translate into vehicle price increases, and the consumer might not pay for them. Also, not accounted for here is the vehicle resale value and a discount rate, both of which can play into net costs. Cellulosic ethanol and biodiesel also come with a moderate cost or even a savings, depending on petroleum price.

LD VEHICLE CONCLUSIONS

This section reviewed the status of fuel consumption and CO_2 reductions in the light-duty sector. Most of the potential savings comes from improvements in the powertrain. Some of the technologies are favored for light-load or city operation, like hybridization and engine downsizing, and some do better in high-load or highway operation, like diesel. Both the gasoline and diesel engines can improve, on the order of 20-30% from today's engines. These technologies are generally cost effective and will save the consumer money over the life of the vehicle. The greatest CO_2 saving comes for the diesel hybrid. To meet IPCC goals of 80% reductions by 2050, electrification of the vehicle and a green grid are necessary. The PHEV is on the vehicle pathway to accomplish this, but it is a relatively expensive option.

<figure 21 here>

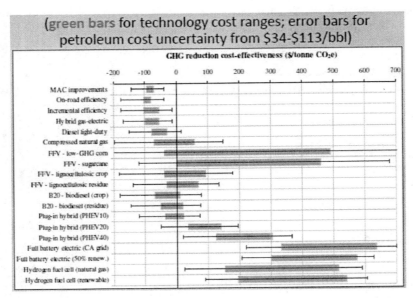

Figure 21. The cost effectiveness of GHG reductions in $ per tonne of CO₂ (eq.) reduced on a well-to-wheel basis. Improvements in powertrains commercially sold today have the lowest costs (48).

HEAVY DUTY POWERTRAIN TECHNOLOGIES

About 6-8% of GHG emissions come from the heavy-duty (HD) truck sectors in the US, Europe, and Japan (51, 52, 53). In the US, about 75% is emitted by the large Class 8 trucks. The HD truck sector is strongly driven to drop fuel consumption, as fuel costs represent 20 to 30% of the total life cycle cost of the truck (second only to wages), and can represent 2 to 2.5X the cost of the truck itself (54). However, many of the fuel-saving technologies have not been introduced to the market because the industry is quite risk adverse, operates on a tight margin, has to face fuel price volatility, and might not have access to good fuel consumption information. For example, many fleet operators require an 18 to 24 month payback period for new technologies, shorter than many feasible technologies can deliver. As such, regulatory pressures are increasing to force fuel savings technologies.

Converse to the light-duty sector, wherein 70% of potential fuel savings comes from engine improvements (Figure 11), in the long haul truck sector 65% potential savings comes from vehicle and operations improvements and 35% comes from the engine(55).

Some examples of technologies and costs for reducing fuel consumption for long haul trucks are reported (56) in an extensive study commissioned by the Northeast States Center for a Clean Air Future (NESCCAF) and by the International Council on Clean Transportation (ICCT). The study evaluated only technologies that are in production or are emerging but have a design specification in the literature.

Figure 22 shows a summary of various results. Relative to a baseline 2010 truck with a 13 liter engine and 10-speed manual transmission, between 1 and 10% fuel consumption savings can be realized with engine and powertrain modifications; 5 to 21% savings can come from operational measures, like low speed driving and double trailers; but 18 to 28% reductions can come from vehicle modifications, such as aerodynamic streamlining and low rolling resistance tires. Using combinations of technologies that are already deployed on some trucks can save 8 to 18% fuel. These technologies include hybridization, turbo-compounding, and the modest Smartway 1 package. Up to 50% fuel savings might be realized with the most advanced technology combinations. Deploying technologies with a three-year payback period ($2.50/gallon, 120,000 miles/yr) can save 17% of the long haul fleet fuel in 2030, and 39% would be saved using technologies that pay back in 15 years (1.2 million miles).

<figure 22 here>

Regarding progress on improving engine efficiency, Stanton showed that advanced engine measures, can reduce fuel consumption by 14% relative to a 2007 production engine running at the same NOx level of ~1.3 g/kW-hr (57). Technologies employed include combustion optimization (high pressure and multiple fuel injections, bowl design, variable swirl, variable valve actuation); advanced EGR (low pressure drop, high flow, advanced cooling); air management (2-stage boost, electrically assisted turbocharger); and advanced controls (mixed mode combustion, closed-loop control). Note that these additional technologies are not included in the NESCCAF report, except variable valve actuation and advanced EGR. If exhaust emission control

PACKAGE NAME	FUEL CONSUMPTION/ CO₂ REDUCTION (%)	INCREMENTAL VEHICLE COST ($)ᵃ	LIFETIME COST OF OWNERSHIP (15 YEARS, 7%)ᵃ	TIME TO PAYBACK *(YEARS)
HEAVY-DUTY LONG HAUL CO₂ AND FUEL CONSUMPTION REDUCTION AND COST RESULTS FOR ANALYZED PACKAGES				
Baseline	n/a	n/a	n/a	n/a
Building Block Technologies				
SmartWay 2007 (SW1)	17.8%²	$22,930	-$23,600	3.1
Advanced SmartWay (SW2)	27.9%²	$44,730	-$55,800	3.8
Parallel hybrid-electric powertrain (HEV)	10%³	$23,000⁴	$100	7
Mechanical turbocompound	3.0%	$2,650	-$5,500	2.0
Electric Turbocompound	4.5%	$6,650	-$5,500	3.5
Variable Valve Actuation (VVA)	1.0%	$300	-$2,500	0.6
Bottoming cycle	8.0%	$15,100	-$4,800	5.2
Advanced EGR	1.2%	$750	-$2,600	1.4
Operational Measures				
Rocky Mountain Double (RMD) trailers	16.1% (grossed out) 21.2% (cubed out)	$17,500	-$34,100⁵	2.1
60 mph speed limit	5.0%	$0	-$13,900	n/a
Maximum Reduction Combination Packages				
Maximum reduction combination 1 (standard 53' trailer, hybrid, BC, SW2, 60 mph)	38.6% (grossed out)⁶ 40.2% (cubed out)⁶	$71,630	-$27,300⁵	4.8
Maximum reduction combination 2 (RMD, hybrid, electric turbocompound, VVA, SW2, 60 mph)	48.7% (grossed out)⁶ 46.2% (cubed out)⁶	$80,380	-$41,600⁵	4.3
Maximum reduction combination 3 (RMD, BC, hybrid, SW2, 60 mph)	50.6% (grossed out)⁶ 48.3% (cubed out)⁶	$89,130	-$37,200⁵	4.7

Figure 22. Technology assessment for fuel savings from long haul HD trucks. Lifetime cost and payback period estimates are based on high volume technology costs, 1.2 million miles over 15 years (120,000 miles per year for payback time), and fuel priced at $2.50/gallon ($0.67/liter). (56)

achieves 97% deNOx efficiency instead of 80% to attain US2010 NOx standards, an additional 10% fuel consumption reduction might be achieved. This is due to the general inverse relationship between fuel consumption and NOx.

The above technologies are relatively advanced but nevertheless are deployed or moving towards deployment. They are incremental. To look at the longer term, Schmidt did a thermodynamic analysis of engine efficiencies and thus opportunities for improvements (58). Improving combustion efficiency improvements can potentially save 5% fuel; friction reduction and improved gas handling - 2% each; and reducing heat loss to the wall 6 to 18%. Heat losses to the cooling water and exhaust combine for ~55% loss of efficiency.

As with light-duty engines summarized above, heavy-duty waste heat recovery systems based on the Rankine cycle (evaporation of organic working fluid, expansion, condensation, return) are being evaluated (59). Heat is taken out of the EGR loop and exhaust system (after emission control system) and converted to mechanical energy for a 6 to 7% fuel savings, depending on level of EGR. A pathway to achieving a potential 9.5% fuel savings was itemized. Figure 23 shows a proposed timeline for implementation of heat recovery systems (or bottoming cycles) in the heavy-duty sector (60).

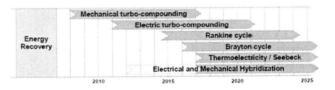

Figure 23. Estimate of implementation timeline for waster heat recovery system for heavy-duty engines. (60)

Combustion research is shifting from diesel- to gasoline-fueled engines with high levels of EGR and premixed combustion, similar to the HEDGE concept summarized in the light-duty section. Reitz, et al. (61), ran a 2.4 liter single-cylinder engine up to 11 bar BMEP at 1300 RPM using 80% multiport injected gasoline and 20% diesel to ignite the charge. They reported 53% indicated thermal efficiency (ITE, no friction or pumping losses) compared to 44% for the diesel baseline. Almost all the indicated fuel savings (~20%) came from reduced thermal losses to the cooling water and exhaust. The NOx emission was nominally 20 mg/kW-hr, and PM emission was 8 mg/kW-hr. Johansson adapted a 12 liter 6-cylinder HD engine to burn gasoline, and reported a brake thermal efficiency (BTE; all engine losses) of 48% at 18 bar IMEP and 1300 RPM (62). This represents perhaps a 10-15% fuel consumption reduction. However, NOx and PM emissions were much higher than reported by Reitz.

HD HYBRIDIZATION

Compared with light duty applications, heavy-duty hybridizaton is just beginning. It is mainly focused on urban vocational applications, but can save much fuel. For example, as shown in Figure 24, a medium duty utility truck can save 8 to 27% fuel depending on drive cycle (63). When used for stationary work, 80% of fuel is saved. Similarly, a medium duty box truck can save 24 to 32% of fuel, and a PHEV urban bus can save from 35 to 65% of fuel. For local courier delivery trucks, about 30% fuel is saved using a hydraulic hybrid design (64). These values compare with 10% savings for long haul applications, including idle reductions, and 5% savings when traveling (Figure 22).

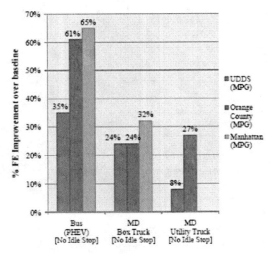

Figure 24. Fuel savings for various types of hybrid heavy duty trucks in different use patterns (63).

Historically, the economics of the systems restricted their use to heavily subsidized applications. However, this is changing. For example, in the courier truck application the additional $7000 for the hydraulic hybrid system has a two-year payback period at fuel prices of about $0.80 per liter ($3/gallon). For electric hybrid applications, the $16,000 incremental cost (65) of the system might be recovered in four years. As such, some projections show 40,000 hybrid trucks being sold in the US by 2015 (65), with 20% of them being hydraulic hybrids.

Hybrid configurations are even beginning to show in the nonroad construction sector. Caterpillar introduced the D7E diesel electric hybrid bulldozer, which reduces fuel consumption by 20% (66). The $100,000 incremental cost on the $600,000 machine has a payback period of 2.5 years (67), largely because it is 10% more productive.

HD VEHICLE CONCLUSIONS

The HD truck sector emits about 6% of the GHGs in the US, Europe, and Japan. In the US, about 70% of this comes from the long haul sector. Contrary to the LD sector, wherein most of the fuel savings comes from engine improvements, in this segment, 65% of the opportunity is in the vehicle (e.g., aerodynamic design) and operations. Looking at the whole long haul truck, 8 to 18% of fuel savings can come from wider use of technologies that are already on some trucks. In 2030, 18% of the segment's fleet-wide fuel can be saved by utilizing technologies that have a three year payback period. Considering only the engine, research engines are delivering 14% lower fuel consumption using largely incremental advancements in current engine technology. An additional 10% reduction might be gained by increasing emission control system deNOx efficiency from 80 to 97%. Waste heat recovery systems (bottoming cycles) have demonstrated 6 to 7% fuel savings in preliminary tests. Further out, HD engine research is migrating towards gasoline engines with high amounts (40-50%) of EGR. Steady state BTE values of 48% at 18 bar IMEP and 1300 RPM have been reported, for a roughly 15% fuel savings and a 25% CO_2 reduction. Hybrid HD trucks are emerging in the vocational market, with payback periods of 2 to 4 years becoming possible.

Black carbon reductions from vehicles can have a profound effect on GHG impact, accounting for upwards of ~20% of CO_2 reductions proposed by the IPCC by 2050.

BLACK CARBON REDUCTIONS

As mentioned earlier, black carbon particles in the air retain heat and can cause warming. As it is a short-lived emission, staying in the atmosphere for weeks instead of hundreds of years like CO_2, early reductions can have immediate impacts. Black carbon is a significant fraction of the particles emitted by diesel engines without filters. Aside from the climate benefits, the health benefits alone justify PM reductions via the use of diesel particulate filters.

Figure 25 shows how worldwide vehicle black carbon emissions from on-road vehicles are projected to vary over time (68). About 60% of the emissions are from trucks. The base case assumes PM regulations that are currently planned. Emissions increase after about 2025 as the developing countries grow. The bars represent emissions if, by 2015, Euro VI HD and Euro 6 LD regulations are implemented in China, India, and Brazil; Euro IV HD and Euro 4 LD standards are implemented in Africa and the Middle East; and Euro 3 motorcycle regulations are implemented in Africa, the Middle East, and Latin America. By 2050, these advanced regulator initiatives remove 19 million tonnes of black carbon, or the equivalent of 38 billion tonnes of CO_2. This is ~20% of the total CO_2 reductions the UN proposes between now and 2050.

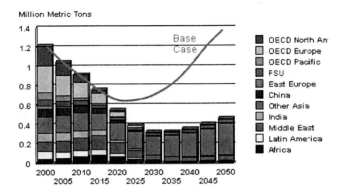

Million Metric Tons

Figure 25. Black carbon reductions from on-road vehicles under current regulations (line) compared to those if developing countries adopt tighter PM standards by 2015. About 60% are for the HD sector. The difference represents about 20% of the total CO_2 reductions proposed by 2050 (68).

OVERALL SUMMARY/ CONCLUSIONS

Driven by scientific consensus of climate change impacts by anthropogenic sources and the desire to diversify energy sources, regulations governing vehicle emissions of CO_2 and fuel consumption are just now emerging in the light-duty sector, but are poised to be a long term trend and involve other vehicle sectors as well. Like criteria pollutant regulations, CO_2 regulations will force technologies on the vehicles that might not go commercial otherwise. This is a paradigm shift.

Many of the goals on emissions and fuel diversity are met with new fuels. As such, governments are implementing biofuel mandates. The US will be requiring that about 20% of the transportation fuels have >50% of the GHG emissions reductions (in neat form) versus conventional fuels by 2022. Europe will require 10% penetration of similarly effective fuels by 2020. Short term, the best approaches involve utilizing the cellulosic portion of plants or using waste products. The long term trend is towards integrating biofeedstocks into refinery operations to expand applications and reduce specific application issues.

For light duty vehicles, most of the potential savings comes from improvements in the powertrain. Some of the technologies are favored for light-load or city operation, like hybridization and engine downsizing, and some do better in high-load or highway operation, like diesel. Both the gasoline and diesel engines can improve, on the order of 20-30% from today's engines. These technologies are generally cost effective and will save the consumer money over the life of the vehicle. The greatest CO_2 saving comes for the diesel hybrid.

The HD truck sector, most of the energy saving opportunity is in the vehicle design and truck operations optimization. Although low fuel consumption in this segment is critical to a successful product today, 8 to 18% of fuel savings can come from wider use of technologies that are already on some trucks. Considering only the engine, research engines are delivering 14% lower fuel consumption using largely incremental advancements in current engine technology. Additional reductions can come from higher deNOx emission control efficiency and the use of bottoming cycles. HD engine research is migrating towards gasoline engines with high amounts (40-50%) of EGR. These engines might have 25% lower CO_2 emissions than today's best commercial diesel engines. Hybrid HD trucks are emerging in the vocational market, with payback periods of 2 to 4 years becoming possible.

Black carbon reductions from vehicles can have a significant effect on GHG impact, accounting for upwards of ~20% of CO_2 reductions proposed by the IPCC by 2050.

REFERENCES

1. Smyth, J. G., "New Directions in Engines and Fuels", presentation at US Department of Energy Directions in Engine Efficiency and Emissions Research (DEER) Conference, Dearborn, Michigan, August 2009.

2. International Energy Outlook 2009, US Department of Energy, Energy Information Administration, www.DOE.EIA.gov

3. BP Statistical Review of World Energy, June 2009.

4. Sheikh, I. Rocky Mountain Institute,. Data Source: BP data as graphed by USDoD JASON, "Reducing DoD Fossil-Fuel Dependence" (JSR-06-135, Nov. 2006, p. 6, www.fas.org/irp/agency/dod/jason/fossil.pdf), plus IEA's 2006 World Energy Outlook estimate of world demand and supply to 2030, plus RMI's coal-to-liquids (Fischer-Tropsch) estimate derived from 2006-07 industry data and subject to reasonable water constraints. Graph courtesy of Shell.

5. Climate Change 2007: Synthesis Report, UN Intergovernmental Panel on Climate Change.

6. Huntington, H. P., "Update on Selected Climate Change Issues of Concern", Arctic Monitoring and Assessment Program, Oslo 2009.

7. Smith, K. R., "Household Energy, Black Carbon, Climate and Health", 2009 International Workshop on Black Carbon in Latin America, sponsored by the International Council on Clean Transportation, Mexico City, October 2009.

8. Hansen J., Sato M., Kharecha P., Russell G., Lea D.W., Siddall M., "Climate Change and Trace Gases", Philosophical Transactions of the Royal Society A 365:1925-1954, 2007.

9. European Commission press release, "Commission plans legislative framework to ensure the EU meets its target for cutting CO2 emissions from cars", IP/07/155, Brussels, 7 February 2007.

10. Charmley, W., "EPA's Recent Advance Notice on Greenhouse Gases", presentation at US Department of Energy Diesel Engine Efficiency and Emissions Research (DEER) Conference, Dearborn, Michigan, August 2008.

11. International Council on Clean Transportation, Annual Review of Public Health, 29, 2008. www.TheICCT.org

12. An, F., et al., "Passenger Vehicle Greenhouse Gas and Fuel Economy Standards: A Global Update", the International Council on Clean Transportation, June 2008, updated November 2009. www.TheICCT.org

13. Regulation of the European Parliament and of the Council setting emission performance standards for new passenger cars as part of the Community's integrated approach to reduce CO2 emissions from light-duty vehicles, Brussels, 25 March 2009, 2007/0297 (COD).

14. Cackette, T., "Reducing Vehicle Emissions to Meet Environmental Goals", presentation at US Department of Energy Directions in Engine Efficiency and Emissions Research (DEER) Conference, Dearborn, Michigan, August 2009.

15. Wani, K., "Fuel Efficiency Standard for Heavy-Duty Vehicles in Japan", presentation at International Workshop: Fuel Efficiency Policies for Heavy-Duty Vehicles, International Energy Agency, Paris, June 2007.

16. "Regulating Greenhouse Gas Emissions under the Clean Air Act", US Environmental Protection Agency, 40 CFR Chapter I, [EPA-HQ-OAR-2008-0318; FRL-8694-2], RIN 2060-AP12, July 2008.

17. Final Regulation Order to Reduce Greenhouse Gas Emissions from Heavy-Duty Vehicles, California Air Resources Board, December 9, 2009.

18. Agryropoulos, P., US EPA, "The Energy Independence and Security Act - A new Road to Renewable Fuels", Energy Frontiers International, Emerging Energy Technology Forum, Miami, February 2008.

19. "EPA Proposes New Regulations for the National Renewable Fuel Standard Program for 2010 and Beyond", EPA-420-F-09-023, May 2009

20. "Rulemaking to Consider the Proposed Regulation to Implement the Low Carbon Fuel Standard", April 2009, http://www.arb.ca.gov/regact/2009/lcfs09/lcfs09.htm

21. Flach, B., et al., "EU-27 Biofuels Annual Report 2009", Global Agricultural Information Network (GAIN), UDA Foreign Agricultural Service, GAIN Report NL9014, June 15, 2009.

22. Smyth, J. G., "New Directions in Engines and Fuels", presentation at US Department of Energy Directions in Engine Efficiency and Emissions Research (DEER) Conference, Dearborn, Michigan, August 2009.

23. Coordinating Research Council, Inc. (CRC) Report E-87-1, "Mid-Level Ethanol Blends Catalyst Durability Study Screening", June 2009.

24. Modlin, R., Raney, D., "Assessing Effects of Mid-Level Ethanol Blends", presentation to the US EPA Mobile Sources Technical Review Subcommittee, May 2008.

25. Coordinating Research Council, Inc. (CRC) Report AVFL-17, "Investigation of Biodistillates as Potential Blendstocks for Transportation Fuels", June 2009.

26. Wright, K., "New Directions in Fuel Technology", US Department of Energy Directions in Engine Efficiency and Emissions Research (DEER) Conference, Dearborn, Michigan, August 2009.

27. Hillebrand, D.G., "Alternative Fuels: Does the Model Work? What are the Powertrain-Related Technical Challenges?" Panel Discussion at SAE 2008 World Congress, April, 2008. (Figure courtesy of Diesel Fuel News, May 12, 2008.)

28. West, B., "Effects of Intermediate Ethanol Blends on Legacy Vehicles and Small Non-Road Engines, Report 1", US Department of Energy National Renewable Energy Laboratory report NREL/TP-540-43543, October 2008.

29. Tanaka, H., Matsumoto, T., Funaki, R., Kato, T. et al., "Effects of Ethanol or ETBE Blending in Gasoline on Evaporative Emissions for Japanese In-Use Passenger Vehicles," SAE Technical Paper 2007-01-4005, 2007.

30. Rueger, J.-J., "Clean Diesel - Real Life Fuel Economy and Environmental Performance," Presentation at SAE 2008 Government and Industry Meeting, May, 2008.

31. Rueger, J.-J., "Powertrain Trends and Future Potential", presentation at US Department of Energy Directions in Engine Efficiency and Emissions Research (DEER) Conference, Dearborn, Michigan, August 2009.

32. Tate, E.D, Harpster, M.O., and Savagian, P.J., "The Electrification of the Automobile: From Conventional Hybrid, to Plug-in Hybrids, to Extended-Range Electric Vehicles," *SAE Int. J. Passeng. Cars - Electron. Electr. Syst.* 1(1):156-166, 2008.

33. Fulbrook, A., "Evolution of European Powertrains", European Environmental Policy Conference, London, CSM International, March 2009.

34. "Proposed Rulemaking To Establish Light-Duty Vehicle Greenhouse Gas Emission Standards and Corporate Average Fuel Economy Standards - Part II", Proposed Rule 49 CFR Parts 531, 533, 537, et al., US EPA and DOT Notice of Proposed Rulemaking, September 28, 2009.

35. Körfer, T., Lamping, M., Kolbeck, A., Pischinger, S. et al., "Potential of Modern Diesel Engines with Lowest Raw

Emissions - a Key Factor for Future CO_2 Reduction," SAE Technical Paper 2009-26-025, 2009.

36. Koerfer, T., "The Future Power Density of HSDI Diesel Engines with Lowest Engine out Emissions - A Key Element for Upcoming CO_2 Demands", F2008-06-015, FISITA Conference, Munich, September 2008.

37. Alger, T., "SwRI's HEDGE™ Concept - High-Efficiency Dilute Gasoline Engines for Automotive, Medium Duty and Off-Road Applications," Panel Presentation at SAE 2009 Powertrains, Fuels, and Lubricants Meeting, November, 2009.

38. Obieglo, A., et al., "Future Efficient Dynamics with Heat Recovery", presentation at US Department of Energy Directions in Engine Efficiency and Emissions Research (DEER) Conference, Dearborn, Michigan, August 2009.

39. Fairbanks, J. W., "Vehicular Thermoelectric Applications Session DEER 2009", presentation at US Department of Energy Directions in Engine Efficiency and Emissions Research (DEER) Conference, Dearborn, Michigan, August 2009.

40. Ayres, S., et al., "Development of a 500 Watt High Temperature Thermoelectric Generator", presentation at US Department of Energy Directions in Engine Efficiency and Emissions Research (DEER) Conference, Dearborn, Michigan, August 2009.

41. Moawad, A., et al., "Impact of Real World Drive Cycles on PHEV Fuel Efficiency and Cost for Different Powertrain and Battery Characteristics", EVS24 International Battery, Hybrid, and Fuel Cell Electric Vehicle Symposium, Stavanger, Norway, May 2009.

42. www.Chevy-Volt.net

43. Brandt, K., "Li-Ion Batteries for PHEV and EV Applications", presentation at Plug-In 2009 Hybrid Electric Vehicle Conference, Long Beach, California, August 2009.

44. Kalhammer, F., "Battery Workshop Summary", presentation at Plug-In 2009 Hybrid Electric Vehicle Conference, Long Beach, California, August 2009.

45. Nelson, P., "Modeling Lithium Ion Battery Manufacturing Cost for PHEVs", presentation at Plug-In 2009 Hybrid Electric Vehicle Conference, Long Beach, California, August 2009.

46. Gaines, L., "Lithium Ion Batteries: Material Demand and Recycling", presentation at Plug-In 2009 Hybrid Electric Vehicle Conference, Long Beach, California, August 2009.

47. Mock, P., "Assessment of Future Lithium Ion Battery Costs", presentation at Plug-In 2009 Hybrid Electric Vehicle Conference, Long Beach, California, August 2009.

48. Yeh, S., "Reducing Long-Term Transportation Emissions: Electricity as a Low Carbon Fuel", presentation at Plug-In 2009 Hybrid Electric Vehicle Conference, Long Beach, California, August 2009.

49. Ryan, T., Pirault, J. P., "Cost Effectiveness of Technology Solutions for Future Vehicle Systems", presentation at the US Department of Energy Diesel Engine Energy and Emissions Research Conference, Dearborn, August 2008.

50. Leonhard, R., "Clean Diesel Technology - Efficient Emission Reduction", presentation at the Near Zero Emission Vehicle Conference, Jeju, Korea, June 2009.

51. "Inventory of US Greenhouse Gas Emissions and Sinks: 1990-2007", US Environmental Protection Agency, April 2009.

52. European Commission, Mitteilung der Kommission an den Rat und das Europäische Parlament, {SEK(2007)60}, Brüssel, 2007.

53. Hoshi, A., "Integrated Approach - Pass for the Sustainable Future of Transport", presentation at The Automobile Industry and CO_2, The Road Ahead", Copenhagen, COP15, December 14, 2009.

54. White Paper on Life Cycle Costs, Kenworth Truck Corporation, www.Kenworth.com, 2004.

55. Wall, J., "The Right Technology Matters - The Importance of Public-Private Partnerships for Engine Technology Development", presentation at US Department of Energy Directions in Engine Efficiency and

56. Cooper, C., et al., "Reducing Heavy-Duty Long Haul Combination Truck Fuel Consumption and CO_2 Emissions", Miller Paul, editor, NESCCAF, October 2009.

57. Stanton, D., "Technology Development for High Efficiency Clean Diesel Engines and a Pathway to 50% Thermal Efficiency", presentation at US Department of Energy Directions in Engine Efficiency and Emissions Research (DEER) Conference, Dearborn, Michigan, August 2009.

58. Schmidt, K. D., "Emissionsminderung in Nutzfahrzeugen", presentation at the International CTI Forum Selective Catalytic Reduction (SCR) Systems, Karlsruhe, May 2007.

59. Nelson, C., "Exhaust Heat Recovery", presentation at US Department of Energy Directions in Engine Efficiency and Emissions Research (DEER) Conference, Dearborn, Michigan, August 2009.

60. Pinson, J., "New Directions in Engines - The Road Ahead", presentation at US Department of Energy Directions in Engine Efficiency and Emissions Research (DEER) Conference, Dearborn, Michigan, August 2009.

61. Reitz, R. D., et al., "High Efficiency Ultra-Low Emission Combustion in a Heavy Duty Engine via Reactivity Control", presentation at US Department of Energy Directions in Engine Efficiency and Emissions Research (DEER) Conference, Dearborn, Michigan, August 2009.

62. Johansson, B., "Partially Premixed Combustion, PPC, for High Fuel Efficiency Engine Operation," Panel Presentation at SAE 2009 Powertrain, Fuels, and Lubricants Meeting, November, 2009.

63. Lei, N., "HD Truck and Engine Fuel Efficiency Opportunities and Challenges Post EPA2010", presentation at US Department of Energy Directions in Engine Efficiency and Emissions Research (DEER) Conference, Dearborn, Michigan, August 2009.

64. Burnsed, B., "High Hopes for a New Type of Hybrid Vehicle", Business Week magazine, January 8, 2009.

65. Kar, S., "Class 6-8 Truck Hybrid Powertrain Systems Market", presentation at the Hybrid Truck User's Forum, Atlanta, October 2009.

66. Goff, T., "Efficiency Improvement Pathway", presentation at US Department of Energy Directions in Engine Efficiency and Emissions Research (DEER) Conference, Dearborn, Michigan, August 2009.

67. Scanlon, J., "Caterpillar Rolls Out Its Hybrid D7E Tractor", Business Week magazine, July 20, 2009.

68. Analyses by Walsh Michael, international regulation consultant.

CONTACT INFORMATION

Timothy V. Johnson
JohnsonTV@Corning.com

ACKNOWLEDGMENTS

Charles Sorensen of Corning Incorporated conducted the GREET analyses of well-to-wheel CO_2 emissions. The following experts provided the author with valuable comments on this paper;

Wendy Clark, National Renewable Energy Laboratory

Kevin McMahon, Martec

Jack Peckham, Hart Diesel Fuel News

Michael Walsh, International Regulatory Consultant

Ken Wright, ConocoPhillips

Uwe Zink, Corning Incorporated

AUTHOR DISCLAIMER

FUEL ECONOMY IMPACT EVALUATION OF HYBRID VEHICLES IMPLEMENTATION IN THE BRAZILIAN FLEET

David Queiroz Luz; Gilmar de Paula Junior
Ford Motor Company - Brazil

ABSTRACT

With the turmoil of the finance crisis in 2008 and 2009 and increasing volatility of petroleum prices at US gas stations, the automobile industry has been forced to review its product cycle plan and to expedite implementation of fuel economy driven technologies such as Hybrid and Battery Electrical Vehicles. For the Brazilian market, due to its independent fuel supply base with adequate infrastructure and the established demand for Flex Fuel Vehicles, hybrid technologies are not currently in the scope of short term developments of local industry. However, driven by upcoming regulations with more restricted levels of emissions, increasing market interest on fuel economy and more global synergies, hybrid technology is in the strategy perspective for the following five to eight years as a bridge action to BEVs. Based on the fuel economy provided by models in production in 2009 and on market projections of vehicles sales in Brazil, the objective of this study is to forecast fuel economy associated to the ramp-up implementation of Hybrid vehicles in the local market.

INTRODUCTION

This study brings into discussion the negative effects caused by the excessive fuel consumption of the vehicles in use today, and also proposes an estimative of the fuel economy caused by the future fleet, when the hybrid technology will be available for the Brazilian market. The Hybrid Vehicles use the conventional internal combustion engine (ICE), with the assistance of an electric engine, in order to allow a better fuel economy and, then, reduced levels of emission of harmful gases in the atmosphere.

PRO-ALCOHOL PROGRAM AND FLEX FUEL ENGINES

Due to petroleum crisis in 1973 and 1979, the Brazilian government financed in 1975 a well-succeeded program of substitution on a large scale of the oil derivatives called Pro-Alcohol. Approximately 5.6 million ethanol vehicles were produced from 1975 to 2000. The substitution of the gasoline for the alcohol in the 1976-2004 period represented a US\$ economy of 61 billion[1] or US\$ 121 billion[2]. (BERTELLI, 2009)

The program also substituted a fraction of pure gasoline (from 1.1 to 25%) for alcohol. For a fleet of more than 10 million vehicles, it prevented the importation of approximately 550 million barrels of oil and provided an economy of the order of 11.5 billion dollars. The Pro-Alcohol is, therefore, a historical example of introduction of a new fuel alternative to the costumer.

According to Figueiredo (2006), three factors contributed to the mindset change in relation to the ethanol as a fuel: the growing market demand related to the transportation sector; the increasing oil prices; and the world forecast of exhaustion of petroleum in the next 30 years.

In the 90's, with the prices reduction of petroleum, gasoline became more attractive and the Alcohol demand decreased significantly forcing the automakers

---

[1] Dollar of December, 2009
[2] With the interests of the external debt.

practically to abandon production of ethanol-powered vehicles.

In the beginning of 21st century, ethanol and gasoline prices variation let both competitive depending on the season or market demand variation (FIGUEIREDO, 2006).

Added to the government incentives for ethanol vehicles, which would be extended to Flex Fuel Vehices (FFVs), and the evolution of researches on FFVs using cost competitive solutions, in 2003 the first FFV was launched. The vehicle could now be powered by any blend of gasoline or ethanol using the same fuel tank and with no human interference required (ALVES, 2007).

These two important milestones in history of the transportation sector are shown on Graphic 1.

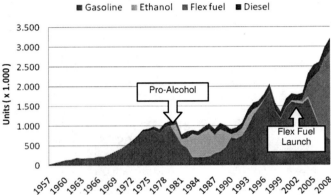

Graphic 1 – Absolute vehicle production in Brazil by fuel type (including cars, light commercials, trucks and buses).

HYBRID VEHICLES

A hybrid vehicle is an automobile that has more than one propeller engine, which uses different types of feeding. Hybrid-Electric vehicles (HEVs) combine the benefits of internal combustion engines and electric motors and can be configured to obtain different objectives, such as improved fuel economy, increased power, or additional auxiliary power for electronic devices and power tools.

TYPES OF HYBRIDS - There are three different configurations for a hybrid automobile:

- Parallel Hybrid - The internal combustion engine is responsible for the movement of the automobile and the electric one is an extra aid to improve the performance of the same.
- Series Hybrid - The electric engine is responsible for the movement of the automobile, being that the internal combustion engine only puts into motion a generator responsible for generating the energy necessary for the automobile to move itself and to load the batteries.

- Series-Parallel Hybrid – It is the combination of the characteristics of the system in series with the parallel system, with the objective to maximize the benefits of both. This system allows the supply of energy to the wheels and to generate electricity simultaneously, using a generator, differently of what occurs in the simple parallel configuration. It is possible to use only the electrical system, depending on the load conditions, or to use both systems at the same time.

Some of the advanced technologies typically used by hybrids include:
Regenerative Braking - The electric motor applies resistance to the drivetrain causing the wheels to slow down. In return, the energy from the wheels turns the motor, which functions as a generator, converting energy normally wasted during coasting and braking into electricity, which is stored in a battery until needed by the electric motor.

Automatic Start/Shutoff - The engine is automatically shut off when the vehicle comes to a stop. It is restarted when the accelerator is pressed. This prevents wasted energy from idling.

HYBRIDS SALES AND FUEL PRICES – The first hybrid electric vehicle in the US market was the Honda Insight (61 mpg city, 70 mpg highway) in 1999, achieving a total global sale of only 18,000 approximately. The Toyota Prius (52 mpg city, 45 mpg highway) was launched in 2000, and the Honda Civic Hybrid (46 mpg city, 51 mpg highway) went on sale in 2002. The first full-size pickup hybrids, the Chevy Silverado and Dodge Ram hit the streets in 2004, as did the first SUV hybrid, the Ford Escape. (HYBRIDCARS.COM). Graphic 2 shows the US hybrid vehicle sales from 1999 to 2009[3].

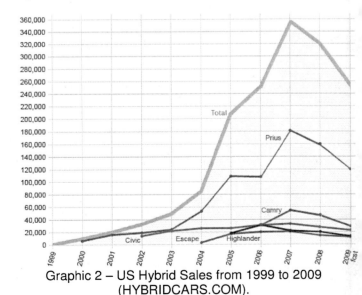

Graphic 2 – US Hybrid Sales from 1999 to 2009 (HYBRIDCARS.COM).

---

[3] 2009 forecast

According to a survey conducted by Klein (2007) with people who have bought a hybrid vehicle, the reason why people buy it can be divided into five segments. Four of these are related to financial motivations.

Segment	% of Sample
Early Adopters	27%
The Prius Cost Less than the Alternative	40%
Buyer Calculated the ROI Differently than the Experts	16%
Buyer Bought the Prius to Drive in the Carpool Lane	12%
Buyer Bought the Prius as an Inexpensive 'Fun' Car	5%
Total	100%

Table 1 – Survey showing the reasons why people buy Hybrid vehicles (KLEIN, 2007).

These data suggest that Hybrid sales may be correlated to fuel prices. Graphic 3 shows the Gasoline and Diesel prices variations collected from Jan/2006 to May/2009 when compared to the hybrid sales.

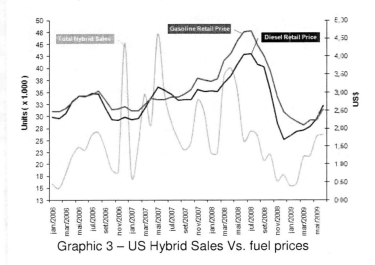

Graphic 3 – US Hybrid Sales Vs. fuel prices

BRAZILIAN MARKET

According to the National Energy Balance of 2008 (BEN, 2008), the transportation sector in Brazil was responsible for 27% of the final consumption of energy in 2007 and, consequently, is the most important segment among the consumers of oil and derivatives, being responsible for 50.5% of this consumption, as can be seen in Graphic 4

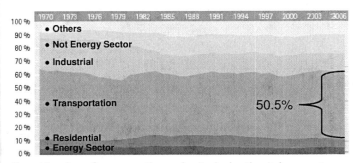

Graphic 4 – Consumption of oil derivatives by sector. (BEN, 2008)

FLEX FUEL FLEET AND FUEL SALES – By the end of 2008, 74.7% of cars and light commercial vehicles sold in Brazil were Flex Fuel capable. In 2009, from January to June, 85% of the selling transportation vehicles were Flex-Fuel Vehicles (FFVs), and considering only cars, the percentage increases to 97% (ANFAVEA, 2009).

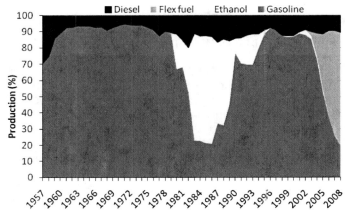

Graphic 5 – Percentual Vehicle production in Brazil by fuel type (including cars, light commercials, trucks and buses).

This change of mix in the Brazilian fleet causes a variation on the type of fuel demand. The growing production of ethanol can be realized in graph 6.

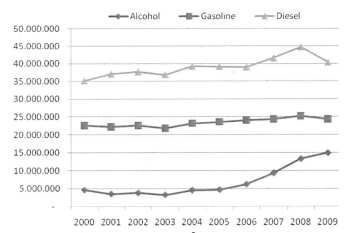

Graphic 6 – Consumption in m^3 of Alcohol, Diesel and Gasoline in Brazil from 2000-2009[4] (ANFAVEA, 2009)

REGULATORIES - Evidenced the severity of the pollution generated by the vehicles during the decade of 80, the CONAMA - National Council of the Environment, established the PROCONVE – Automotive Vehicles Air Pollution Control Program. With this program, the international methodologies had been adapted to the Brazilian necessities and new techniques had been developed to reduce the pollution generated by the vehicles. The pollutants emission homologation is

---

[4] 2009 linear consumption projected based on data from Jan-Apr.

mandatory for all the new models of vehicles and national and imported engines. For such, the engineering parameters relative to the vehicle and to the engine's pollutants emission are analyzed, being also submitted to rigid laboratory tests, where the exhaust pipe emissions are quantified and compared with the maximum limits in vigor. (CETESB, 2009)

Since its implementation in 1986, the Program reduced the emission of pollutants of new vehicles in about 97%, by means of the gradual limitation of the emission of pollutants. This contributed with the advance of the automobile industry through the introduction of technologies such as catalyst converter, electronic fuel injection and improvements in automotive fuels. (CETESB, 2009)

The major restriction regarding pollutants emissions is directly correlated to the vehicle's fuel consumption. Better fuel economy means improved emissions performance.

Since the 80s decade, Brazil stimulates the energy consumption rationalization. One of the initiatives that contributes to the utilization of more efficient products by the society is the Brazilian Labeling Program, which makes it possible to the customer to select more economic products by the use of informative labels affixed to the equipment. Since April 2009, the program began to label vehicles[5] commercialized in the country with fuel economy information. The world-wide experience shows that these programs induce to the manufacture of more economic vehicles and with lesser emission, benefiting the local consumers and the environment. (CONPET, 2009)

According to the CONPET program, the search for vehicular energy efficiency brings a technological evolution to the Brazilian automobiles, increasing the competitiveness of this industrial segment and improving the economic, social and environmental sectors.

These two main drivers, emissions and fuel economy, make hybrid vehicles technology very favorable alternative for Brazil, which is world-widely recognized for the spread use of bio-fuels and flex fuel engine technology.

BRAZILIAN FLEET ESTIMATION FOR 2012-2018

To make a quantitative evaluation of the potential fuel economy that can be generated with the implementation of hybrid vehicles in the domestic market, it is necessary to make a projection of the national vehicle fleet for the period of interest (from 2012 to 2018).

Graphic 7 – Cars, Trucks and Buses – Brazilian Fleet from Jan/2000 to Jan/2009 (DENATRAN, 2009)

To estimate the future fleet of vehicles in Brazil, first it is necessary to evaluate the behavior of the historical data and to define a function that better describes its tendency, using the Simple Linear Regression method. For such, it was considered that the sales curve follows a trend of the type Y(t)=AX+B, which follows the evolution of the Brazilian fleet curve for the last years, from 2000 to 2009. Graphic 7 presents the historical data evolution for the Brazilian fleet of vehicles from Jan/2000 to Jan/2009.

SIMPLE LINEAR REGRESSION: LEAST SQUARE METHOD - The coefficients A and B can be found by a applying the least square method in the historical data of the Brazilian vehicle fleet. This principle affirms that, to adjust a straight line to the data values, it is necessary to find a straight line in which the sum of the squares of the vertical distances of each point to the straight line is the minimum possible. This rule is arbitrary, but of great efficiency, and is only one form to describe a straight line that passes through the middle of the data (HILL, 2003). The formulas for the calculation of the least squares for A and the B are:

$$A = \frac{N \sum X_t Y_t - \sum X_t \sum Y_t}{N \sum X_t^2 - \left(\sum X_t \right)^2}$$

$$B = \overline{Y} - A\overline{X}$$

Determination Coefficient – The R^2 measure is called Determination Coefficient. The closest the value of R^2 is to 1, the better will have been the work to explain the variation of Y in function of X. The R^2 measures the linear association, or the quality of the adjustment, between all the sample data and its predicted values.

---

[5] This initiative is optional for each automaker

The coefficient is given by:

$$R^2 = \frac{\sum(\hat{Y}_t - \overline{Y})^2}{\sum(Y_t - \overline{Y})^2}$$

SIMPLE LINEAR REGRESSION ANALYSIS - Table 2 demonstrates the values found for the linear and exponential functions by vehicle type. Graphic 8 shows the trend line plotted over the real historical data. It can be observed that, for the three segments of vehicles, the exponential functions are the ones that better describe the behavior of the curves, since the values of the determination coefficient R² are closer to 1 than the values of R² for the linear functions.

	Linear (Y=AX+B)		Exponential (ln(Y)=AX+B)	
	Function	R²	Function	R²
Cars	Y=1E+06x + 2E07	0,9651	Y=2E+07e^{0,0506x}	C,9828
Light Trucks	Y=205599x + 3E06	0,9557	Y = 3E+06e^{0,0516x}	C,9769
Bus + Trucks	Y = 58277x + 2E06	0,977	Y = 2E+06e^{0,0286x}	C,9851

Table 2 – Functions and R² values per vehicle type.

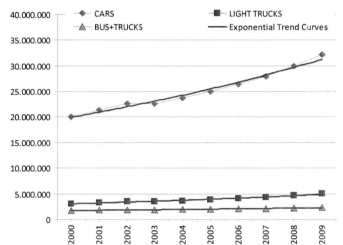

Graphic 8 – Brazilian fleet data showing the exponential trend curves.

BRAZILIAN FLEET PROJECTION - Based on the regression projection, the set of data from 2010 to 2018 was generated considering an exponential growth, as depicted in Graphic 9.

- In 2012, it's forecasted a total of 36.38 millions of cars, 5.73 millions of light trucks, and 2.50 millions of buses and trucks. This represents a total fleet growth of 12,5% related to 2009.

- For 2018, it's forecasted a total of 49.27 millions of cars, 7.80 millions of light trucks, and 2.96 millions of buses and trucks. This represents a total fleet growth of 51,4% related to 2009.

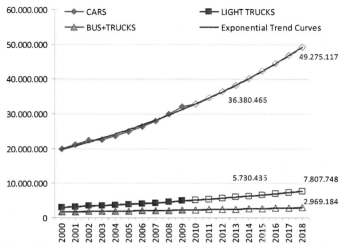

Graphic 9 – National fleet estimation based on Simple Linear Regression method.

VEHICLE PRODUCTION BY FUEL TYPE – As seen on Graphic 10, in 2008 Flex Fuel Vehicles were responsible for 69% of fleet sales. In 2009, a linear projection based on January to April sales indicates FFVs as responsible for 83% of fleet production.

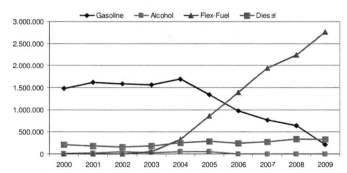

Graphic 10 – Vehicle production in Brazil by fuel type (ANFAVEA, 2009).

Considering that the introduction of Hybrid vehicles would have the same growth trend of FFVs, the following estimations take into account Hybrids starting-on-production in 2014.

BRAZILIAN FLEET PRODUCTION PROJECTION - Applying the same methodology of the Brazilian Fleet Projection to the Brazilian Fleet Production, the following results can be observed:

- In 2012, it's forecasted a total of 4,0 millions of vehicles will be produced. FFVs will be responsible for 89%. This represents a production growth of 20% related to 2009.

- In 2018, it's forecasted a total of 5,4 millions of vehicles will be produced. FFVs will be responsible for 50%. Hybrids will be 36%. This represents a production growth of 62% related to 2009

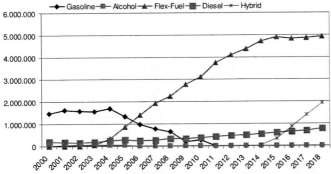

Graphic 11 – Vehicle production volumes per fuel type – projection of 2010-2018

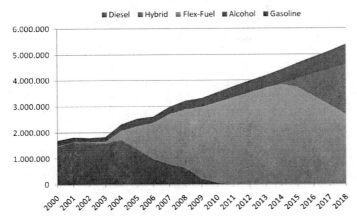

Graphic 12 – Vehicle production volumes and participation per fuel type – projection of 2010-2018

BRAZILIAN FLEET PROJECTION PER TYPE OF FUEL - Applying the same methodology of previous sections, the projection to the Brazilian Fleet Production per Fuel presented the following results:

- In 2012, it's forecasted Gasoline as 44%, FFVs as 44%, Diesel as 11% and Hybrids not available yet. 1% remaining from Alcohol vehicles.

- In 2018, it's forecasted Gasoline as 25%, FFVs as 55%, Diesel as 12% and Hybrids as 8%.

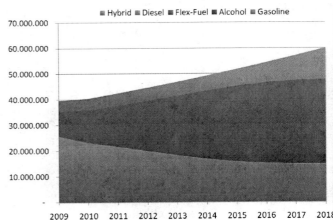

Graphic 13 – Brazilian fleet per fuel type – projection of 2010-2018

FUEL ECONOMY OF HYBRIDS - Applying the same methodology of the Brazilian Fleet

HYBRID Models	Km/L COMBINED (50% City and 50% Highway)	CONVENTIONAL Models	Km/L COMBINED (50% City and 50% Highway)	Difference
Honda Civic Hybrid 4 cyl, 1.3 L, Automatic (variable gear ratios), HEV, Regular	18,07	Honda Civic 4 cyl, 1.8 L, Manual 5-spd, Regular	12,75	-29,41%
Nissan Altima Hybrid 4 cyl, 2.5 L, Automatic (variable gear ratios), HEV, Regular	14,45	Nissan Altima 4 cyl, 2.5 L, Automatic (variable gear ratios), Regular	11,48	-20,59%
Toyota Camry Hybrid 4 cyl, 2.4 L, Automatic (variable gear ratios), HEV, Regular	14,24	Toyota Camry 4 cyl, 2.4 L, Automatic 5-spd, Regular	11,05	-22,39%
Ford Escape Hybrid FWD 4 cyl, 2.5 L, Automatic (variable gear ratios), Regular	13,82	Ford Escape FWD 4 cyl, 2.5 L, Automatic 6-spd, Regular	10,20	-26,15%
Mazda Tribute Hybrid 2WD 4 cyl, 2.5 L, Automatic (variable gear ratios), Regular	13,82	Mazda Tribute FWD 4 cyl, 2.5 L, Automatic 6-spd, Regular	10,20	-26,15%
Mercury Mariner Hybrid FWD 4 cyl, 2.5 L, Automatic (variable gear ratios), Regular	13,82	Mercury Mariner FWD 4 cyl, 2.5 L, Automatic 6-spd, Regular	10,20	-26,15%
Chevrolet Malibu Hybrid 4 cyl, 2.4 L, Automatic 4-spd, Regular	12,75	Chevrolet Malibu 4 cyl, 2.4 L, Automatic 4-spd, Regular	11,05	-13,33%
Saturn Aura Hybrid 4 cyl, 2.4 L, Automatic 4-spd, Regular	12,75	Saturn Aura 4 cyl, 2.4 L, Automatic (S6), Regular	11,69	-8,33%
Saturn Vue Hybrid 6 cyl, 3.6 L, Automatic (variable gear ratios), Regular	12,12	Saturn Vue FWD 6 cyl, 3.6 L, Automatic 6-spd, Regular	8,72	-28,07%
Saturn Vue Hybrid 4 cyl, 2.4 L, Automatic 4-spd, Regular	12,12	Saturn Vue FWD 4 cyl, 2.4 L, Automatic 4-spd, Regular	9,57	-21,05%
Ford Escape Hybrid 4WD 4 cyl, 2.5 L, Automatic (variable gear ratios), Regular	11,90	Ford Escape 4WD 4 cyl, 2.5 L, Automatic 6-spd, Regular	9,35	-21,43%
Mazda Tribute Hybrid 4WD 4 cyl, 2.5 L, Automatic (variable gear ratios), Regular	11,90	Mazda Tribute 4WD 4 cyl, 2.5 L, Automatic 6-spd, Regular	9,35	-21,43%
Mercury Mariner Hybrid 4WD 4 cyl, 2.5 L, Automatic (variable gear ratios), Regular	11,90	Mercury Mariner 4WD 4 cyl, 2.5 L, Automatic 6-spd, Regular	9,35	-21,43%
Toyota Highlander Hybrid 4WD 6 cyl, 3.3 L, Automatic (variable gear ratios), HEV, Regular	11,05	Toyota Highlander 4WD 6 cyl, 3.5 L, Automatic (S5), Regular	8,50	-23,08%

TABLE 3 – Fuel efficiency difference between hybrid and conventional vehicles. (HYBRIDCARS.COM)

With the forecast available, an average fuel efficiency difference between a hybrid and a conventional internal combustion engine vehicle is needed to estimate the total fuel economy generated. Table 3 shows the fuel efficiency of hybrid vehicles and its conventional internal combustion engine versions, and calculates the difference between their efficiencies[6].

- Average Hybrid Fuel Consumption: 13,19 Km/L

- Average Conventional Engine: 10,25 Km/L

- Fuel Economy Hybrid / Conventional: 22,07%

The consumption of fuel per type was defined based on Graphic 6 and Brazilian fleet in 2009. The same methodology is applied to forecast ethanol, gasoline and diesel consumption up to 2018.

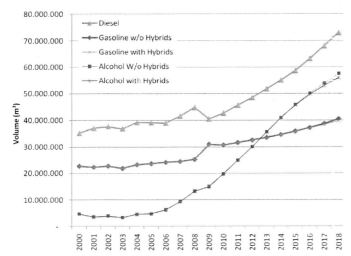

Graphic 14 – Fuel consumption variation caused by Hybrid Vehicles implementation.

To calculate fuel economy provided by hybrics implementation, the following assumptions were stated:

- Hybrid vehicles will affect only FFVs production and fleet. Therefore, fuel economy will be related to gasoline and alcohol.

- The same consumption of liters per year, based on 2009 consumption, is assumed from 2010 up.

Fuel Economy calculation resulted in a reduction of total fuel production from 2014 to 2018. These calculations can be found on graphic 15.

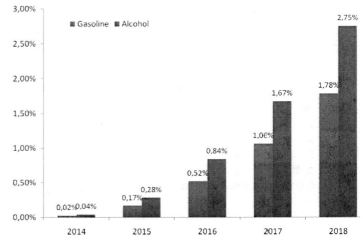

Graphic 15 – Fuel economy provided by Hybrid Vehicles implementation

CONCLUSION

Because of FFVs increase in participation, dominating the market over gasoline vehicles, according to the projections, the alcohol will become the main fuel for automotive vehicles. This is expected to happen in 2012.

Diesel, for being used on a large scale in the services sector, is responsible for the biggest consumption in volume. However, technologies that substitute the powertrain of trucks and buses still are more embryonic than of vehicles.

The verified fuel economy impact takes into consideration a macro scenario (the total volume production of ethanol and gasoline). A more detailed investigation considering the fuel economy generated vs. the technology cost needs to be done.

Up to 2018, the fuel economy is estimated to be 1,78% for gasoline and 2,75% for alcohol. In matters of money, this means approximately a R$ 3,5 billions saving[7] for gasoline and alcohol in 2018.

Even though the projection had a sped up implementation curve for Hybrid vehicles, the fuel economy generated seems to be small for a short term. However, the high 22% average better efficiency, the increase in market participation for a long term, and the promising technology improvements and cost reduction, are all favorable for the spread utilization of hybrid technology.

---

[6] Toyota Prius was not considered because it doesn't have na equivalent conventional ICE version.

[7] Considering January 2009 fuel prices.

REFERENCES

1. ALVES, M. L., "Carro Flex Fuel: Uma Avaliação por Opções Reais", PUC-Rio, Rio de Janeiro, 2007.

2. ANFAVEA - Associação Nacional dos Fabricantes de Veículos Automotores

3. Associação Nacional dos Fabricantes de Veículos Automotores, available at: <http://www.anfavea.com.br/tabelas/autoveiculos/tabela11_vendas.pdf>, accessed in August 01, 2009.

4. BEN – Balanço Energético Nacional – Ministério de Minas e Energia, 2008.

5. BERTELLI, L.G., 2009, "A Verdadeira História do ProÁlcool", O Estado de S. Paulo, available at: <http://www.biodieselbr.com/proalcool/historia/proalcool-historia-verdadeira.htm>, accessed in July 7, 2009

6. CETESB - Companhia Ambiental do Estado de São Paulo, available at: <http://www.cetesb.sp.gov.br/Ar/emissoes/proconve.asp>, accessed in July 7, 2009

7. CONPET - Programa Nacional da Racionalização do Uso dos Derivados do Petróleo e do Gás Natural, available at: <http://www.conpet.gov.br/projetos/pbeveicular_01.php?segmento=corporativo>, accessed in June 5, 2009

8. DENATRAN - Departamento Nacional de Transito, available at: <http://www.denatran.gov.br/frota.htm>, accessed in July 21, 2009

9. FIGUEIREDO, S., "O Carro a Álcool: Uma Experiência de Política Pública para a Inovação no Brasil", UnB-CDS, Brasília, 2006.

10. FUELECONOMY.ORG, available at: <http://www.fueleconomy.org>, accessed in July 17, 2009.

11. HILL, R. C., GRIFFITHS, W. E., JUDGE, G. G., Econometria, Saraiva, São Paulo, 2003.

12. HYBRIDCARS, available at: <http://www.hybridcars.com>, accessed in February 17, 2009.

13. KLEIN, J., "Why People Really Buy Hybrids", General Partner, Top Line Group, April, 2007.

14. Le Monde Diplomatique, 2006, "Cronologia da OPEP", available at: <http://diplo.uol.com.br/2006-05,a1304>, accessed in February 17, 2009.

15. ProÁlcool - Programa Brasileiro de Álcool, Available at: <http://www.biodieselbr.com/proalcool/proalcool.htm>, accessed in July 7, 2009.

CONTACT

Gilmar de Paula Junior – Ford Motor Company
E-mail: gpaulaju@ford.com

David Queiroz Luz – Ford Motor Company / Universidade Federal da Bahia
E-mail: dluz4@ford.com / david_queiroz@hotmail.com

SAE Paper No. **2009-26-075**

Solution for India Towards Clean and CO_2 Efficient Mobility

Ravi Kishore, Patrick Leteinturier and Wong Sou Long
Infineon Technologies India Pvt. Ltd, India

ABSTRACT

In the recent times in India the importance of environment, energy and CO_2 reduction are being discussed and debated by all the government environmental agencies, automotive manufacturers and Tier-1 suppliers. With the advent of Tata Nano and introduction of other low end vehicles into the Indian automotive space, it is estimated that the number of on road vehicles would definitely cross over 40 Million by 2012. With this expected surge in the vehicle population on Indian roads the air quality has become a major concern to the government which in turn puts the pressure on the local OEM's. This also calls for the enforcement of the strict emission norms (like BSIII and BSIV) and CO_2 reduction regulations as already announced in EU and USA. All the aforesaid statements definitely pose a greater challenge on the vehicle design aspects which include the efficient energy management system, implementation of stop/start and hybridization, electronically enabled engine management systems and CO_2 reduction in power train systems etc. Furthermore all the techniques mentioned above could only be achieved by incorporating efficient and powerful semiconductor systems like high performance microcontroller based EMSs, IGBT and microcontroller based hybrid systems etc.

The scope of this paper discusses the detailed analysis of the technologies that could be implemented in all the vehicles which would enable CO_2 reduction and achieve the enforced emission standards for greener future. It also provides an insight into some technological challenges and solutions with semiconductors which could be successfully implemented with out compromising the safety and reliability aspects of the vehicle design. In extension to this, it also covers the importance of introducing semiconductors in the Indian automotive space today, which is the best solution for a greener tomorrow.

INTRODUCTION

The recent variations in the barrel prices has shown that crude oil will not be the energy source forever. On one hand we have to conserve it as much as possible while on the other hand preparing for its eventual replacement. The challenge is not only energy availability, but also the effect of exhaust gases on our environment. It is not only vital to reduce emissions such as CO, NOx, HC, PM, but we also have to combat CO_2 in order to limit global warming. The Kyoto treaty, the ACEA (Association des Constructeurs Européens d'Automobiles in English: European Automobile Manufacturers' Association) and the CAFE (US Corporate Average Fuel Economy) regulations are all working in this direction. Engineers from all countries are working on this challenge, and different steps are required: improve engine efficiency, improve transmission for fuel economy, save further fuel by using hybrid solutions, save energy during braking, and introduce high-performance battery management systems. The ultimate step will be to migrate toward the hydrogen economy. In order to achieve this powertrain revolution, the electronics that will have to control these functions will need a tremendous increase in innovation.

CO_2 IMPACT ON THE ENVIRONMENT

CO_2 is one of several gases in the Earth's atmosphere which absorb infrared radiation and is therefore categorized as a greenhouse gas. Gases with this property are essential to maintaining the temperature of the Earth, but in excess, can contribute to global warming. CO_2 is produced by many natural and industrial processes, but it is primarily the industrial sources which have been linked to the rise of CO_2 concentration over the past 150 years. Automobiles are responsible for almost 25% of CO_2 emissions in certain areas of the world, and are therefore prime targets for increases in fuel efficiency and reduction of CO_2 emissions.

CO$_2$ REGULATION AND TAXES

At the end of 2007 the target for US CAFE has been established at 35 MPG in 2020 (155 gCO$_2$/km). The ACEA puts its target for European CO$_2$ emissions at 120 gCO$_2$/km in 2012 with some possible negotiation up to 130 gCO$_2$/km by using bio-fuel or other measures to reduce CO$_2$ in the traffic. Europe also has the ambition to further reduce the CO$_2$ emissions to 95 gCO$_2$/km by 2020 (Fig. 1). In order to drive the market in the right direction three measures can be taken. The first is the price of the fuel, the second is carbon tax for the consumer, and the third one is a carbon tax on the OEM. These regulations may be also applied to the Indian OEM's sooner or later. Table-I Provides the regular gasoline gCO$_2$/km, l/100km and MPG.

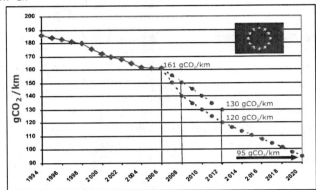

Figure 1 : gCO$_2$/km Emissions Vs. Time

gCO$_2$/km	155	140	130	120	110	100	90
L / 100km	6.72	6.08	5.65	5.21	4.78	4.34	3.91
MPG	35.00	38.69	41.66	45.13	49.24	54.16	60.18

Table- I : For Regular Gasoline gCO$_2$/km, l/100km, MPG

CO$_2$ COST-BENEFIT ANALYSIS : Reduction in CO$_2$ emission in the past, was based only on the reduced fuel but the situation looks somewhat different:

- The customer benefits from saving fuel, which is getting more expensive all the time.

- The customer benefits from reduced CO$_2$ taxation (in preparation).

- The manufacturer benefits from complying with fleet fuel economy standards and avoids penalty payments (in preparation).

- The manufacturer benefits from having to grant fewer discounts required to sell unattractive cars.

Taking all these factors into account, the question is: What value or benefit does a reduction of 1 gram of CO$_2$/km represent for the consumer as well as for the automobile manufacturer?

To find an answer to this question we want to use the oft-quoted relationship between the consumption of fuel and the consumption of electrical power or the carrying of additional weight.

Fig. 2 shows that the consumption of 100 watts of electrical power or carrying 50 kg of additional weight equals the consumption of 0.1 liters of fuel per 100 km. Since burning 1 liter of gasoline emits 23.6 grams of CO$_2$/km (26.5 grams of CO$_2$/km per liter of diesel fuel), emitting 1 gram of CO$_2$/km is the equivalent of consuming approximately 40 watts of electrical power or carrying 20 kg of additional weight.

100 W el.	⇔	0.1 l/100km
50 kg	⇔	0.1 l/100km
1 l/100km Fuel	⇔	23.6 g CO$_2$/km
1 l/100km Diesel	⇔	26.5 g CO$_2$/km
1 g CO$_2$/km	⇔	40 W el.
1 g CO$_2$/km	⇔	20 kg

Figure 2 : Relationship Between CO$_2$ Emissions and the Consumption of Electrical Power or the Carrying of Additional Eight

If we manage to reduce the vehicle's electricity consumption by 400 watts or more, we could save at least 10 grams of CO$_2$/km or more, which is a huge amount. Implemented such measures will undoubtedly lead to higher development and manufacturing costs, but these can ideally be compensated for by yet-to-be-determined CO$_2$ savings. If this is the case, such improvements would also make sense financially and would not have to be enforced via governmental regulations. Below we will take a closer look at the potential benefits of CO$_2$ savings for vehicle buyers as well as vehicle manufacturers. Fig. 3 lists some of the benefits of cutting down on CO$_2$ emissions. First of all, a vehicle with high CO$_2$ emissions will be harder to sell in the future than will be the case with competing models. To be able to sell a € 20,000 vehicle with a higher emission of 50 grams of CO$_2$/km, the dealer may have to grant a discount of 5% or 1,000. Scaled to the CO$_2$ emission, this would cost the manufacturer € 20 per vehicle and per gram of additional CO$_2$/km emission. To put it another way, the manufacturer could invest € 20 into his vehicle to achieve a reduction of 1 gram of CO$_2$/km without reducing his profit. The same type of calculation can be used for any penalties or compensation payments which may be instituted in the future if the manufacturers fleet average exceeds the maximum fuel economy. Since such models are currently being developed by the EU Commission, we have to do with our own rough estimates for the purposes of this paper.

A starting point is the proposal made by automobile business scientist Prof. Ferdinand Dudenhöffer in January 2007 [2], who wants to subsidize small cars like the Smart with € 720 and tax larger vehicles like the Porsche Cayenne Turbo S with € 7,100. Considering the Smarts CO_2 emission of 116 grams per km and the Porsche Cayennes of 378 grams per km, we arrive at benefit € 29.80 per gram of CO_2 saved per vehicle. The amount of potential savings would be similar at € 37.50 per gram of CO_2/km if you take the additional costs into account which it takes to save one ton of CO_2 according to the ACEA study .An Integrated Approach to Reducing Passenger Car-Related CO_2 Emissions, of January 2006 [3].

A third approach would be the current rate of €22 per ton of CO_2 quoted for emission certificates covering the years 2008 through 2012. Assuming that a vehicle lasts for 150,000 kilometers, this is the equivalent of 3.30 per gram of CO_2/km. For a manufacturer, the financial benefits from being able to sell the vehicle for a better price and having to pay lower fleet fuel economy penalties (Approach 1) come to €49.80 per vehicle and per gram of CO_2/km (see Fig. 4). A surprisingly high number that may be based on vague assumptions, but nevertheless shows that there is some financial room for investing in CO_2 reduction efforts also under commercial aspects.

For the vehicle buyer, the calculation looks similar. Assuming a gasoline price of €1.40/l and 15,000 km driven per year, the potential savings from reducing the CO_2/km emission by 1gram come to 8,90 per year (see Fig. 3). In addition, lower CO_2 emissions will also lead to less CO_2 taxation in the future.

The sum of these two potential savings comes to €10.23 per gram of CO_2/km saved, as shown in Fig. 4. Extrapolated to 15,000 kilometers driven per year, reducing emission by 1gram of CO_2/km therefore results in savings of 6,8 €-cents per 100 km for the vehicle owner. The potential savings or cost benefits are significant for manufacturers as well as for buyers and justify the additional costs of producing vehicles with lower CO_2 emissions. Automobile manufacturers have recognized this, too. Uwe Michael, head of development of electrical and electronic systems at Porsche AG [4] Optimizing these systems (power electronics, electrical motors and battery systems) contribute to fuel savings We will fight for every tenth of a liter.

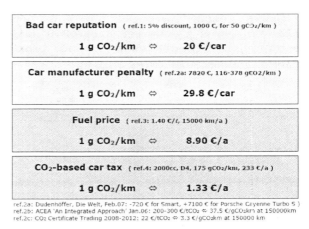

Figure 3 : Automotive Mechanic Fundamentals

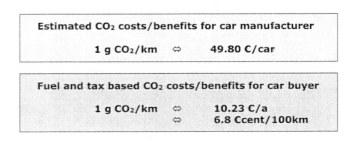

Figure 4 : Cost Benefits or Potential Savings for Vehicle Manufacturers and buyers Resulting from Reducing Emissions by 1 Gram of CO_2/km

HOW CAN ELECTRONICS REDUCE CO_2?

To understand the use of energy in the car, we have to come back to the fundamentals of vehicle mechanics. The car is moving on the road and of course has to run against the air resistance, the road resistance and the potential change of altitude. The last one is in theory neutral, as the energy you use to go uphill is coming back when you go downhill.

It is only in theory, as most of the time we use the brakes when going downhill. We don't have an efficient system for regenerative braking today. The scenario of starting at sea level and going at 3000 m altitude and then going back to sea level will require a huge amount of batteries.

The equations in Fig. 5 give the first lesson. In order to reduce fuel consumption it is required to reduce car mass, air drag, car front surface, tire resistance, road resistance and of course the vehicle speed. Also, we can never forget the fuel consumption increase with the square of the vehicle speed.

The next question is to go through the energy transformation chain and identify the losses. One liter of gasoline has about 10kWh. Only 20% of the energy is used, the rest is just an accumulation of losses: engine cooling, exhaust losses, idling, braking, mechanical and electrical losses (Fig. 6).

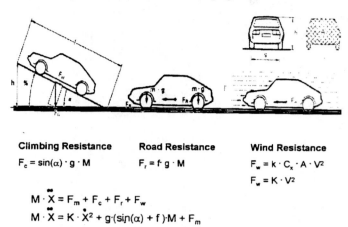

Climbing Resistance

$F_c = \sin(\alpha) \cdot g \cdot M$

Road Resistance

$F_r = f \cdot g \cdot M$

Wind Resistance

$F_w = k \cdot C_x \cdot A \cdot V^2$

$F_w = K \cdot V^2$

$M \cdot \overset{\bullet\bullet}{X} = F_m + F_c + F_r + F_w$

$M \cdot \overset{\bullet\bullet}{X} = K \cdot \overset{\bullet}{X}^2 + g \cdot (\sin(\alpha) + f) \cdot M + F_m$

Figure 5 : Automotive Mechanic Fundamentals

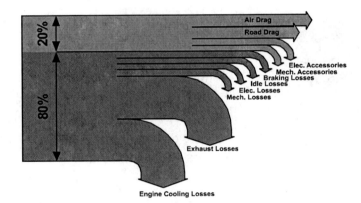

Figure 6 : Vehicle Energy Utilization

In most car dealerships and automotive magazines the information about fuel consumption is given in gCO_2/km. The regulation does not take into account the energy used by certain accessories such as: air conditioning, light system, infotainment, etc. This leaves room for lot of inefficiency. This also gives the idea for improvement of the regulation to further reduce CO_2 (Table-II).

Today in Europe the average car is producing 160 gCO_2/km, the ideal target will be to reach 95 gCO_2/km in 2020. This represents 40% reduction. Is this target achievable at reasonable cost or just a dream?

The table above gives the potential fuel savings we can achieve in the long term, starting from a 160g CO_2/km car. As diesel engines are already very efficient the future saving is less than the gasoline engine. Of course all the values cannot be summed, but all together 45% improvement can be made (see Fig. 7).

	Saved CO_2
Gasoline Engine	30%
Diesel Engine	10%
Transmission	2%
Micro Hybrid (Stop/Start)	10%
Mild Hybrid	20%
Full Hybrid	30%
HVAC	1%
Light system	1%
EPS	3%
Infotainment	1%
Engine Cooling system	3%
Fuel Pump	1%
Total above accessories	10%

Table- II : CO_2 Potential Saving for 160g CO_2/km Car

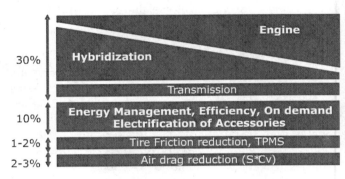

Figure 7 : CO_2 Potential Saving for 160g CO_2/km Car

The target is achievable but the price to pay will be not negligible. The first g CO_2/km will cost about 30€, but the last will be more around 150€. To reduce further the CO_2 emissions will definitely require a change of technology and a migration to either electric propulsion or the hydrogen economy. It will also be very important to count the total CO_2 over complete life, from manufacturing to recycling of the car with the complete well to wheel consumption during the vehicle life.

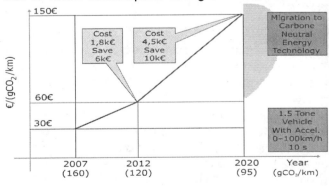

Figure 8 : System Cost to Reduce Emissions by 1g CO_2/km

IMPROVED ENGINE

The principle of the internal combustion gasoline engine was first patented by Eugenio Barsanti and Felice Matteucci in 1854. It then became conceptualized by Alphonse Beau de Rochas in 1862 and independently by Nicolaus Otto in 1876. Rudolf Diesel's patent was filed in 1893. Since that time many generations of engineers have worked on improving the technology but the core principles have remained constant.

The diesel engine has taken benefit from several advanced technologies such as downsizing, turbo charging, and common fuel rail. Further improvements have been developed in the after treatment systems to bring emissions levels down.

The gasoline engine has not been as successful and is still very primitive. For a very long time it has been a simple and low cost technology but the efficiency is now coming into question. The gasoline engine is following its brother, the diesel engine, in the way of direct injection and turbo charging. The challenge for both engines will be to change the combustion principle and use Homogeneous Charge Compression Ignition (HCCI). This target could be achieved by 2015.

The field of automotive electronics is making a large contribution to enable the control, sensing and actuation of the engine. The critical components, such as the injectors and the valves are carefully controlled to reach the maximum efficiency. To reach a better control, the pressure in the combustion chamber will be measured and processed in real time. The expectation is to reduce by 30%, the fuel consumption of the gasoline engine and by 10%, the diesel engine.

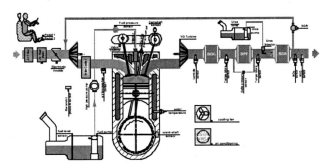

Figure 9 : EURO6 Diesel Engine Schematic

ELECTRIFIED ENGINE ACCESSORIES

There are several engine accessories which are typically controlled by systems which, in the past, have been designed with the lowest possible cost, not energy efficiency, as a priority. This includes systems such as the fuel pump, oil pump, water pump, alternator, starter and cooling fan. These accessories are now moving from belt-driven, low electronic content systems to electrically controlled, demand based actuation.

Driving these engine accessories in a more efficient manner has a direct impact on the fuel economy of the vehicle. Each individual improvement in efficiency has only a small impact on the overall efficiency of the vehicle, but when several systems are improved, the sum of all the efficiencies can provide fuel economy improvements in the range of 5% - 7%.

IMPROVED TRANSMISSION

The drive towards higher fuel efficiency is bringing about the rapid adoption of more fuel efficient transmissions. There is a clear trend to replace 4-speed transmissions with 6-speed transmissions and introduce 7-speed and 8-speed transmissions in high-end vehicles. These added gear ratios can increase fuel efficiency by as much as 9% when compared to typical 4-speed transmissions. Other technologies are being introduced, such as double clutch transmissions, which are projected to see significant growth in the next 5 years in both Europe and the United States. These new transmission technologies have the ability to bring both increased fuel efficiency and enhanced drivability to the vehicle.

These new automatic transmission designs primarily use hydraulics to shift gears. Electronically controlled actuators, or solenoids, are used to precisely manage the pressure of fluid to shift gears and actuate the clutches. The precise control of pressure and timing in engaging and disengaging the clutches affects the drivability and efficiency of the transmission. The need for precision has brought about the wide use of Variable Force Solenoids (VFS), which offer designers greater accuracy in timing and pressure control. The number of VFS in transmissions has risen from one in a typical 4-speed transmission to as many as seven in some 6-speed transmissions.

The ability of semiconductor manufacturers to integrate more functionality into smaller form factor devices which can operate in increasingly harsher environments at cost effective price points has enabled the proliferation of VFS in automatic transmissions. The sophisticated electronics to control these transmissions is a combination of high-end microcontrollers, intelligent power devices, and speed sensors.

HYBRIDIZATION

Hybridization has, for a long time, been a very questionable solution. With the recent progress of battery technologies together with advances in power electronics, hybrid is becoming an unavoidable path towards the car of the future.

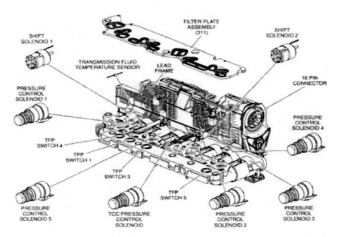

Figure 10 : Illustration of the Valve and Solenoid Assembly for a General Motors Hydra-Matic 6L80 6-Speed Automatic Transmission, which Incorporates Six VFS

Hybridization has multiple levels of implementation, starting from micro-hybrid, followed by mild and full hybrid. The micro-hybrid is a very promising technology in Europe. The merge of two functions starter and alternator into one machine opens new horizons. On one hand this offers the possibility to turn off the engine when the vehicle is stopped, on the other hand this allows to improve the generator efficiency by active rectification and control of the excitation. This machine of 2 kW allows a kind of regenerative braking system. All together we can reach 10% fuel saving: 5% idle losses, 2% alternator efficiency improvement and 3% regenerative braking.

This solution also requires a better battery than the traditional lead-acid technology. New generations such as Adsorbed Glass Mat (AGM) or Gelled Electrolyte (GEL) are being investigated. This advanced system must work together with a smart battery sensor to determine the state of charge and state of health of the battery and manage the energy distribution within the vehicle.

The mild and full hybrids have pushed the development of high power modules for 20 kW to 80 kW motors. The technology used is 3rd generation IGBT trench running up to 600 V. The inverters are controlled by coreless transformer drivers. These power modules are specially designed to satisfy the power dissipation and the temperature cycling requirements of hybrid vehicles. The solution will definitely not be complete without a Li-Ion battery. This battery technology is the most promising for its power density and the number of charge /

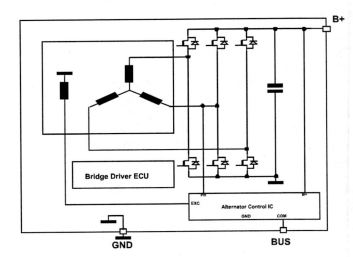

Figure 11 : Micro-Hybrid Schematic

discharge cycles. Two critical points should be nevertheless solved: battery management and the smart power switch to connect and disconnect the battery to the load. The Li-Ion cell requires a very accurate control for charging and discharging. A cell cannot tolerate to be overcharged or undercharged without taking the risk to shorten the life time of the battery. Only a very accurate control system allows balancing the energy between the cells. The management is also monitoring the state of charge, the state of heath, and the temperature of the battery pack.

A 2 kWh battery at 450 V does not switch on and off without a certain challenge. The switch also plays a role in the safety of the vehicle in case of an accident and should guarantee very high impedance. Here electronics has a critical function to avoid any arcing during turn on or off.

Figure 12 : Hybrid Power Modules

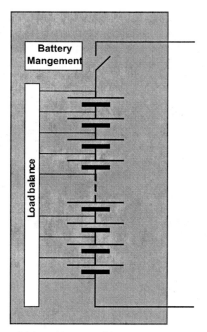

Figure 13 : Li-Ion Battery Management

ENERGY MANAGEMENT

As vehicles become increasingly electrified the management of energy being produced and consumed has a critical impact on the fuel economy and range of the vehicle. The energy management system is tasked with organizing the power consumption of the vehicle. All requests for electrical energy must be prioritized in order to use the available energy in the most efficient manner.

The simple strategies for energy management of current vehicles will no longer suffice in the fuel efficient vehicles of tomorrow. For example, in today's vehicles the headlights are not turned on until the engine has gone through its crank cycle and is running. In future stop / start systems, this energy management tactic will not be used because the engine will go through many more crank cycles within a single drive. The state of charge of the battery system has to be monitored to determine when the engine must be restarted.

In electric and hybrid vehicles, efficient charging and discharging of the battery is a key task of the energy management system. When excess energy is available, such as during a regenerative braking cycle, the electrical system should be in a state that allows for the immediate use or storage of that energy so that it is not wasted.

The management of energy is an area where new and complex strategies will be required to improve the efficiency of vehicles.

MASS REDUCTION

The calculation of the mass effect on CO_2 emissions is very critical. We can save 1g CO_2/km for every 20kg we remove from the car. Electronics can also play an important role. The first is merging functions, for example, the starter and alternator in one machine will save some mass. Electronic modules are also getting smaller and smaller over time. The size of the PCB and module housings will also get smaller. The modern car has up to 80 ECUs (electronic control units) and consequently it is logical to merge ECUs into fewer but more capable super-computer ECUs. Electronics can also reduce the mass of the wiring harness and connector by means of a smart networking system.

DRAG REDUCTION

Small improvements in the coefficient of drag can have significant positive effects on the fuel efficiency of vehicles. Most improvements in aerodynamics are done through vehicle body design. However, actively controlled aerodynamic improvements, which require electronics in their implementation, will see growth in the future. Some examples of actively controlled devices include air dams which are lowered at high speeds to prevent air flow, rear spoilers which are raised at high speeds to reduce lift, and active ride height which raise and lower the vehicle depending on speed.

DRIVER ASSISTANCE

In the modern car the cruise control is already taking over responsibility from the driver to regulate the vehicle speed. Driving without strong acceleration and deceleration also reduces the CO_2. Future cruise control systems will be adaptive with the use of radar and cameras. The assistance will not always be sophisticated; sometimes just a lamp on the dashboard can be enough to inform the driver if the right gear ratio is selected. Another example is the enhancement navigation systems. Today's navigation systems are displaying the routing for the shortest or the fastest way, but none are giving the route for the lowest CO_2 emissions. Navigation and hybrid will be working together to create the best battery charging strategy according to the road topology.

TRAFFIC MANAGEMENT

Managing traffic and lessening the average travel time per trip can have a large effect on fuel consumption. A 2002 study from the US Department of Energy's Oak Ridge National Lab attributed an annual delay of 296 million vehicle hours to poor signal timing alone. This statistic is leading to development in traffic light synchronization initiatives and traffic monitoring technologies. On the vehicle level, technologies such as Vehicle to Vehicle (V2V) communication are also being developed as a means of traffic management.

EMERGING APPLICATIONS

Using an internal combustion engine creates heat. Today 75% of the energy is lost to heat, either in the engine or in the exhaust gas. Thermal recuperation systems are now being envisioned. This should not stop us from being even more imaginative; why not place solar cells on the roof or an auxiliary fuel cell to recharge batteries.

IMPACT ON E/E ARCHITECTURE

In today's vehicles there are as many as 80 ECUs, which can result in a completely overloaded network with several bottlenecks. In this architectural strategy the behavior has largely been that when a new function was needed a new ECU was added to the vehicle and placed on the network. There is typically one or more gateway ECUs which are dedicated to multiplexing the different networks within the vehicle. This distributed function architecture is increasingly difficult to test and verify at the vehicle level.

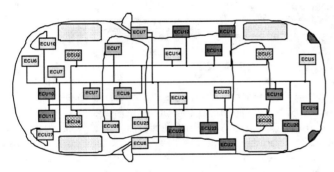

Figure 14 : Distributed Function Architecture Connected Via Several Networks

The proliferation of functions in a vehicle which are implemented with electronics is causing a clear trend away from the distributed network architectures to domain architectures connected via a high-speed, robust backbone network. These architectures will consist of a low number of powerful domain ECUs which are connected via the backbone network. Each domain ECU may then have a set of smaller, less capable ECUs, smart sensors, smart actuators, and mechatronic modules connected via a local domain network. The first implementations of domain architectures have the domain ECUs allocated by function, where there is a main body controller, powertrain controller, safety controller, etc. There is also the possibility to allocate the domain ECUs by location within in the vehicle, where there would be a rear controller, front controller, middle controller, etc.

This is a major departure from today's strategy in that each domain ECU could be responsible for tasks which are not necessarily related from a functional point of view, but which are linked by their distance from the ECU.

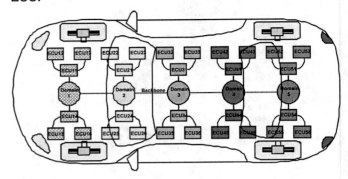

Figure 15 : Location Based Architecture Connected via Backbone

IMPACT ON SENSORS

Multiple sensors are used in the vehicle to measure position, speed, acceleration, pressure, and temperature. The previous passive sensors (e.g. VRS, potentiometers, etc.) are finding limits in most fields. The applications are very often asking for better accuracy over the product lifetime, and improved power-up strategies. The challenge is also to reduce the hardware components to keep sensors and ECUs as small as possible. To reduce system cost, the sensor information must be distributed to avoid a wrong sensor data proliferation. All these features are now available on active sensors. These active sensors also have the possibility to improve electronic safety such as: self-diagnosis, self repair, self calibration, and the implementation of digital communication to avoid problems of losses and noise. A new generation of sensors is coming. TPMS (Tire Pressure Monitoring System) benefits from a high system integration. Radars are now using SiGe 77GHz technology to replace current GaAs. This offers better signal strength together with a huge system cost reduction.

IMPACT ON MICROCONTROLLERS

Electronic systems are consistently requiring better performance, bigger memories, low power consumption, and low price. The CPU performance and memory size is currently doubled every 3-4 years. This performance increase has mainly been met by increasing the clock frequency of the microcontrollers. In 2000, the typical clock frequency in engine control was 40 MHz. Today's microcontrollers in the same applications can reach CPU frequencies up to 200 MHz and can still grow up to 330MHz. However, this value represents a limit that is difficult to cross with existing technologies (90 nm with embedded Flash) for automotive embedded control. Increasing frequency requires higher supply current for a device. This generates more heat, which must be dissipated. Technical solutions exist at the packaging level to handle this situation, such as extra substrate layers, thick copper, heat slugs and heat spreaders, and enhanced BGA packages. However, the current cost of those innovations challenge their implementation in the near future. This will give enough time for the automotive industry to migrate software to dual core architecture. Nevertheless, dual core solutions are already supplied to accelerate the learning curve within the supplier chain.

IMPACT ON POWER DEVICES

In the power domain the drivers are energy efficiency, integration of functions with more control, safety features, diagnosis, and integration of more power to actuate stronger loads. The increased power demand can lead toward lower Rdson at 12-14 V or an increase of the voltage level. The car requires energy management and energy on demand. As mechanical components are migrating towards electric drives, power supplies with linear regulators are being replaced by DC/DC converters. Loads are driven in PWM to control the power based on demand.

The demand on hybridization has pushed the development of power modules for 20kW to 80kW motors. This technology used is IGBT trench 3rd generation running up to 600V.

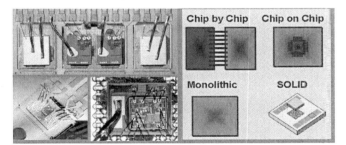

Figure 16 : Semiconductor Integration Methods

CONCLUSION

It is clear that the political, economic, and environmental pressures to increase the fuel efficiency of vehicles all around the world has gained enough momentum that the next 10 to 20 years will yield a wealth of changes in the vehicles which are produced. There is not one large technical innovation which will allow vehicles to become more fuel efficient. Instead the innovations will be many, each one bringing only incremental improvements in efficiency, but when summed together will reach the goals which have been set.

The role of electronics in the fuel efficiency race is paramount. Almost every improvement which is now being designed or considered involves using electronics where before there were none or less capable versions. The key challenge to the semiconductor and electronics makers is to use their innovations to solve the problems in a manner which is cost effective. On the other hand, the opportunities to innovate are plentiful and touch every vehicle manufactured globally.

DEFINITIONS, ACRONYMS, ABBREVIATIONS

Association des Constructeurs Européens d'Automobiles in English: European Automobile Manufacturers' Association

AGM	-	Adsorbed Glass Mat
BGA	-	Ball Grid Array
CAFE	-	Corporate Average Fuel Economy
CNG	-	Compressed Natural Gas
CPU	-	Central Processing Unit
ECU	-	Electronic Control Unit
EPS	-	Electronic Power Steering
GEL	-	Gelled Electrolyte
HCCI	-	Homogeneous Charge Compression Ignition
HVAC	-	Heating, Ventilation and Air Conditioning
IGBT	-	Insulated Gate Bipolar Transistor
LNG	-	Liquefied Natural Gas
OEM	-	Original Equipment Manufacturer
PWM	-	Pulse Width Modulation
TPMS	-	Tire Pressure Monitoring System
V2V	-	Vehicle to Vehicle (communication)
VFS	-	Variable Force Solenoids
VRS	-	Variable Reluctance Sensor

CONTACT

Prita Peter
Head, Corporate Communications
Infineon Technologies India Pvt. Ltd.
13th Floor, Discoverer Building, International Tech Park, Whitefield Road, Bangalore - 560066
Tel.   : + 91 80 4139 2001,
Fax   : +91 80 4139 2333
E-mail : Prita.Peter@infineon.com

SUSTAINABLE MOBILITY

FCV's for a More Sustainable Mobility in 2050	2010-01-0850
	Published
	04/12/2010

Eduardo Velasco
UAEM & GMM

ABSTRACT

Today we are facing an accelerated vehicle technology development, it has progressed so much that the propulsion systems have been diversified; vehicle electrification is a reality, and we already are in a technological transition to the mobility for the future. Then, technological trends vs. a technological vision are addressed in order to understand how to create a technological momentum for a more Global Sustainable Mobility (GSM). Fuel Cell Vehicles (FCV's) are undergoing extensive research and development because of their potential for high efficiency, low emissions and no petrol demand; However, Is this technology that will dominate the roads in 2050? An investigation and analysis of the Future Vehicle Propulsion Systems (FVPS's) are presented to meet the vision for a GSM by the year 2050 as a long-term thinking that might produce better short-term decisions.

Internal combustion engines (ICE's) developments and their major changes may be truncated while technology is growing on batteries, fuel cells and the efficient use of energy. Hybrid Electric Vehicles (HEV's) have the potential to meet the short term stringent regulations as well as Fuel Economy (FE) and CO_2 goals, it can be considered as the first step forward into GSM, therefore there is an urgency for a global energy policy that promotes a real efficient use of energy as the cheapest and fastest way to move towards a global future sustainable mobility. GSM is possible, a change of mindset is necessary for sustainable growth through education of our future generations.

INTRODUCTION

The main driver for the present investigation is to support the expanding vision of GSM and this paper provides a better understanding of how current technologies in the present may survive or evolve on 2050 year toward this vision.

Our planet is under unprecedented ecological and environmental strains because of the way we manage our economies as a result of an unsustainable model for global growth. It is clear today how the environmental & biodiversity challenges accompanying our economic development are not in the same equation of the model. A growing sustainable model through the innovation research at the intersection of the different disciplines, no science borders and a sustainable change of mindset over the present and future generations in the next decades, appears to be the better short-term decision.

The growing numbers of Non Governmental Organizations (NGO's) around the globe are coinciding to establish the year 2050 for setting global targets. Until now, a reasonable start for setting numbers would be 50% of the primary energy production and 80% of world's vehicle propulsion from Renewable Energies (RE's) by 2050. This assumes the mindseting change of present and future generations as well as a technological innovation through the contribution and benefice of all disciplines envisioning sustainable energy economy. Energy efficiency becomes the first step; improvements of it can deliver some of the immediate largest and cheapest CO_2 reductions. Doing all of this in an environmental responsible manner some important & multidisciplinary international NGO's like IEA, IPCC, UNFCCC and the G8 leaders agreed to seriously consider a global 50% CO_2 minimum reduction target for 2050 [1].

Trends of how the future could be unfolded are easy to indentify; scenarios are images of alternative futures, these are neither predictions nor forecasts, but for sure they show how future could change by giving a more precise picture of the behavior of complex systems. Therefore technological trends without a solid and global vision will not ensure a sustainable growth.

Figure 1 represents the transition to a renewable energy future model following the vision of a global sustainable energy economy, and at the same time by explaining itself the direction to evolutionary mobility development as a technological momentum of improvement over the coming years of energy efficiency as the GSM. Today one of the strongest drivers of it is the climate change which is forcing humankind to modify our energy consumption patterns by a significant reduction of burning fossil fuels. As a consequence, automakers have started the fuel diversity and their search for FVPS's.

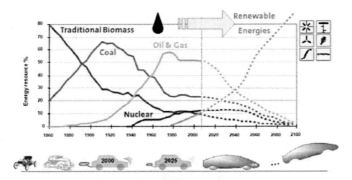

Figure 1. Energy transition to RE's [2].

Going into 2050 a strong link between energy production and vehicle propulsion is very clear as a dependence of each other's energy and mobility, efficiency and performance. Paradigms such as levitation and wireless energy transmission to re-invent the mobility again should be broken. Nature as an alternative guide of design for modern technology has the potential to improve the way we do everything; current design challenges may probably already had solved by nature. Designs where the required rigidities are achieved with the lowest possible weight and less use of materials with the highest precision and efficiency are those desired features for the future propulsion technologies. Biomimicry, as a new discipline that studies nature's best ideas imitating these designs and processes as biologically inspired design, might help to solve our global unsustainable issues; energy and transportation for example.

Jules Verne's novels have been noted for being startlingly accurate anticipations of modern times and in terms of energy he wrote his vision of the fuel of the future: "water" in their novel "The mysterious island, Part 2 The abandoned in Chapter 11" first published in 1874.

"Yes, my friends, I believe that water will one day be employed as a fuel, that hydrogen and oxygen, which compose it, used alone or together, will furnish an inexhaustible source of heat and light and with intensity that coal cannot provide".

Given the progress we have already made with the technology and by adding GSM vision, we can foresee what kind of automobiles should probably be in the roads by 2050: no question about safety, free of harmful emissions, completely recyclable, built from intelligent materials, powered by energy from renewable sources and biological sensors to assist the human-vehicle interface. Nowadays, current hybridization has shown how different combined technologies provide better fuel economy, lower emissions and efficient energy conversion than a single technology. In fact, they even have seen to solve each other's weaknesses. Fuel cells and batteries combined for example are therefore expected to be an important energy technology for the future, ultracapacitors may also do too.

A GSM VISION

Based on the celebrated report from the United Nations in 1987, "Our common future", (Brundtland report); Global Sustainable Development (GSD) is the development that allows people to meet their needs of the present, without compromising the ability of future generations to meet their own needs. Furthermore, GSM definition is implicit in the above description; development that allows people to meet their transportation needs of the present without compromising the ability of future generations to meet their own transportation needs. In other words, GSM should become the central guiding principle of the government and private institutions, organizations and enterprises or any regular people by adding today to the business as usual equation some parameters to resolve and decide always with a Continuous Positive Sustainable Impact (CPSI) delivery.

Figure 2. Global message in Front of City Hall, Seoul, South Korea, April '07.

Figure 2 interpretation as a message for every person in the planet, a change in mindset is necessary for sustainable growth and the best way to do it is to teach our children at home and in classrooms around the globe, keeping in touch with nature about their wonders but also telling them about nature limitations of sources availability from our planet. This early training can produce a tremendous impact on their impressionable brain and gradually a generation of responsible citizens will decide the future of our specie. These citizens will then become the new customers with requirements of mobility more in line with the GSD and GSM.

Recently automotive companies believe that its own long term viability depends on achieving sustainable mobility, they should start wrestling with complex challenges that must be addressed and achieve global goals for economic growth CPSI and social progress simultaneously, in a GSM context. Then challenges and opportunities are global; the initial direction will depend on co-operation among governments, industries, consumers and other elements of civil society. In fact, all disciplines of science need to come together to establish new direction to follow envisioning a GSM. Meanwhile a new generation of responsible citizens grows and evolves to take their role in the future as decisions takers to continue the GSM vision.

Currently an accurate assessment of how propulsion systems are compared in terms of efficiency considering the environment impact is commonly called a Well-to- Wheel (WTW) analysis. It can infer about the future fuel/propulsion system alternatives as a complete vehicle fuel-cycle study. WTW analyzes energy use and emissions associated with fuel production activities (also known as Well-to-Tank activities). The other part of the analysis to close the loop is the energy use and emissions associated with vehicle operation (or Tank-to-Wheel) activities. As a reference the WTW efficiency of a gasoline vehicle is 14% for the average car (2005 year), while a gasoline - electric hybrid vehicle (where engine is off when traction forces are zero negative or low) receives approximately 26% and a FCV can receive up to 42% WTW efficiency [3]. It looks like the broad industry goal is moving vehicle technology towards greater efficiency, lower air pollutant and Green House Gases (GHG's), through the use of simpler and more reliable vehicle designs. But there are no specific targets and it is not clear their commitment for a GSD and GSM vision.

DEVELOPMENT OF FVPS'S

Road blocks such as fuel prices and the urgency for sales are moving automakers toward smaller, lighter and more fuel efficient vehicles. At the present vehicle mass production is one of the important requisites only if it meets and exceeds customer expectations. In addition the energy diversity and development of alternative fuels are creating a new transitional hybrid technology for the vehicle propulsion system. However, this doesn't warranty GSM environment although; it is evident that customers are looking for cleaner and more energy-efficient cars at affordable prices. Such technologies are generally being described Advanced Internal

Combustion Engines (AICE's), hybrids, fuel cells and batteries as follows:

The AICE dominant role in the automobile propulsion looks like it will be maintained in the next 20 years, although, the polarization on this unique form of propulsion appears as unrealistic for the future. Hybrids, where a minimum of two energy sources (fossil fuels, H_2, Batteries, capacitors, flywheels, etc.) are combined to deliver power to the wheels. A hybrid vehicle is a combination or blending of more than one propulsion system, the market share of hybrids will further increase as the fuel and energy diversity increases. Finally the vehicle electrification via fuel cells and/or batteries has been a progress absolutely remarkable but not enough to be considered today an independent source of energy available in the market.

What drives automakers today to develop new propulsion technologies? Well, first we start with the main contributors; leadership and successes over the competitors as a believer of potential mechanism of vehicle sales increase. Second the increasing pressure from society to have greener technologies as a contribution to reduce GHG effects. And the third one is from some countries to avoid dependence from foreign oil. At the end of the day the balance shows that none of them are driven to meet any level of sustainability or at least future common goals. The evolution of the car based on above automakers drivers (see figure 3) shows how the car transformation may happen over the next years. Bridges between fossil fuels to full electrification are the hybrids with fuel diversification.

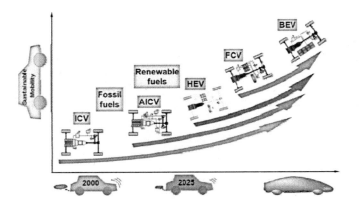

Figure 3. The most general automakers accepted transformation model to FVPS's path [2].

Current diversification of fuels & propulsion systems has established a number of alternatives for FVPS's by combining fuels and drive trains (see figure 4). Over the next 5 to 10 years automakers are planning investments on R&D activities focused mainly on FE improvement and CO_2 emissions reduction by decreasing vehicle mass, improvements on combustion, reduction on friction, the usage of biofuels, the implementation of idle-off capabilities and certain level of electrification. The final idea is to attract the attention for additional sales. After all every day is heightened FE & CO_2 pressure worldwide.

At the end of the path of the <u>figure 4</u> we classified the FVPS's in four viable potential fuel/vehicle technologies which already exist or are under extensive development.

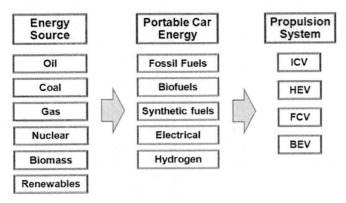

Figure 4. Current powering automobile diagrams.

ICEV - Internal Combustion Engine Vehicle; has fossil fuels diversified; gasoline, diesel, gas, biofuels, synthetic fuels, etc and their environmental impact will never reach zero levels, however today is the predominant technology on the roads. Additionally the ICE developments and their major challenges are in a process to be truncated while technology growth on hybrids, fuel cells, batteries and ultracapacitors due to their lower irreversibility's and energy conversion efficiencies.

HEV - Hybrid Electric Vehicle; It is considered the next generation of propulsion systems for automobiles. This technology can be diversified depending of the hybridization level from (ICE) Idle-off capability to full electrification using Fuel Cells (FC's).

BEV - Battery Electric Vehicle; assuming affordable batteries are developed along with low recharge times ($\sim$ 30 min); they would allow to achieve quasi free oil technology and meet 2050 targets.

FCV - Fuel Cell Vehicle; this technology also can contribute to 2050 targets however a number of technological challenges need to be solved first. They can benefit from an increased understanding of nano technology, development of fuel cell variable functions and the support of multidisciplinary research. Biology is one of the most promising knowledge fields to develop and create synthetic materials and devices for fuel cells that can carry out tasks with the precision and efficiency of biological systems.

ENERGY SOURCE

Energy consumption has been growing exponentially since recorded data is available, energy consumption patterns are both physically and socially unsustainable. This growth is partially attributable to the needs of an exponentially growing population and therefore mobility.

The last decade has brought significant development on more efficient renewable energy technology in other words; a more sustainable way to generate and use energy to drive societal development. Further developments are focusing the attention to the recent promising research on nanostructured materials, which have demonstrated significantly enhanced energy conversion and storage efficiency in solar cells, batteries and catalysis due to their large surface to volume ratios and favorable transport properties [4].

<u>Figure 5</u> shows the total primary energy supply in which for 2010 year the estimated value of 510 EJ was based on growth rates as well as the key world energy statistics 2009 from the IEA. By 2050 reasonable future energy requirements may be increased up to 3 times from today.

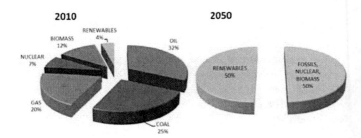

Figure 5. Total primary energy supply.

The share of renewables in the global energy mix could go up from its existing very low base to about 50 per cent by the middle of the century. This assumes that the hunt for technological breakthroughs to make renewables cheaper will be successful.

PORTABLE CAR ENERGY

Vehicle transportation has two modes of fuel supply to their propulsion system; portable and non portable. For cars portable fuel that is the only mode used and looks like it will remain at least for the following decades until the paradigm may be broken with wireless energy transference or other mechanism.

Today the strong link between energy production and vehicle propulsion play an essential role in the automotive industry, fuel diversification is destined to emerge in the near future, see <u>figure 6</u>. Various alternative fuels from biomass are possible as potential substitute to fuels from fossil sources, and will play an increasingly important role in the future, making an optimization of the engine with the amount of fuel necessary. However in order to meet GSM, zero emissions is a prerequisite along with energy efficiency and the alternative is electro-mobility (powertrain electrification) in which it uses different kinds of fuels; the batteries and fuel cells. Therefore fuel diversity may work well during certain period of time as the bridge for electro-mobility.

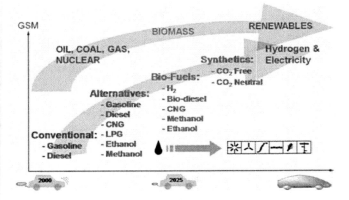

Figure 6. Primary energy source and vehicle fuels change toward GSM [2].

Fuels can be produced from a number of different primary energy sources, such as a fossil feedstock, or directly extracted from renewable energy and biomass. In this study, we have included 6 primary energy sources and 5 general fuels that appear relevant for the foreseeable short terms & medium term future. The number of conceivable fuels and fuel production routes is very large, however the combinations were considered less relevant have been left out from the analysis see figure 7.

ENERGY SOURCE	FUEL						PROPULSION TECHNOLOGY
	PORTABLE					NON-PORTABLE	
	FOSSIL	SYNTHETIC	BIO-FUEL	HYDROGEN	ELECTRICITY		
OIL	Gasoline, Diesel, LPG	--	--	--	--	--	ICE, Hybrid
COAL	DME, Methanol	D100	--	Compressed & Liquid	Electricity	--	ICE, Hybrid, Fuel Cell, Electric
GAS	DME, Methanol, CNG	D100, D 5/95	--	Compressed & Liquid	Electricity	--	ICE, Hybrid, Fuel Cell, Electric
NUCLEAR	--	--	--	Compressed	Electricity	--	Hybrid, Fuel Cell, Electric
BIOMASS	--	Diesel, D95	E100, E95, Compressed gas, D100, D5/95	Compressed & Liquid	Electricity	TBD	Hybrid, Fuel Cell, Electric
RENEWABLES	--	--	--	Compressed & Liquid	Electricity	TBD	Hybrid, Fuel Cell, Electric

Figure 7. Primary energy sources and cars portable fuels.

FVPS'S

If the GSM vision, hybridization and electrification trends continue, the world of 2050 and onward will be more biologically diverse and many of today's decisions as the near term will have long-term positive consequences having clear that the FVPS's will move people around the globe in a

sustainable manner. Long-term thinking as GSM & GSM vision has the power to incorporate sustainable short-term decisions in parallel to the gradual change of mindset thought education of our current and future generations. In the case of future sustainable technology development may be reached with the contribution from the different science disciplines giving on this way the result of a CPSI to our environment and biodiversity. Result of current research and development for future vehicle propulsion systems suggests the idea to find technical solutions on nature follow evolutionable designs and create multivariable functions.

Energy	Fuel	Technology	Propulsion system
OIL	Diesel	Hybrid DICI + battery	HEV
		Hybrid DICI-PF + battery	
		Hybrid FC-R + battery	
		DICI	ICV
		DICI-PF	
	Gasoline	DISI	ICV
		Hybrid DISI + battery	HEV
		Hybrid DICI-R + battery	
	LPG	Hybrid FC-R + battery	HEV
COAL	Dimethylether	Hybrid DICI + battery	HEV
		DICI	ICV
	Electricity	Na-NiCl	BEV
		Li-ion	
		Ni-MH	
	H₂ Compressed	Hybrid FC + battery	HEV
		Fuel cell	FCV
	H₂ Liquid	Hybrid FC + battery	HEV
		Fuel cell	FCV
	Methanol	Hybrid FC-R + battery	HEV
	Synthetic Diesel	Hybrid DICI + battery	HEV
		Hybrid DICI-PF + battery	
		DICI	ICV
		DICI-PF	
NATURAL GAS	H₂ Compressed	Hybrid FC + battery	HEV
		Fuel cell	FCV
	CNG	PISI	ICV
NUCLEAR	Electricity	Li-ion	BEV
		Na-NiCl	
		Ni-MH	
	H₂ Compressed	Hybrid FC + battery	HEV
		Fuel cell	FCV
RENEWABLE	Electricity	Li-ion	BEV
		Na-NiCl	
		Ni-MH	
	H₂ Compressed	Fuel cell	FCV
		Hybrid FC + battery	HEV

Figure 8. Future cars tend to be fueled by a diversity of energy forms (Biomass not included).

The table of figure 8 shows the potential propulsion technologies for the next decades; note that H_2 has the highest feedstock flexibility of any available energy carrier and electricity as second place. H_2 is also able to provide a sustainable storage media and facilitate the transition from today's fossils economy into a renewable energy development. Hydrogen used as an automotive fuel in hybridized fuel cell powertrains provides superior Well-to-Wheel (WTW) efficiencies compared to today's fuels and thus reduce primary energy consumption and greenhouse gas emissions.

Figure 9 shows the energy conversion efficiency from the primary energy sources (except Biomass) to the vehicle wheels where the message is clear on renewables as the highest efficiency to produce H_2 and electricity over fossil fuels. In the case of the renewable as primary energy source fuels produced are H_2 and electricity for FCV - HEV and BEV respectively. The longer bar for renewable energy to produce electricity in figure 9 depends on where the energy is produced and where it will be applied, therefore in order to get the highest efficiency for BEV electricity produced needs to be as close as possible and the rest of it produced to be used for other local application different from transportation.

The case for GHS's specifically CO_2 is addressed in figure 10 shows also the no impact by renewable considering the full chain; from renewable to wheels. Coal is the worst case for primary energy source not only for their lower efficiencies but also for their higher GHG's. So we can conclude that this type of energy source and the fuels produced are made in an unsustainable way that in the short to medium term might be truncated by the rest on figure 10.

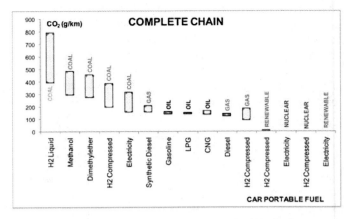

Figure 10. Estimated CO_2 Generated while fuels are produced and processed by the different propulsion technologies (ref. figure 8).

For the 2050 year, following its targets and a GSM vision, the economic model of mobility will be less dependable on fossil fuels for the energy and transportation sectors as shown on figure 11. This model indicates that the best options for GSM in terms of fuels are the H_2 and electricity produced by renewables. A number of combinations of AICE's, FC's and batteries for hybridization over 50% of electrification might dominate the roads in the following decades as part of the transition to fully electro-mobility drive trains. Then the roles for hybrids are an important technology, after all they easily achieve increased fuel economy by reducing the amount of inefficient operation of the ICE increasing system efficiency.

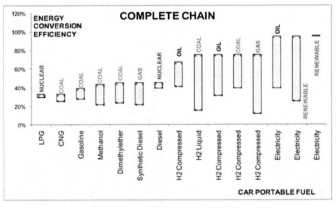

Figure 9. Estimated efficiency to produce fuels for a variety of propulsion technologies (ref. figure 8).

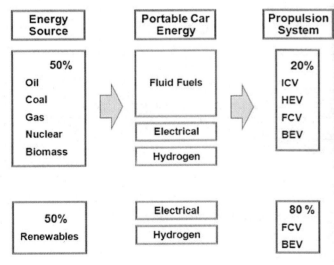

Figure 11. Model for predominant powering automobile by 2050 in a sustainable energy development.

THE FUTURE OF FCV'S

Future cars tend to be fueled by a diversity of energy forms. The FCV's have the potential to meet future mobility requirements in harmony with the biodiversity if the technological development is aligned with GSM vision.

FCV's are undergoing extensive research and development because of their potential for high efficiency, low emissions and no petrol demand [5]. It is a fact that H_2 has to be produced using primary energy but electricity too and electricity is by no means the only path to generate H_2. Therefore BEV's will always be more efficient than FCV's, but the combination and optimization of both technologies may meet future mobility requirements while experiencing the lower GHG emissions, higher FE, the maximum positive sustainable impact and support the GSM vision.

Today is clear how the road transport has started by looking at the fuel and powertrain diversity, but based on this paper the long term mobility has to converge slowly at first, but certainly the electric car will conquer the roads. Battery and fuel cell complement each other, pure battery car will not be the general solution for our future mobility needs, but the fuel cell on the other hand cannot perform well without a battery. All fuel cell cars operating or being developed now has a buffer battery to complement variable fuel cell functions depending on power demand. Due to the universal application and diversity of users along with many environmental conditions to deal with, propulsion systems of future cars are evolving to perform multivariable operational functions. Variable fuel cell functions will allow this technology to improve energy conversion efficiency, therefore a good recommended design direction to follow in the next decades should be the implementation of variable fuel cell functions.

SUMMARY/CONCLUSIONS

For the analysis considered in this study, the following conclusions are drawn.

A vision to revert the negative trends & projections is possible via GSM by taking right decisions for a long term benefit. According to the results COAL as one of the primary energy source is most pollutant as well as the less efficient for fuel production route and usage on drive trains. Contrary renewables are the promising primary source of energy to allow vehicle electrification.

Hydrogen with the highest feedstock flexibility of any available energy carrier can provide a clean storage media and facilitate the transition from our today's global unsustainable energy economy into a sustainable world. Hydrogen therefore allows extending significantly the use of renewable energy sources in the transportation sector.

Our future propulsion system will be a part of a complex development models but is clear that GSM vision can change our world in a positive impact on environmental and biodiversity. Global efforts and mindseting change of present and future generations might be the difference between allows or not allows people to meet their transportation needs of the present without compromising the ability of future generations to meet their own transportation needs.

Battery and fuel cell complement each other by solving each other's weaknesses. Assuming a progress toward GSM, the propulsion system for the car of 2050 has to be intelligent, predictable and capable of multivariable functions for the highest precision and efficiency of their main components and system itself; the electric motor, the battery and probably even most of those will also have a hydrogen tank and a fuel cell.

REFERENCES

1. G8 SIRACUSA 22, 23, 24 Apr 2009. Ensuring Green Growth in a Time of Economic Crisis: The Role of Energy Technology. International Energy Agency, OECD/IEA, 2009.

2. Velasco E. "Mirando Hacia el Automóvil del 2050", SOMIM paper CAA5-48, XII SOMIM annual Congress, Acapulco, Mexico, September 2006.

3. Smith K. K., Powering Our Future, iUniverse, 2005 by Alternative Energy Institute. N. Y. (2005). p122.

4. Wang Ying and Cao Guozhong, "Nanostructured Materials for Advanced Li-Ion Rechargeable Batteries," IEEE Nanotechnology Magazine, 14-20 (June 2009).

5. Delorme, A., Rousseau, A., Sharer, P., Pagerit, S. et al., "Evolution of Hydrogen Fueled Vehicles Compared to Conventional Vehicles from 2010 to 2045," SAE Technical Paper 2009-01-1008, 2009.

CONTACT INFORMATION

Eduardo Velasco can be contacted at eduardo.velazco@gm.com.

ACKNOWLEDGMENTS

Author would like thank Carlos Esquivel, Noe Dominguez and Alberto Sierra for their time and recommendations on this paper.

DEFINITIONS/ABBREVIATIONS

GSM

Global Sustainable Mobility

FCV

Fuel Cell Vehicle

FVPS

Future Vehicle Propulsion System

ICE

Internal Combustion Engine

HEV

Hybrid Electric Vehicle

FE

Fuel Economy

NGO

Non Governmental Organization

RE

Renewable Energie

IEA

International Energy Agency

IPCC

Intergovernmental Panel on Climate Change

UNFCCC

United Nations Framework Convention on Climate Change

GSD

Global Sustainable Development

CPSI

Continuous Positive Sustainable Impact

WTW

Well to Wheel analysis

GHG

Green House Gases

AICE

Advanced Internal Combustion Engine

ICEV

Internal Combustion Engine Vehicle

HEV

Hybrid Electric Vehicle

BEV

Battery Electric Vehicle

DICI

Direct Injection Compression Ignition

PF

Particle Filter

R

Reformer

DISI

Direct Injection Spark Ignition

PISI

Port Injection Spark Ignition

CNG

Compressed Natural Gas

FURTHER WORK

There is a personal committed to continue working on more detailed data, analysis of efficient energy conversion and meet the PhD Program.

2009-01-0598

Is Mobility As We Know It Sustainable?

Philip Gott
IHS Global Insight, Inc.

ABSTRACT

Economic growth, quality of life and mobility are inextricably intertwined in virtually every culture of the globe. With only 1/5 of the world's population living in "Industrial Countries," and with those countries responsible for some 60% of the demand for transport energy and the associated environmental challenges, what will happen by 2035 with almost universally higher standards of living? This paper explores the extent to which known transport technologies can reduce energy demand and greenhouse gas emissions, and identifies opportunities for actions that should be taken to mitigate the impact of the huge growth of the fleet by 2035.

KEYWORDS

sustainable, mobility, motorization, population, demand, political, regulatory, economic, solution, fuel cell, battery, energy conversion, consequence

INTRODUCTION

Sustainable mobility provides for the safe freedom of movement for people and goods in a way that uses renewable sources of energy while creating no adverse impacts on the earth and its environment As such, Sustainable Mobility is today on everyone's lips, yet we have a relatively long history of concern about the sustainability of modern modes of transport. Indeed, there have been doubts about the long-term availability of petroleum ever since the 1920s, if not before. For most of automotive history, the availability of petroleum has been the major factor seen to limit the use of the automobile. Yet the oil industry has continued to provide a seemingly limitless supply, causing some analysts to consider oil as "an infinitely available finite resource."

Because we have so far been able to prove as false such predictions of impending doom, we as a society have become more or less hesitant to take such forecasts too seriously. Only relatively recently have concerns over climate change begun to raise everlasting questions about the mobile lifestyle of modern developed nations and the implications of the global spread of the motor car. In January 2008, the announcement of the microcar "Nano" by India's Tata Group was greeted with expressions of grave concern by environmental groups worldwide. Giving access to a very low-cost car to a market of a billion people is thought certain to have a profound effect on not only demand for fuel, but also on an already CO_2-sensitive global environment.

This paper attempts to quantify the likely impact of the motorization of developing markets, put our current conservation efforts in context of what is needed in the long term, and propose some long-term solutions to the challenge of sustainable mobility.

THE MODERN MODEL OF MOBILITY

Mechanized mobility has become essential for our modern way of life. Economic prosperity grows on the back of goods transport modes. The more goods we can move faster to market, the greater the wealth resulting from the value created by those goods. Globalization, an economic concept based on the specialization of nations or regions producing those goods for which they are best suited, is an extreme case, but is not an essential prerequisite for high levels of motorization.

A strong correlation exists between per capita income and the number of 4+ wheeled vehicles (both cars and trucks) per capita (motorization rate). This correlation appears in Figure 1 and can be characterized by Equation 1.

$$Vehicles/1000\ People =$$
$$-0.5413(PC\ GDP)2 + 37.375(PC\ GDP) - 131.53$$

Equation 1

Increasing wealth is accompanied by an increase in the motorization rate that begins in earnest at about $5,000 per person and levels off in the range of $35,000 per person, on a globally normalized basis of Purchasing Power Parity (PPP – meaning that the value of the dollar is normalized such that in each country it has the same purchasing power).

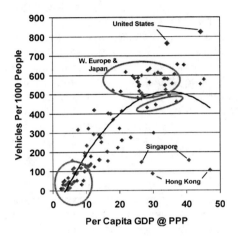

Figure 1: Historical relationship between individual wealth and motorization rate (data from 2002 and 2007)[1]

At the $5,000 level, most of the vehicles are Light or Medium commercial vehicles used for goods transport. As populations become wealthier, the mix of vehicles includes more Heavy commercial vehicles for goods transport, and passenger cars begin to dominate the in-use fleet. As populations become even wealthier, the markets become saturated and are limited by the number of licensed drivers. As a result, the motorization rate levels off. Another important factor ultimately limiting motorization is the movement of populations into cities where a combination of congestion, high ownership costs and convenient access by foot or public transport to most essential destinations (shopping, work, and leisure) can be accomplished without a private motor vehicle. In extreme cases, very high levels of taxation, outright limitations of motor vehicle numbers, congestion or other external factors cause the motorization rate to be very low, despite high levels of per capita income, as in Singapore and Hong Kong.

GROWTH OF THE IN-USE FLEET

At the moment, there are globally an estimated 800 million 4+ wheeled vehicles in-use today. Many estimates of the future growth of this fleet expect this number to double by 2050, with the associated consequences in terms of air quality, fuel consumption and climate change. Most policy makers, in turn, base regulatory targets on the expected impact of this growth in the local or global vehicle parc and establish performance targets for vehicles such that the impact of this larger vehicle fleet will be reduced to acceptable levels.

In reviewing the estimated impact of these future fleets on fuel consumption and CO_2 emissions, the author began to question the basis of the forecast for overall fleet size. This questioning gave rise to the following analysis.

Economic forecasts expect increased prosperity in all regions of the world. These forecasts indicate that by 2035, national economies will grow such that every region of the world will have a per-capita GDP above $5,000 at PPP. Populations will grow significantly in Asia and Africa (see Figures 2, 3 and 4). Highly populated regions of the world will be climbing the rising side of the motorization curve of Figure 1. China and the combination of the U.S. and Canadian economies will grow by about the same amount in terms of total GDP.

Changes in purchasing power of the various currencies present a somewhat different view than might be first apparent from the overall wealth of nations. There are relatively large differences in overall change in wealth of the various countries or regions (Figure 2). Differences in per capita purchasing power of the consumers in the various countries are further impacted by stagnation or decline in the population of the developed markets. As a consequence, there is much more purchasing power in the hands of the Japanese, Europeans and Americans as well as the Chinese than the relative growth in GDP would at first suggest.

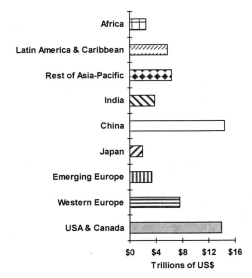

Figure 2: Regional gains in GDP, 2000 to 2035

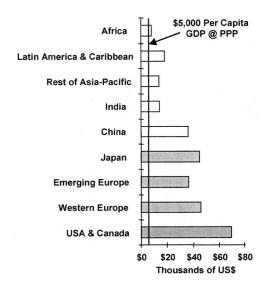

Figure 4: Per Capita GDP at PPP

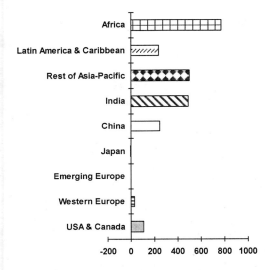

Figure 3: Africa, Asia and India Outstrip China in terms of population growth, 2000 to 2035

Forecasts of Per Capita GDP at PPP enable us to project the motorization rate in each of these countries or regions. Equation 1, previously developed, is the model of mobility as we know it today. Applying equation 1 to the PPP GDP, and then multiplying by the population gives us an overall estimate of the economic demand or potential for the size of the regional or national and global in-use fleet by 2035.

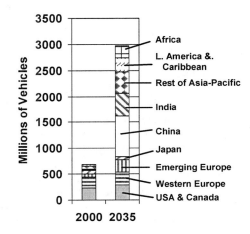

Figure 5. By 2035, the economic potential for the size of the in-use fleet will be about 3 billion vehicles, an increase of almost 4 times that of the current in-use fleet.

This model projects that the global in-use vehicle fleet will be on the order of 3 billion vehicles by 2035. This is an in-use fleet that is 3.75 times the size of the current fleet. That is, almost three to four times the demand for physical space, raw materials and fuel than exists today, and 3 to 4 times the input to the atmosphere of toxic and climate-changing emissions (assuming no change in vehicle characteristics and similar annual distances traveled).

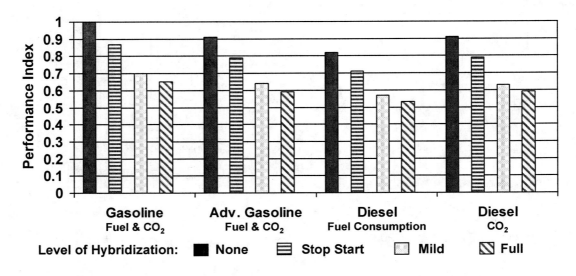

Figure 6: Indexed fuel consumption and CO_2 performance of advanced Diesel, gasoline and hybrid powertrains[2]

THE CONSEQUENCES

Is this possible? No-one expects this fleet size to be achieved simply because the implications of such an occurrence are staggering. The significant increases in vehicle population would occur in regions where the roadway and parking infrastructure is least capable of accommodating large numbers of vehicles. The demand for fuel and natural resources is enormous. Clearly, these are improbable factors, by themselves, never mind the consequences of the resultant manufacturing and tailpipe emissions. Still, if demand for mobility follows current patterns, the economic potential is headed in this direction, and the consequences are clear, present and onerous.

Why make this calculation? As stated at the outset, no other forecaster has publicly entertained such a massive in-use fleet. The sheer numbers are staggering, and the consequences boggle the mind. However, no other forecaster has been able to clearly state how and why such an economic demand will NOT occur. Projections of a fleet only half this size by 2050 are more readily accepted, and the consequences deemed manageable, albeit with much challenge.

The danger comes in not recognizing the plausibility of this scenario, and taking the necessary steps to ensure that it does not occur. Every person with access to modern communications sees the freedom and extent of mobility available to people in the developed nations.

Every one in emerging and developing nations aspires to similar levels of mobility. Clearly, applying our contemporary model of mobility to the rest of the world is not sustainable, yet most of the world clamours for just that.

"OUT OF THE BOX" THINKING NEEDED

There are tremendous technical advances available to us today to get us part of the way towards minimizing the impact of the future in-use-fleet. Powertrain technology alone can reduce CO_2 and fuel consumption by about 40 and 50% respectively (Figure 6) compared to conventional gasoline engines.

Alternative fuels can displace more of the conventional fuels, but many are alternative forms of fossil fuels which are ultimately in limited supply and add carbon sequestered naturally from previous generations into our biosphere. Fuels made from vegetable stock are more benign, and once the transition disruptions are addressed, represent a viable alternative to replace at least 15% of our transportation energy demand before we run into a food-fuel conflict.[3]

A significant shift towards smaller vehicles that fundamentally require less energy to propel them will also help, of course. However, without measures that many consumers would deem oppressive, a massive shift is unlikely. We have modelled the implementation of relatively severe circulation taxes (taxes paid every year based on the CO_2 output of a vehicle) and assessed their impact on the sales of vehicles of different size classes. An extreme case is one in which the costs for transportation as a percentage of disposable income in Western Europe is similar on average to the current situation in Italy. On that basis, the market shares of the different vehicle segments in Europe would shift by 2025 to be similar to that of Italy today as shown in Figure 7. Impacts in other regions would be similar.

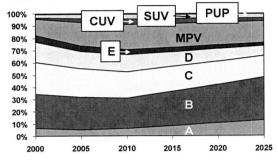

Figure 7: Possible vehicle segment market share shifts in Western Europe due To high levels of CO_2-based taxation

Such an apparently substantial change in the market mix actually results in only a modest change in net transportation energy demand. If everything else (including the market shares of different powertrains in each segment) is held constant, possible shifts in the markets of W. Europe, the U.S. and Japan result in a net decrease in the fuel consumption of each region by 4%, 12% and 5%, respectively. The very low impact in Europe is due to the predominance of gasoline engines in the segments of smaller vehicles, and the popularity of Diesels in the larger, heavier vehicles. As the market moves from large to small vehicles, the consequential shift (in this analysis) from diesel to gasoline offsets the gains obtained from reduced vehicle size and mass. Clearly, a shift to smaller vehicles must be accompanied by a shift to advanced propulsion systems.

The net benefit, then, of deploying technical changes combined with plausible CO_2- or fuel-conscious changes in the market mix of otherwise conventional vehicles is a net reduction on the order of 40% on average per vehicle in W. Europe (given that a little more than half the new vehicles are already Diesel) and about 60% elsewhere. With the forecast quadrupling of the vehicle in-use fleet and a 3 to 4 fold increase in net transportation energy demand, these gains, as significant as they may be, are nonetheless insufficient to achieve even a net status quo in terms of CO_2 and fossil fuel demand. If the current mobility model remains in place, we will still see an approximate doubling of land transport-related energy demand and fossil-related CO_2 emissions even with full deployment of our most advanced powertrain technologies and a downsizing of the fleet.

Clearly, non-conventional solutions are needed.

POSSIBLE SOLUTIONS

The search for non-conventional solutions needs to embrace the full spectrum of options.

Most attractive in the list of possibilities is the fuel cell. The attraction is not so much due to its use of hydrogen (the well-to-wheels benefits of which are often questionable) but more because it promises freedom from the constraints of the Carnot cycle, and is expected to yield efficiencies that are almost double those of combustion technologies. While this doubling will be insufficient to offset the almost fourfold potential increase in the number of in-use vehicles globally, it will be a part of the solution, but not the entire solution.

Combinations of technologies can enable further reductions, as we've already seen with the combination of electric and internal combustion engines in today's hybrids. A move to smaller, lighter vehicles will also help. Substantial weight- and size-related improvements will come only when the vehicle is significantly smaller.

Fuel-cell powered 2-wheelers have been demonstrated, but they are not for every purpose. Indeed, much of the developing world wants to move away from 2-wheels and get on 4! Do we tell the developed markets they must move from 4-wheels to 2 (Figure 8)?

Figure 8: Daimler demonstrated a fuel-cell-powered scooter at the 2007 Challenge Bibendum in Shanghai, the epicenter of China's shift from 2- to 4-wheeled personal transport.

Or, do we:

- Force everyone into Nano-sized vehicles?
- Focus on pure electric vehicles charged from an emissions-free electric grid?
- Change the model of mobility away from physical mobility and towards virtual mobility via the internet or some other means?
- Prevent by some means the natural increase in the global fleet?

- Fundamentally change everyone's expectations of the necessary attributes of a personal motor vehicle, moving towards perhaps something like Toyota's i-swing (Figure 9)?

If one considers the fundamentals, it makes the most sense to:

- Create energy conversions (fuel to electricity, for example) on a large scale, as these are usually the most efficient.

- Adopt cogeneration principals to make the best use of all energy released in any given process

- Minimize the number of times we convert from one form and/or state of energy to another (electricity to hydrogen gas then to compressed gas or liquid and back again, for example)

- Utilize the smallest possible (least energy intensive) vehicle for any given mission

Figure 9: Toyota's i-swing, intended for personal mobility, is conceived for equal ease and safety on sidewalk and motorway alike.[4]

Is the answer, then, to develop and build large stationary fuel cells, powered by natural gas, and in combination with other sources generate power for the electrical grid, and then charge batteries in electric vehicles from the grid?

CONCLUSION

Is mobility as we know it sustainable? Clearly, No!

Instead, the global model of mobility will change to embrace all these solutions and more. A wholistic approach must adopt political, market, regulatory and fiscal measures to evolve our current model towards one that minimizes the fundamental energy intensity of personal and goods transport.

What we must do is view the current regulatory standards for the next decade as the first steps, not the final objective. These are the foundation upon which we will build a new era of mobility.

The challenge before us is to recognize on a global scale that these are the necessary choices, and act soon enough such that the transition to this wholistic approach is managed as an opportunity rather than reacted to as a crisis. Left unabated, the natural economic demand for mobility will overwhelm the ability to sustain it, and we will enter into an era of profoundly disruptive change: a crisis of resources, of pollution, of mobility, with all the resultant political and economic consequences.

ACKNOWLEDGEMENT

The author is grateful to his many colleagues at Global Insight and the automotive industry at large. Their insights and direction have been critical to the evolution of this paper.

REFERENCES:

[1] Source: IHS Global Insight, Inc.

[2] *The Biofuels Boom*, copyright 2007, Global Insight, Inc.

[3] Source: *Future Powertrain Technologies*, 2010 to 2025, May, 2007, Global Insight, Inc. and TIAX LLC

[4] Picture courtesy of Toyota

2009-01-1187

Sustainable Green Design and Manufacturing Requirements and Risk Analysis Within A Statistical Framework

Paul G. Ranky
New Jersey Institute of Technology

ABSTRACT

A well-tested, integrated approach and toolset is introduced to sustainable green design and manufacturing engineering requirements and risk analysis within a statistical framework with practical, industrial case studies. The emphasis is put on the integration of advanced process modeling, customer requirements analysis, statistical methods and risk analysis. The objective of the effort presented is to create a generic and systematic green design and manufacturing architecture that is sustainable, as well as complies with set guidelines by the green mobility industry, following international quality standards, and extended lean six-sigma methods.

INTRODUCTION

The fundamental purpose of the greening effort is to help to increase quality of life. Furthermore, it is to improve the quality of green compliant products and processes, and simultaneously reduce cost, pollution, the carbon and the environmental footprint of all product design, manufacturing and related activities. This includes raw material processing, warehousing, transportation, logistics, remanufacturing, recycling, and even reuse. The goal is that we can all stay compliant with USA, European, and other valuable international 'green' principles and laws that govern global trade and supply chains, whilst maintaining crucially important IP (Intellectual Property) rights.

What makes green design and manufacturing (in other words green engineering) very exciting is that it is an interdisciplinary subject. It should attract a flexible person with an open mind, who is ready to think laterally, structure, reason and integrate quality information, and then turn it into new knowledge to help mankind and all living entities on our planet. Consider the fact that factory pollution created in one continent can now be measured in another... therefore it is not a local issue anymore... pollution and toxicity changes everybody's life on Earth... not just those who are polluted, but also those who are polluting! Maybe it is time for mankind to wake up and realize that we are all in the same boat and it is our common interest to change our polluting, toxic products, processes, factories and systems for sustainable, energy efficient green solutions... and as we'll see later in this presentation, greening makes excellent business sense too.

The other very important driving force towards green, is that governments enforce compliance and IP, and consumers in the USA, as well as in Europe, and increasingly in Asia (China, Hong Kong, Taiwan and Japan) demand green products, made in non-polluting sustainable green factories... Consider this interesting fact: over 92% of young graduates in the USA want to work for a 'green' enterprise... also, 9 out of 10 new venture capital applications in California relate to some kind of greening invention... and this is just the beginning. Greening will become a bigger revolution than what the Internet has created for all of us!

In a free society consumers have a lot of power, and can change entire industries by purchasing only environmentally friendly, green products... designed and made in green, sustainable factories... Shouldn't we all be ready to drive this major transition?

A SOLUTION FOR GREEN ENGINEERING

The tested solution integrates object-oriented process modeling, requirements and risk analysis, statistical methods, design of experiments, and 3D interactive multimedia methods and tools, and it is 100% web-compatible. Furthermore, the methods and software tools are generic, in that they can be applied to a large variety of different industries and systems, from green mobility, automobile, aerospace, computer and other manufacturing, assembly, de-manufacturing and disassembly, and even service industries, supporting sustainable green systems.

In this paper, due to the limited space, the focus is on providing an introduction to the core concepts of the method, the tools, as well as some practical case studies. During the live demonstration of the tool-set several validated, practical examples will be shown, using the active code programs and interactive 3D CAD, green PLM, requirements and risk analysis, and statistical models. Sustainable green design and manufacturing, and the related product lifecycle management challenges are considered to be one of the most complex information systems and engineering architectures, due to the large number of attributes, processes and dynamic changes such projects go through during their lifecycle.

Following the integrated approach, developed by Ranky (as outlined in **Figure 1**, Refs. Ranky, 2001-2009), every green design and manufacturing project is treated as a system built of components and objects, and classes of objects. Then the interaction of these system components is analyzed. Once the behaviors and the interaction of the objects are understood, the system is analyzed as a set of layered processes in which each process step should satisfy green customer requirements and also reduce risks. The model is embedded into a statistical analysis and 3D interactive CAD / PLM / multimedia framework. The primary focus with statistical methods is to capture processes before they go out of control, and therefore become non-sustainable. For visualization, communication and education purposes 3D interactive multimedia, video-conferencing and 3D visualization methods are deployed over the web. This itself is a green approach because it cuts down travel time, cost and the carbon footprint when global team members communicate and collaborate. Fundamentally sustainable green design and manufacturing means waste reduction, and optimization, therefore they advocate data-driven, quantifiable lean six-sigma concepts and methods. It is very important because customers request green products and services.

In terms of green design and manufacturing sustainability one should consider an eco-friendly state of a product, or a process that can be maintained over time, for a very long time. Also, the ability of an ecosystem to maintain ecological advanced manufacturing processes and functions, biological diversity, and productivity over time.

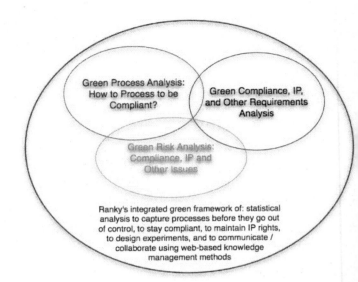

Figure 1. Ranky's integrated approach to sustainable lean and green design and manufacturing

WHAT ARE THE SUSTAINABLE GREEN DESIGN AND MANUFACTURING REQUIREMENTS?

Sustainable green encompasses the concept of meeting present manufacturing needs without compromising the ability of future generations to meet their needs. Without green design, there is no green manufacturing. Just as in the case of quality, eco-friendliness / green sustainability must be designed into the product. In other words, the characteristic of green, sustainable advanced design and manufacturing is being able to coexist with another system indefinitely, without either system being damaged. Sustainable green advanced manufacturing concept also emphasizes that the creation of wealth within the community considers the wellbeing of both the human as well as natural environments, and is focused on the more complex processes of development rather than on simple growth or accumulation only.

Advanced, sustainable green design and manufacturing is used in all areas of manufacturing, including product and process design, control, fabrication, test, assembly, disassembly, and remanufacturing / recycling. According to the U.S. Department of Commerce: *'...Sustainable manufacturing is defined as the creation of manufactured products that use processes that are non-polluting, conserve energy and natural resources, and are economically sound and safe for employees, communities, and consumers... As companies look for new ways to make more efficient use of resources, ensure compliance with domestic and international regulations related to environment and health, and enhance the marketability of their products and services, one area that has grown in importance to U.S. businesses is the implementation of sustainable manufacturing practices... Many U.S. firms have demonstrated that being environmentally sustainable can also mean being profitable...'*

In terms of a practical sustainable green design and manufacturing requirements rule-set, consider the following 18 'monozukuri' principles (Ranky, 2007):

Principle-1: Design and simulate in the digital domain, meaning on the screen first, before anything is built on the factory floor, following eco-friendly, 'monozukuri-focused' product, assembly system and factory design rules. 'Monozukuri' means eco-friendly, sustainable design, manufacturing and assembly for the purpose of reducing waste, helping the environment, the communities and in the long term increasing profits and product quality. The other key concept to notice here is the extensive use of advanced digital manufacturing / assembly / packaging and flexible manufacturing / assembly system design and simulation tools. These tools help to design and optimize in the digital domain first. This represents huge savings since changes are easier and quicker in the digital domain, than in the physical domain.

Principle-2: Design products for automated assembly, but start with mostly human operated assembly systems and gradually introduce the appropriate level of automation. 'Monozukuri' here means stay as lean, agile, reconfigurable and flexible as you can by employing your best associates along the manual, or semi-automated assembly lines, because they'll help you to iron out issues at an early stage. Furthermore by being able to assemble several different types of product on the same balanced line, pulled by the kanban system will not suffer due to major order fluctuations. This principle should not be misinterpreted though, by saying, that 'don't invest in automation anymore...' The contrary is the truth. In successful green-focused companies engineers, management and associates work together, and learn together, because they are motivated to gradually learn together the areas where automation offers clear benefits.

Principle-3: Management decisions should reflect long-term sustainable thinking, even if this means initially hard-to-accept financial returns. 'Monozukuri' here means developing a culture for eco-friendly, lean, agile, reconfigurable, green and flexible production with a long-term view. It reflects a philosophy that supersedes short-term decision-making. It means work hard, grow with your customers, and align the entire organization that is greater than just making profit. It also means continuous urge to generate value for customers, and the society in a responsible manner.

Principle-4: Develop outstanding leaders who fully understand and support the company's green philosophy, and wisdom. This means identifying and growing leaders from within the organization who truly understand and share the wisdom, as well as are seen as role models for others to follow. The greening philosophy should be taught at all levels by a competent chief engineer and his/her team, who not only understand the science, technology, project management and human aspects, but also the financial ROI issues. (Please refer to the Appendix for Ranky's Green Engineering Program outline.)

Principle-5: Create a strong company culture that does not tolerate poor quality work and focuses on continuous improvement and waste reduction methods and tools at all levels. 'Monozukuri' here means continuous waste reduction, continuous improvement, and a strong and stable culture in which company values are respected and the benefits are widely shared. In terms of greening, his includes the elimination of toxic materials, in every process step, both in terms of input, as well as output. Furthermore, the company culture should guide designers, procurement, manufacturing, environmental, quality and other engineers and managers, to ensure that all material and energy inputs and outputs are as inherently non-hazardous as possible. As the old principle states: It is better and less expensive to prevent waste, than to treat or clean up waste when it is too late... On the same token, separation and purification operations should be a component of the design framework, and system components should be designed to maximize mass, energy and temporal efficiency, most importantly focusing on sulfur dioxide (SO_2) emission and the carbon footprint.

Principle-6: Good leaders will spend time at all levels of the company to thoroughly understand all aspects of a greening decision ('genchi genbutsu'). The best managers and leaders know their people, their systems, and their strengths and weaknesses, just as a captain of a ship; therefore continuously strive for improvement. The best practice here is to personally observe greening challenges and green improvement processes and data BEFORE any major decisions are made in respect of the associates, the product, the line, or the factory. (Note, that the preventive attitude, as with all sustainable greening decisions, is critical.)

Principle-7: Your company's supply-chain network plays a key part in designing green products, processes and systems, therefore help them to continuously improve. Suppliers are crucial, in particular when companies increasingly rely on complex global supply chains. Since in many cases major product recalls have their roots in poor quality supplier products (unfortunately seatbelts are excellent examples for major recalls by Nissan in 2004, and Toyota in 2009, costing hundreds of million US Dollars), it is essential to have respect and offer greening help to supply chain companies too. This of course does not exclude continuously challenging business partners to grow, improve quality, and reduce energy use, the carbon footprint and cost. Supply-chain 'marriages' can be long term, but don't have to be forever... therefore if a supplier is not willing to become lean and green, and non-toxic (see the Chinese lead in paint, or poisoned milk scandals) then it must be dropped (else it will drag you down and cause major, maybe fatal damage to the company's brand name).

Principle-8: Leaning and sustainable greening decisions should be achieved by means of team consensus. 'Monozukuri' here means, that for every important decision several alternative decisions should be considered by an able team, and these decisions should include the environment too. (The Japanese call this decision process the 'nemawashi'.) This means discussing problems and solutions with everybody involved, and every realistic option and outcome considered. The key here is to maintain a strong direction towards waste reduction and eco-friendly design and manufacture without sidetracking due to short-term gains.

Principle-9: Design continuous process flow with built-in flexibility and agility to be able to produce several (e.g. 5-8, or even more) different product types on the same line. 'Monozukuri' here means stay as lean, agile, reconfigurable and flexible as you can, because order fluctuations can happen. The key here is good product and process design, that focuses on reusable, well-tested objects. Note, that reusability is a very important aspect of greening. Other key elements are flow production in which waiting time is not tolerated, and integrated workflow with visual factory controls (see also Principle-17), so that every associate on the line, as well as managers can immediately see waste, and act. In this sense quantifiable, data-driven, real-time controlled lean and green processes are essential.

Principle-10: Balancing the workload is essential to avoid high WIP (work – in –progress) and buffer size fluctuations ('heijunka'). The key is not to overload associates, but to achieve a well-balanced, steady flow of operations during which high quality can be maintained.

Principle-11: The production control system should always focus on producing based on the market's needs, not the factory's maximum capacity. The kanban, or 'pull' production system shouldn't be limited to individual lines inside the factory, but should also be well integrated into the supply chain of suppliers and customers. This literally means just-in-time production throughout the entire system with inventory holding times as short as 3-3.5 hrs! (Easier said than achieved...)

The benefits are huge since all supply chain and customer chain factories work in balanced harmony, therefore waste and cost is down. The risks of course include running out of parts and stopping the line... (In such very rare cases, as an example DENSO in the USA will immediately turn the effected line into a 'learning factory' and start 'kaisen' activities: see Principle-15. These are great opportunities for introducing green engineering concepts and solutions to the workforce; also see the Appendix for a proposed course structure.)

Principle-12: Follow reusable, standardized green processes. Reusable objects and well tested, standardized green, non-toxic processes pay off on the computer screen in the digital design domain, as well as in real production because they help save funds, improve quality and support the greening effort.

Principle-13: It is 'OK' to stop the line if the associate recognizes a quality issue, a sustainability challenge, that needs to be fixed immediately ('jidoka'). This principle underlines the important fact, that if any of the associates detect an error that can or will effect product quality / sustainability, can and should stop the line without penalty. Obviously the goal is to prevent such cases by means of quality product and process design, well-educated associates, statistical quality control and variation trend analysis, by means of statistical methods, visual factory controls, and by building quality culture into the product, the people, the systems, and even the users. 'Jidoka' literally means machines with human intelligence. 'Smart' machines, such as automated robotized inspection stations with vision are typical implementation examples for 'jidoka'.

Principle-14: Technology used should be the 'appropriate, tested level of sustainable green technology' versus the latest, for the sake of using the latest technology. The latest technology does not always represents the best, nevertheless should be explored. Technology should support associates on the lines, not replace them. In Japanese factories processes are ironed out by the most experienced operators first and automation is brought in only after that.

Principle-15: Design continuous leaning and greening improvement techniques, or Kaisen methods into every process. There is no doubt that our Japanese colleagues are the masters of gradual improvement and we should collectively learn from each other. As they define it in Japan *"Improvement is an attempt to breakthrough from the present and control is an effort to prevent slip-off from an improved stage. These two complement each other as two wheels of the same cart to create a better workshop."*

Principle-16: Design products and processes in the digital domain first, to avoid MUDA, MURA, and MURI.

MUDA

MUDA means waste. It means many different types of waste, including: unbalanced workload, long waiting times and uncoordinated action plans between departments, as well as waste in terms of, *"is the right man doing the right job?"*

MURA

MURA means irregularity, or differences and variability caused by men.

MURI

MURI means stress and strain. It reflects on problem areas such as: is there a shortage of manpower? Is there an opportunity to simplify the operation unloading strain of the operators?

Principle-17: Visual factory. Visual factory often means simple, nevertheless effective visual information aids, including signs, charts, pictures illustrating processes, color coding, machines and workstations with red, yellow and green lights, scoreboards, and others. At an advanced level, visual factories deploy real-time interactive multimedia support systems, advanced statistical control methods and networks to make everybody aware of how the factory works at the time. It is truly amazing how even simple visual factory controls can improve a line or factory. The key here is the culture of willingly and openly sharing versus hiding and protecting information, in particular if the information indicates a process going out-of-control.

Principle-18: Strive towards the wisdom of a learning organization through analysis and reflection ('hansei') and continuous improvement ('kansei'). The key 'monozukuri' drivers and achievable gains with the introduction and execution of this principle include performance advantage through improved green organizational capabilities, and alignment of improvement activities at all levels to an organization's strategic intent.

- Performance advantage through improved organizational capabilities.
- Alignment of improvement activities at all levels to an organization's strategic intent.
- Flexibility to react quickly to opportunities.
- A consistent organization-wide approach to the continual improvement of the organization's greening performance.
- Employees should be offered training in the methods and tools of continual improvement.
- Continual improvement of products, processes and systems should be an objective for every individual in the organization.
- Employees and management should establish goals to guide, and measure to track continual improvement.
- Management should recognize and acknowledge improvements.

- As a result of this, large organizations, in particular those dealing with mission critical applications and systems, such as health care, aerospace, automotive and transportation product manufacturers, energy generation companies, and others have adopted continuous quality principles and practices with great success.

Companies that continuously improve their green engineering quality have gained international recognition, respect, productivity and major profits, as a result of their focus on continuous quality improvement.

THE BENEFITS AND RISKS OF SUSTAINABLE GREEN ENGINEERING

The diagram below clearly illustrates the essence of the benefits of green design and manufacture. As it can be seen, it is every company's aim to innovate, create new products, processes and services that are required by customers, that are high quality, lean and green, and can be introduced to the market at a competitive price. **Figure 2** clearly illustrates the extremely high cost of product design changes. This is typical without deploying green and sustainable concurrent engineering / PLM (Product Lifecycle Management) methods... clearly, one need methods, tools and technologies, that will lead to innovate, and minimize the number, and therefore reduce the cost of changes during the entire lifecycle of a product (hence the terminology: 'Sustainable Green PLM').

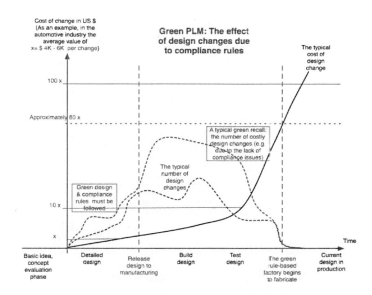

Figure 2: The benefits of 'Sustainable Green PLM'

Furthermore, as illustrated below, it is important to understand, that Green PLM, Digital Design and Digital Manufacturing, Total Quality Management, Decision Analysis Methods and Project Management offers major product life-cycle cost reductions. As a simplified justification, consider the following: In engineering systems the term 'life-cycle cost' means the sum of all the costs, both recurring as well as non-recurring, related to product, structure, system, process, IT, and/or service during its life span. (A typical life-cycle is illustrated below in Figure 3.)

As you can see from this graph, life-cycles typically start by identifying a customer need, a requirement, a want, a desire, and/or a sound business opportunity, and end with product/process retirement, conversion, re-engineering, recycling and eventual disposal activities. (Note, that in the case of a modern, environmentally friendly, lean and green, concurrently engineered product/process design system, 85-100% of products are eventually re-used and recycled. Since by understanding the underlying drivers and processes of sustainable lean and green design and manufacture, the opportunities for saving valuable product/process design and development cost and time are huge. Besides others, the graph in **Figure 3** underlines the fact, that sustainable green engineering economic studies, process-by-process cost and gain assessments are essential part of the big picture, because they help to analyze and compare alternatives in virtual environments, costing a lot less than real-world experimentation on the shop floor when the product/process is already in full swing production and the committed costs are already high.

factory) will have to understand and master green PLM and NPI&I (New Product & Process Innovation & Introduction) methods and apply them at all stages of the business to be able to successfully compete. It is also important to recognize, that as shown in Figure 3, the committed cumulative life-cycle cost curve increases rapidly during the *basic product/process idea to detailed design* transition phase. In general typically 80% of life-cycle costs are 'locked in' at the end of this phase and at the beginning of the 'release design to manufacturing' phase by the decisions made during the green requirements analysis, preliminary and detailed design phases. As illustrated by the cumulative, actual life-cycle cost curve, only about 20% of actual costs occur during the acquisition phase, in other words around the *basic idea to detailed design* transition phase, with about 80% being occurred at the *product construction, operation* phase.

The conclusion is, as shown below in our process model in **Figure 4**, that the lean and green NPI&I activity must understand the potential for product / process life-cycle savings in all processes, with their requirements and risks involved. This should include their time and cost demands too. The key then is to simulate these issues in advanced virtual (digital) environments, and take decisions as early as possible (i.e. 'front loading') to avoid dead-end paths and costly changes downstream... easier said than done! (Note, that this statement is true for every product/process, IT, and service product development activity!)

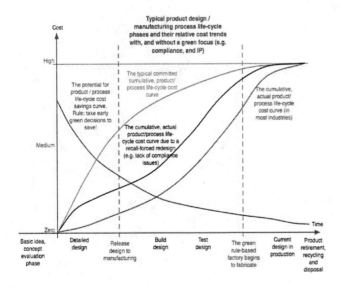

Figure 3: Some of the major cost saving opportunities of sustainable green and lean design and manufacturing, based on Ranky's research and analysis

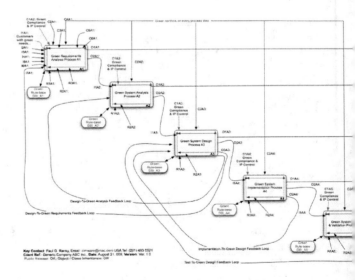

Figure 4: Ranky's object-oriented lean six-sigma process model of sustainable green and lean design and manufacturing. (Note the feedback control loops in the model, allowing continuous improvement processes.)

We can furthermore see, that the greatest savings occur at the acquisition phase, in other words around the basic idea to detailed design transition phase. This is why a green factory with a future (i.e. a digital, networked

Risk-based Sustainable Product Design and Manufacture, Satisfying Green Requirements

Risk-based design and process-oriented failure risk analysis (PFRA) is an extended Failure Mode and Effects Analysis (FMEA) method designed to be deployed during the entire product lifecycle (PLM). It represents a systematic computational method developed by the author to identify and minimize potential failure risks / failures of products, processes, sub-assemblies, objects and components, and their effects on the customer (meaning internal and external customers). PFRA is aimed at minimizing dissatisfaction, and financial loss. In order to maximize ROI (Return On Investment), PFRA is typically applied during the planning stages of a green engineering design process, and then updated on a regular basis to document changes throughout the entire Green PLM cycle. It addresses negative quality and is primarily concerned with potential events, that can make the process, or the embedded components fail.

For the purpose of creating an integrated green design and manufacturing methodology several types of the Weibull distributions have been designed from the standard one. In this paper, we focus on Type I analysis that is itself composed of five models. One of the most important applications of such a distribution is the study of a product's reliability over its life cycle. This allows getting first information about the distribution by computing product sustainability-focused statistics, such as maximum, minimum, mean, and variance values. This approach helps the engineering team to work through an analytical, quantitative, open-source computational model, to understand and reduce the risks product design engineers, process planning engineers, line managers and operators face, then see the solutions they have come up with, and then apply the learned problem solving skills to other, new challenges by understanding and then controlling the various risk-factors. In other words, the methodology represents a systematic engineering product and process design method to identify and minimize potential failure risks, meaning failures of products, sub-assemblies, objects and components, as well as various manufacturing / assembly processes, and their effects on the customer.

This approach helps the engineering team to work through an analytical, quantitative, open-source computational model, to understand the following sustainable green design and manufacturing concerns:
- What could go wrong with the product or the processes involved, or one or more of its components?
- How badly might it go wrong, and what could the financial loss be?
- Which are the highest risk processes / operations when working on the product / process, and even service, such as maintenance?

- What needs to be done to prevent failures?
- What tools and fixtures are required to prevent failures, and reduce the risk?
- What education is needed for designers, process engineers, line management and operators to reduce, or prevent failures?
- Others.

In order to support the industry-wide rule-based risk rating standardization process, there are three important files (that are hyperlinked to the spreadsheet too in our system). These being the rating rules, regarding Severity, Detection and Occurrence.

Severity is a rating corresponding to the seriousness of the effect(s) of a potential failure mode. Severity applies only to the effect of a failure mode. As a greening example, the purpose of establishing a value of disassembly severity is to be able to quantify the seriousness of failing during one or more disassembly process steps. For simplicity, and in compliance with basic statistical rules, the value range for severity rating is set between 1 and 10, where 1 is least severe, and 10 is most severe.

As a generic statement, one should realize, that a severity rating system is as good as the data it is based upon, as well as that the ratings we provide here could/ should be standardized by industry groups for conformity. As an example, for disassembly failure risk analysis, let us share some of the rules (shown in the traditional IF... THEN... rule based, versus table format) for calculating severity rating in our programs: Example: If Rating = 1, then this means, that due to the disassembly process performed, there is No Effect on the part/ component, and/or on system performance, and/or on subsequent disassembly process, or operation. Good news, this should be the default value, etc.

The value entered in the spreadsheet for severity, is used for calculating the Risk Priority Number, the prime indicator for risk detection. When analyzing the RPN value, the disassembly manager, or operator should focus first on the highest values, trace back the causes and effects and try to eliminate them. Then, as a secondary optimization level, the medium and low RPN values should be analyzed and eliminated in the same manner as the highly rated values.

As mentioned earlier, our methods and use-cases are object-oriented and self-contained. They can be integrated or grouped into different classes of objects in a lean and flexible way, just as a modern software program, or a modern manufacturing /assembly system can be integrated into different environments. This enables learners as well as instructors and managers to 'plug-and-play' our cases in ways they choose to ([2], and [16] to [20]). The methodology we follow is highly interactive, collaborative, and enables large groups as well as individuals to analyze their greening risks.

Interactive Multimedia for Visualization

The benefits of introducing problems for students and professionals to solve using risk analysis cases in a browser-readable 3DVR interactive multimedia format are manifold. The entire learning / problem solving process becomes more user/ student- versus lecture- or tutor-centered. Students and users can learn by exploring versus being told, and can have as many goes at solving a problem, or exploring an idea, taking as much time as desired or is available. Mistakes made can be corrected without penalties. Multimedia tools, or a subset of such technology and a variety of media, are available during the learning process.

Some Statistical Aspects of Green Design and Manufacturing Our Framework

Besides the extended FMEA models, we also use Weibull distributions with a green focus. This is one of the statistical models that have been created on an empirical basis. Several types of the Weibull distributions have been created from the standard one. In this paper, we focus on Type I, that is itself composed of five models. One of the most important applications of such a distribution is the study of a product's reliability over its life cycle. This allows getting first information about the distribution by computing statistics like maximum, minimum, mean and variance values. This way, the variability of the data set is known:
-if the range is small compared to the mean, the distribution can be modeled by the sample mean,
-if not, the variability needs to be taken in count in the modeling process.

This analysis is aimed to determine if a model is adapted to a data set when it is completed in graphical approaches like frequency histogram or Weibull probability paper plot (WPP). Once the data set has been preliminary analyzed, and if the variability is large enough to be considered, the next step is to select a model. The data set $(t_1,..., t_n)$ becomes an observed value of a set of random variables $(T_1,..., T_n)$ which is modeled by the following probability distribution function:

$$F(t;\theta) = P(T \leq t)$$

where   is the set of parameters for the distribution

Data is often organized in the increasing order and named $(T_{(1)},..., T_{(n)})$. The forth step consists of estimating the set of parameters   using available data. The approaches are either graphical, or analytical, or both. Most of the time, graphical methods are less precise than the analytical ones, nevertheless offer good first-visual estimation.

The last step is aimed to check if the model selected in step 3 fits the data set. The validity can be verified through different numerical methods, depending on different criteria. The following formulas are composed the three-parameter Weibull distribution:

$$F(t;\vartheta) = 1 - \exp[-(\frac{t - \tau}{\alpha})^{\beta}] \qquad t \geq \tau$$

The data set is composed by values for t, which are known for the model. The parameters are:
- the scale
- the shape
- the location τ

They are gathered as the set   = { , , τ } and respect the following constraints:
-    > 0
-    > 0
-   τ >= 0

This is a special case of the former one, it is called the two-parameter Weibull distribution as $\tau = 0$. For simplicity, we will only consider this standard model in this paper:

$$F(t;\vartheta) = 1 - \exp[-(\frac{t}{\alpha})^{\beta}]$$

From this model, seven types of distributions have been created by transforming the variable T into a derived variable Z. The new formulation of the distribution is G (t;). The nature of the function G determines the seven types of the Weibull distribution, as follows:
- The transformation concerns the Weibull variable: Type I models
- The transformation is a modification or a generalization of the distribution: Type II models
- The transformation involves two or more distributions: Type III models
- The Weibull parameters become variables: Type IV models
- The model uses discrete variables: Type V models
- The distribution is multivariate: Type VI models
- The model is stochastic: Type VII models

The Weibull distributions help us during the design, as well as the manufacturing processes to stay on a sustainable green course.

CONCLUSION

As all studied examples clearly indicate, by following the integrated methodology and toolset with the briefly discussed eighteen 'monozukuri-focused' sustainable lean and green design and visual factory management principles offer major long-term benefits to all involved. This win-win formula is clearly visible in the studied Japanese and American DENSO factories, Mori-Seiki, BMW MINI, Bentley, as well as in other factories like Toyota, Honda, Nissan, Apple Computers, Dell, HP, Fanuc Robotics, GE Fanuc, BMW, and many others.

The core message of these principles is to design eco-friendly, quality green products that customers need, want and desire, produce exactly as much as needed, just-in-time, to reduce inventory waste and cost throughout the entire global supply-chain. This is very difficult to achieve in practice.

It is a proven fact that by deploying these generic principles to literally any factory or system, customer satisfaction will increase through quality products and services, and global growth will become reality through anticipation of change. This is because accepting the fact, that continuous improvement is a natural change process, environmental preservation will become a natural integral part of everything we do in harmony with society, including individuals.

The reality is that despite the fact that these principles sound truly very simple, there is a lot of science, engineering and management hiding behind them. This is often misunderstood and even ignored by management, as well as the academic research community.

Implementation is the hard bit here, because it needs collective acceptance and harmony throughout the entire factory, and even in the global supply chain. Analysis performed in General Motors, Delphi, DaimlerChrysler, Ford, and many other companies have clearly shown, that they are all trying to work hard towards implementing these principles, nevertheless what they are missing is the 'collective acceptance', or in other words the 'quality culture' at all levels. Several programs, such as TQM, Six-sigma, Lean, Business Process Improvements, Visual Factory, and others tried to fix this, nevertheless most of them have only isolated impact unless the entire organization is prepared to change for the better. This is why Visual Factory management is a good place to start this culture change because the 'rocks will become visible to everyone', and that might trigger change. After this one should perform integrated process, requirements, and risk analysis, embedded into a statistical framework at all levels. This will bring truly major positive results, but this path is not a 'quick-fix'; it is very hard work for those who truly believe in a sustainable, long-term, continuous improvement process.

ACKNOWLEDGMENTS

The author would like to express his thanks to NJIT, and to his thousands of graduate students in engineering design and quality management, information systems and project management, as well as the US, European, Chinese and Japanese companies he has worked with on a consulting basis during the past 25 years, who have helped to shape and test some of the methods and the tools presented in this paper.

REFERENCES

Ranky, P.G.: A 3D Multimedia Case: Component Oriented Disassembly Failure Risk Analysis, An interactive multimedia publication with 3D objects, text and videos in a browser readable format by www.cimwareukandusa.com, CIMware USA, Inc. and CIMware Ltd., UK, ISBN 1-872631-47-9, 2001-2009. (Note, that there are 6 different volumes available at the time of writing of this program).

Ranky, P.G.: Eighteen 'monozukuri-focused' assembly line design and visual factory management principles, with Denso industrial examples, Assembly Automation, Feature Article, 27/1, (2007) p.12-16, Emerald Group Publising Ltd.

Ranky, P.G.: A 3D Multimedia Case: Component Oriented Disassembly User Requirements Analysis, An interactive multimedia eBook publication with 3D objects, text and videos in a browser readable format on CD-ROM/ intranet by http://www.cimwareukandusa.com, CIMware USA, Inc. and CIMware Ltd., UK, ISBN 1-872631-50-9, 2001-2009 (Published 7 volumes of this main title with different requirements analysis challenges explained).

Ranky, P G., and Ranky R. G.: An Introduction to Concurrent/ Simultaneous Engineering, Green PLM, An Interactive Multimedia Presentation on CD-ROM with off-line Internet support, published by CIMware 1996-2009, Published by www.cimwareukandusa.com

Ranky, P G: An Introduction to Total Quality and the ISO90001 Standard within a Lean Six-sigma Framework. An Interactive Multimedia eBook published by CIMware, 1997-2009, www.cimwareukandusa.com

Throop D R et al: Automated incremental design FMEA, Boeing Co., IEEE Aerospace Conference Proc., 2001, IEEE, May 10-17, 2001, p 73451-73458

Ranky P G: Interactive Multimedia for Engineering Education, European Journal of Engineering Education, Vol. 21, No. 3, 1996, p. 273-293.

Ranky, P. G.: Some Analytical Considerations of Engineering Multimedia System Design within an Object Oriented Architecture, IJCIM (International Journal of CIM, Taylor & Francis, London, New York)), Vol. 13, No. 2, May 2000, p. 204-214

Ranky, P G, Caudill, R. J., Limaye K., Alli, N., Satishkumar ChamyVelumani, Apoorva Bhatia and Manasi Lonkar A Web-enabled Virtual Disassembly Manager (web-VDM) for Electronic Product / Process Designers, Disassembly Line Managers and Operators, a UML (Unified Modeling Language) Model of our Generic Digital Factory, and Some of Our Electronic Support System Analysis Tools, ADAM with IT (Advanced Design And Manufacturing), by: www.cimwareukandusa.com USA, 5 p., Vol. 3., May 2002

Ranky, P.G., Das, S and Caudill, R: A Web-oriented Virtual Product Disassembly and Identification Method for DFE (Design for Environment) and Electronic Demanufacturers, 2000 IEEE (USA) International Symposium on Electronics and the Environment, Organized by IEEE (USA), and the Computer Society, USA, May, 2000, San Francisco, CA, USA, Conference Proceedings.

Ranky, P. G.: A Multimedia Web-based Flexible Manufacturing Knowledge Management Framework, Japan-USA International Symposium on Flexible Automation, ASME (American Society of Mechanical Engineers), July, 2000, Ann Arbor, MI, Conference Proceedings.

Ranky, P G.: An Object Oriented Virtual Concurrent Engineering Model and Product Demonstrator Case Study, Japan-USA International Symposium on Flexible Automation, ASME (American Society of Mechanical Engineers), July, 2000, Ann Arbor, MI, Conference Proceedings.

Ranky, P.G, One-Jeng and Surjanhata, H: Digital Educational Knowledge Assets, International Engineering Education Conference, August 2000, Taipei, Taiwan, Conference Proceedings

Ranky, P G, Herli Surjanhata, One-Jang Jeng, Geraldine Milano: The Design and Implementation of Digital Educational Knowledge Assets (DEKA) with Software Demonstration (An NJIT and Industry Sponsored R&D Project. ASEE (American Society of Engineering Education) NJ Spring Conference, April, 2001 (eProceedings)

Ranky, P G: Virtual Concurrent & Multi-lifecycle Engineering Over the 3D Internet. An invited cluster presentation, INFORMS International Conference, USA, July, 2001

Ranky, P G: 3DVR Component-based User Requirements Analysis Methods and Software Tools for Collaborative Multi-lifecycle Engineering, INFORMS International Conference, USA, July, 2001

Ranky, P G: An Object Oriented Model and Cases of Design, Manufacturing, and IT Knowledge Management Over the 3D- enabled Web and Intranets, INFORMS International Conference, USA, July, 2001, (in the Proceedings).

Ranky, P G: A 3D Web-based Flexible Manufacturing and Demanufacturing Knowledge Management Model, INFORMS International Conference, Miami, November 2001

Ranky, P G: The Design and Implementation of a Case-based Learning Library for Engineering, Management and IT with 3D web-objects, Third Annual Faculty Best Practices Showcase, Invited Paper, NJ Higher Education Network, Proceedings, NJIT, Newark, NJ, Nov 9, 2001

Dunkerley, G., Norton, N. and Ranky, P.G.: A Case-based Introduction to IMI Norgren's Reengineering Project at Kenilworth, UK; An interactive multimedia publication with 3D objects, text and videos in a browser readable format by www.cimwareukandusa.com, CIMware USA, Inc. and CIMware Ltd., UK, ISBN 1-872631-40-1, 2001-2008.

Anastas, P.; Zimmerman, J. "Design through the Twelve Principles of Green Engineering," *Environmental Science and Technology*, 37, 94A – 101A, 2003.

McDonough, W.; Braungart, M.; Anastas, P.T.; Zimmerman, J.B. "Applying the Principles of Green Engineering to Cradle-to-Cradle Design." *Environmental Science and Technology*, 37 (23): 434A-441A, 2003.

Mihelcic, J.; Ramaswami, A.; Zimmerman, J. "Integrating Developed and Developing World Knowledge into Global Discussions and Strategies for Sustainability," submitted to *Environmental Science and Technology*, 2005.

CONTACT

Paul G. Ranky, PhD, Full Tenured Professor, Chartered and Registered Professional Engineer, Department of Mechanical and Industrial Engineering, NJIT, Newark, NJ 07102, USA. E-mail address: ranky@njit.edu, and web address: www.cimwareukandusa.com

APPENDIX: GREEN ENGINEERING EDUCATION PROGRAM OUTLINE

Since green engineering education is very important, not just in schools and universities, but also in industry, Professor Ranky's Green Engineering education program outline is offered below:

1. **Introduction** to sustainable green design and manufacturing, and an overview of the subject. Definitions of important terms, such as sustainability, lean, green design, green manufacturing, and others. Introduction of lean as a data-driven, quantifiable optimization method.

The key compliance requirements, and business drivers in the USA, in Europe and in Asia. Simultaneously acting constraints, including cost reduction, quality improvement and customer satisfaction requirements, compliance and IP issues, waste and risk reduction, carbon footprint reduction, energy reduction, lean and green sustainability, and others. Reasons why we must green our design engineering and manufacturing processes and factories on a global basis. The USA, the international scene, some predictable greening changes in the next 5 to 10 years, and these changes will effect you and me...

2. **The 18 monozukuri principles** to design and maintain sustainable green manufacturing processes with industrial examples and case studies captured in the USA, Europe, Japan, and China. Design For Environment (DFE), Design For Disassembly / Reuse rules and analysis, Disassembly Bill Of Materials (DBOM) analysis, the identification of toxic components, and others.

3. **A set of analytical methods**, computational assessment and statistical tools for evaluating and designing green manufacturing sustainability processes, requirements, and risks. A carbon footprint assessment method and calculator, and an air pollution analysis method and tool. Product Life Cycle Assessment and Analysis standards and some tools. Interactive case studies from a variety of industries and countries.

4. **The sustainable green manufacturing audit process, standards and compliance regulations**, outcomes and assessment methods. How can we evaluate how lean and green are you? International green manufacturing standards and compliance. USA and EU compliance audit standards, and mainland China's recycling law, and what it means for sustainable green manufacturing processes, and export / import activities.

Relevant ISO and USA standards, and compliance issues. International Chamber of Commerce trade rules and regulations, the Rio Declaration on Environment and Development. ISO 14001 and sub-standards. The European born Eco-Management and Audit Scheme (EMAS), and the International Organization of Standards 14001 (ISO 14001) and what it means for green manufacturing and international trade. Worldwide legislative activities and standardization. Interactive case studies from a variety of industries and countries.

5. **Green materials**, including biodegradable materials for green manufacturing processes. The European directive of the Restriction of Hazardous Substances (RoHS), WEEE: Waste Electrical and Electronic Equipment, Framework for setting up Eco-design Requirements for Energy-using Products (EuP), and related compliance issues and constraints that can make a product fail to enter the European market. New USA developments in 'greening' manufacturing and the world. USA compliance issues and restrictions. Also, how can we recycle and reuse materials and make the entire process sustainable?

6. **Industrial ecology** to foster the co-operation among various industries whereby the waste of one production process becomes the feedstock for another... in a similar fashion to the way nature works... to identify ways for industry to safely interface with nature, in terms of location, intensity, timing, value added and non-value added processes, and others, to develop measurable and controllable indicators for real-time monitoring (e.g. 24/7 using sensor networks reporting over the Internet), to reduce the energy requirements when converting materials throughout their entire life-cycle, and others.

Long term land use planning within an ecological framework, that identifies, and protects environmental, cultural and historical values, as well as simultaneously becomes profitable for the community. Interactive case studies from a variety of industries and countries.

7. **Green rapid prototyping** and green rapid manufacturing of product, process, and service systems; Eco-friendly digital design and digital manufacturing engineering principles, methods and analytical tools with several simulation and practical industrial / research case studies. Interactive case studies from a variety of industries and countries.

8. **Green and cleaner flexible automation**, green demand-driven manufacturing and plant design, recipe driven green manufacturing in the pharma industry, disposable, biodegradable manufacturing in the pharma industries, disassembly, demanufacturing and recycling methods and technologies, reuse, and others. Best practice sustainable green manufacturing use-cases based on USA, European, Japanese, Chinese, and other international industrial and R&D examples in a variety of industries.

The concept of ecological industrial parks factories, and industrial facilities designed on the basis of ecologically sound symbiosis models. (This is of great value, since industrial symbiosis yields significant reductions in oil, coal and water consumption, as well as reduces carbon dioxide and sulfur dioxide emissions.) Global reuse of products, systems and sustainable technologies, life cycle thinking and energy efficiency, resource optimization, management & social responsibility, approaches for management and measurement, sustainability drivers and limitations, global market transformation, end-of-life business challenges. Interactive case studies from a variety of industries and countries.

9. **Green manufacturing networking** and communication / collaboration / video-conferencing processes via the Internet. Smart communities and eco-industrial park networking. Eco-industrial parks are emerging as the primary arena for testing and implementing sustainable green manufacturing within industrial ecology. Similar in some respects to standard industrial parks, eco-industrial parks are designed to allow firms to share infrastructure as a strategy for enhancing sustainable green production and minimizing costs.

The distinguishing feature of eco-industrial parks is their use of ecological design to foster collaboration among firms in managing environmental and energy issues. In an eco-industrial park setting, company production patterns, as well as overall park maintenance, work together to follow the principles of natural systems through cycling of resources, working within the constraints of local and global ecosystems, and optimizing energy use. Eco-industrial parks offer firms the opportunity to cooperatively enhance both economic and environmental performance through increased collaboration and networking efficiency, waste minimization, innovation and technology development, access to new markets, strategic planning, and attraction of financing and investment. Interactive case studies from a variety of industries and countries.

10. **Internet / intranet-based collaboration** and documentation / knowledge management methods, tools and technologies with a sustainable green manufacturing focus. Compliance with international IP and other issues and regulations. Interactive case studies from a variety of industries and countries.

11. **International sustainable green manufacturing** methods, processes and R&D case studies focusing on alternative energy resources, applications, and compliant processes. Interactive case studies from a variety of industries and countries.

12. **Fuel-cell, windmill / wind turbine and solar panel manufacturing** processes with USA, and international examples. Alternative energy sources: hybrid and fuel-cell cars, electric automobiles and power generators. Technologies to transition from a fossil fuel economy to a hydrogen economy.

The emission reduction challenge and compliance issues... Oil, gas, coal, and electricity... transportation, agriculture, industrial and residential emission reduction compliance laws, opportunities, methods and technologies. Interactive case studies from a variety of industries and countries. Windmill and solar panel installation and maintenance experiences in New Jersey, in the USA, in Hong Kong and in the Philippines.

13. **China-USA-European-International focus:** Example topics: How do European environmental laws effect design engineering and trade with the USA and Asia? What is the sustainable green design and manufacturing challenge in China? In force as of January 1, 2009, Chinese law states, that governments at all levels should make plans to develop recycling, establish systems to control energy use and pollutant emission, strengthen management on companies with high energy and water consumption, and divert capital into environmentally friendly industries. The recycling law also introduces rewards and penalties for companies, encouraging them to develop recycling by making them responsible for the recycling of their products. What does this mean to green design and manufacturing engineering, and related service industries in Hong Kong, China, and the region? Can China enforce these laws? Will there be a real difference? What are the greening requirements in Japan, India, Russia and Brazil? What are the green design and manufacturing compliance laws and issues countries must follow to be able to trade with Europe and the USA? Interactive case studies from a variety of industries and countries.

14. **Global green design and manufacturing supply chain and logistics** network: integration methods, tools and technologies of sustainable green manufacturing and other systems. Interactive case studies from a variety of industries and countries.

15. **Sustainable green manufacturing project management and communication** challenges, methods and solutions: USA, European, Japanese, Chinese, and other industrial and R&D case studies. Cross-functional greening teams, including research, engineering (materials, industrial, electrical, mechanical, environmental, etc.), finance, marketing and sales, human resources, operations, suppliers, customers, maintenance, recyclers, and others. International teams working on a global basis. Project team design and communication skills with team members. Interactive case studies from a variety of industries and countries.

This is a simultaneously analytical, as well as practical presentation / study with useful knowledge, that you can turn into 'greening' improvement opportunities in almost any factory, institution, or organization, or system, anywhere in the world.

2009-01-3085

Sustainable (Green) Aviation: Challenges and Opportunities

Ramesh K. Agarwal
Washington University in St. Louis

ABSTRACT

Air travel continues to experience the fastest growth among all modes of transportation. Therefore the environmental issues such as noise, emissions and fuel burn (consumption), for both airplane and airport operations, have become important for energy and environmental sustainability. This paper provides an overview of issues related to air transportation and its impact on environment followed by topics dealing with noise and emissions mitigation by technological solutions including new aircraft and engine designs/technologies, alternative fuels, and materials as well as examination of aircraft operations logistics including Air-Traffic Management (ATM), Air-to-Air Refueling (AAR), Close Formation Flying (CFF), and tailored arrivals to minimize fuel burn. The ground infrastructure for sustainable aviation, including the concept of 'Sustainable Green Airport Design' is also covered.

INTRODUCTION

In the next few decades, air travel is forecast to experience the fastest relative growth among all modes of transportation, especially due to many fold increase in demand in major developing nations of Asia and Africa. Figure 1 shows the expected growth in passenger traffic volume in Passenger Kilometers Traveled (PKT) by major modes of transport [1]. Air travel and other high speed transport accounted for less than 1% of world passenger traffic volume in 1950 and for 10% in 2005; they are forecasted to account for between 36 to 40% by 2050. These projections are based on 3% growth in

world Gross Domestic Product (GDP), 5.2% growth in passenger traffic and 6.2% increase in cargo movement. As shown in Figure 2, as the GDP/capita of nations increases, the demand for travel follows [1].

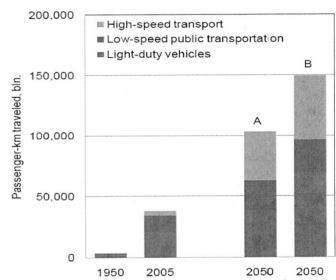

Figure 1 Projected Growth in World Passenger-Kilometers Traveled (PKT) [1]

Based on these demands for air travel, Boeing has determined the outlook for airplane demand by 2025 as shown in Figure 3 [2]. Figure 4 shows various categories of 27,200 airplanes that would be needed by 2025 [2]. The total value of new airplanes is estimated at $2.6 trillion. As a result of three fold increase in air travel by 2025, it is estimated that the total CO_2 emissions due to commercial aviation may reach between 1.2 billion tonnes to 1.5 billion tonnes annually by 2025 from its current level of 670 million tonnes. The amount of nitrogen oxides around airports, generated by aircraft engines, may rise from 2.5 million tonnes in 2000 to 6.1 million tonnes by 2025. The number of people who may be seriously affected by aircraft noise may rise from 24 million in 2000 to 30.5 million by 2025. Therefore there is urgency to address the problems of emissions and noise abatement through technological innovations in design and operations of the commercial aircraft.

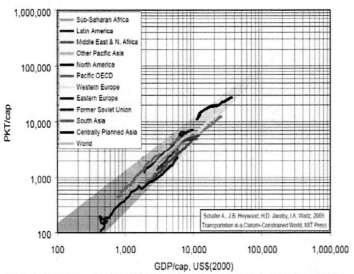

Figure 2: Increase in Travel Demand/Capita with Increase in GDP/Capita [1]

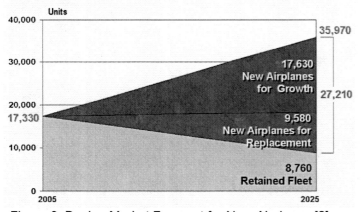

Figure 3: Boeing Market Forecast for New Airplanes [2]

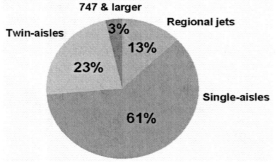

Figure 4: Boeing Demand Forecast for Various Types of Airplanes by 2025 [2]

ENVIROMENTAL CHALLENGES

To meet the environmental challenges of the 21st century, as a result of growth in aviation, the Advisory Committee for Aeronautical Research in Europe (ACARE) has set the following three goals for reducing noise and emissions by 2020; (a) reduce the perceived noise to one half of current average levels, (b) reduce the CO_2 emissions per passenger kilometer (PKM) by 50%, and (c) reduce the NOx emissions by 80% relative to 2000 reference [3]. NASA has similar objectives for 2020 as shown in Figure 5 for N+2 generation [4]. It is expected that the technology readiness level (TRL) of N+1, N+2 and N+3 generation will be between 4 and 6 in 2015, 2020 and 2030 timeframes respectively. The NASA definitions of TRL are given in Reference [5]. TRL 4-6 implies that the key technologies readiness will be somewhere between component/subsystem validation in laboratory environment to system/subsystem model or prototyping demonstration in a relevant environment.

CORNERS OF THE TRADE SPACE	N+1 (2015 EIS) Generation Conventional Tube and Wing (relative to B737/CFM56)	N+2 (2020 IOC) Generation Unconventional Hybrid Wing Body (relative to B777/GE90)	N+3 (2030-2035 EIS) Advanced Aircraft Concepts (relative to B737/CFM56)
Noise (cum below Stage 4)	- 32 dB	- 42 dB	better than -71 dB (55 LDN at average boundary)
LTO NOx Emissions (below CAEP 6)	-60%	-75%	better than -75% plus mitigate formation of contrails
Performance: Aircraft Fuel Burn	-33%***	-40%***	better than -70% plus non-fossil fuel sources
Performance: Field Length	-33%	-50%	exploit metro-plex concepts

*** An additional reduction of 10% may be possible through improved operational capability; metro-plex concepts will enable optimal use of runways at multiple airports within the metropolitan area

Figure 5: NASA Subsonic Fixed Wing System Level Metric for Improving Noise, Emission and Performance Using Technology & Operational Improvements [4]

The achievement of these goals will not be easy; it will require the cooperation and involvement of airplane manufactures, airline industry, regulatory agencies such as ICAO and FAA, R & D organizations, as well as political will by many governments and support of public. However, these challenges can be met with concerted efforts as stated beautifully by the Chairman, President and CEO of Boeing Company, W. J. McNerney, "Just as employees mastered "impossible" challenges like supersonic flight, stealth, space exploration and super-efficient composite airplanes, now we must focus our spirit of innovation and our resources on reducing greenhouse- gas emissions in our products and operations."

A LIST OF NEW TECHNOLOGIES AND OPERATIONAL IMPROVEMENTS FOR GREEN AVIATION

Recently, Aerospace International, published by the Royal Aeronautical Society of U.K., has identified 25 new technologies, initiatives and operational improvements that may make air travel one of the greenest industries by 2050 [6]. These 25 green technologies/concept areas are listed below from Reference [6].

1) "*Biofuels* – These are already showing promise the third generation biofuels may exploit fast growing algae to provide a drop-in fuel substitute.

2) *Advanced composites* – The future composites will be lighter and stronger than the present composites which the airplane manufacturers are just learning to work with and use.

3) *Fuel cells* - Hydrogen fuel cells will eventually take over from jet turbine Auxiliary Power Units (APU) and allow electrics such as in-flight entertainment (IFE) systems, galleys etc. to run on green power.

4) *Wireless cabins* – The use of Wi-Fi for IFE systems will save weight by cutting wiring - leading to lighter aircraft.

5) *Recycling* - Initiatives are now underway to recycle up to 85% of an aircraft's components, including composites - rather than the current 60%. By 2050 this could be at 95%.

6) *Geared Turbofans (GTF)* - Already under testing, GTF could prove to be even more efficient than predicted, with an advanced GTF providing 20% improvement in fuel efficiency over today's engines.

7) *Blended wing body aircraft* - These flying wing designs would produce aircraft with increased internal volume and superb flying efficiency, with a 20-30% improvement over current aircraft.

8) *Microwave dissipation of contrails* – Using heating condensation behind the aircraft could prevent or reduce contrails formation which leads to cirrus clouds.

9) *Hydrogen-powered aircraft* - By 2050 early versions of hydrogen powered aircraft may be in service - and if the hydrogen is produced by clean power, it could be the ultimate green fuel.

10) *Laminar flow wings* – It has been the goal of aerodynamicists for many decades to design laminar flow wings; new advances in materials or suction technology will allow new aircraft to exploit this highly efficient concept.

11) *Advanced air navigation* - Future ATC/ATM systems based on Galileo or advanced GPS, along with international co-operation on airspace, will allow more aircraft to share the same sky, reducing delays and saving fuel.

12) *Metal composites* - New metal composites could result in lighter and stronger components for key areas.

13) *Close formation flying* - Using GPS systems to fly close together allows airliners to exploit the same technique as migrating bird flocks, using the slip-stream to save energy.

14) *Quiet aircraft* - Research by Cambridge University and MIT has shown that an airliner with imperceptible noise profile is possible - opening up airport development and growth.

15) *Open-rotor engines* - The development of the open-rotor engines could promise 30%+ breakthrough in fuel efficiency compared to current designs. By 2050, coupled with new airplane configurations, this could result in a total saving of 50%.

16) *Electric-powered aircraft* - Electric battery-powered aircraft such as UAVs are already in service. As battery power improves one can expect to see batteries powered light aircraft and small helicopters as well.

17) *Outboard horizontal stabilizers (OHS) configurations* – OHS designs, by placing the horizontal stabilizers on rear-facing booms from the wingtips, increase lift and reduce drag.

18) *Solar-powered aircraft* - After UAV applications and the Solar Impulse round the world attempt, solar-powered aircraft could be practical for light sport, motor gliders, or day-VFR aircraft. Additionally, solar panels built into the upper surfaces of a Blended-Wing-Body (BWB) could provide additional power for systems.

19) *Air-to-air refueling of airliners* - Using short range airliners on long-haul routes, with automated air-to-air refueling could save up to 45% in fuel efficiency.

20) *Morphing aircraft* - Already being researched for UAVs, morphing aircraft that adapt to every phase of flight could promise greater efficiency.

21) *Electric/hybrid ground vehicles* – Use of electric, hybrid or hydrogen powered ground support vehicles at airports will reduce the carbon footprint and improve local air quality.

22) *Multi-modal airports* - Future airports will connect passengers seamlessly and quickly with other destinations, by rail, Maglev or water, encouraging them to leave cars at home.

23) *Sustainable power for airports* - Green airports of 2050 could draw their energy needs from wave, tidal, thermal, wind or solar power sources.

24) *Greener helicopters* - Research into diesel powered helicopters could cut fuel consumption by 40%, while advances in blade design will cut the noise.

25) *The return of the airship* - Taking the slow route in a solar-powered airship could be an ultra 'green' way of travel and carve out a new travel niche in 'aerial cruises', without harming the planet."

Some of the ideas listed above require technological innovation in aircraft design and engines, use of alternative fuels and materials while others require operational improvement. Some concepts such as electric, solar and hydrogen powered aircraft are currently feasible but are unlikely to become viable for mass air transportation by 2050. In what follows, we describe the current levels of noise, CO_2 and NOx emissions due to air transportation and possible strategies for their mitigation to achieve the ACARE and NASA goals.

NOISE & ITS ABATEMENT

Historically, the reduction in airplane noise has been a major focus of airplane manufacturers because of its health effects and impact on the quality of life of communities, especially in the vicinity of major metropolitan airports. As a result, there has been a significant progress in achieving major reduction in noise levels of airplanes in past five decades as shown in Figure 6 [7]. These gains have been achieved by technological innovations by the manufacturers in reducing the noise from airframe, engines and undercarriage as well as by making changes in the operations.

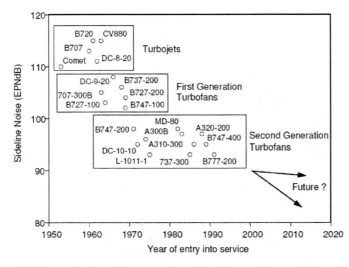

Figure 6: Reductions in Noise Levels of Aircrafts in past Fifty Years [7]

Worldwide, there has been ten fold increases in number of airports since the 1970s that now impose the noise related restrictions as shown in Figure 7 [8]. The airports have imposed operating restrictions and also there has been special attention paid to the planning, development and management of airports for sustainability. Since 1980, FAA has invested over $5billion in airport noise reduction.

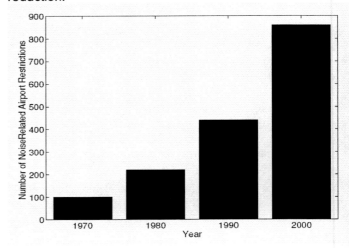

Figure 7: Number of Airports with Noise Related Restrictions in Past Thirty Years [8]

In recent years, the joint MIT/Cambridge University project on "Silent Aircraft" has produced an innovative aircraft/engine design, shown in Figure 8 that has imperceptible noise outside an urban airport [9].

Figure 8: Silent Aircraft SAX - 40 (Joint MIT/Cambridge University Design) [9]

In order to meet the ACARE and NASA goals of reducing the perceived noise by 50% of the current level by 2020, several new technology ideas are being investigated by the airplane and engine manufacturers to both reduce and shield the noise sources as shown in Figure 9 in the chart below by Reynolds [10]. The most promising for the near future are the chevron nozzles, shielded landing gears and the ultra high bypass engines with improved fan (geared fan and contra fan) and fan exhaust duct- liner technology. In addition, new flight path designs in ascent and descent flight can reduce the perceived noise levels in the vicinity of the airports.

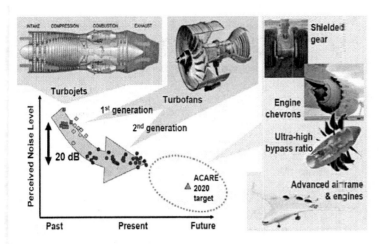

Figure 9: Evolution of Noise Reduction Technologies [10]

EMISSIONS AND FUEL BURN

Aviation worldwide consumes today around 238 million tonnes of jet-kerosene per year. Jet-kerosene is only a very small part of the total world consumption of fossil fuel or crude oil. The world consumes 85 million barrels/day in total, aviation only 5 million. At present, aviation contributes only 2-3% to the total CO_2 emissions worldwide [11] as shown in Figure 10. However, it contributes 9% relative to the entire transportation sector. With 2050 forecast of air travel to become 40% of total PKT (Figure 1), it will become a major contributor to GHG emissions if immediate steps towards reducing the fuel burn by innovations in technology and operations, as well as alternatives to Jet-kerosene are not sought and put into effect.

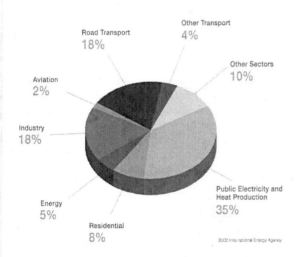

Figure 10: CO_2 Emissions Worldwide Contributed by Various Economic Sectors [11]

Of the exhausts emitted from the engine core, 92% are O_2 and N_2, 7.5% are composed of CO_2 and H_2O with another 0.5% composed of NOx, HC, CO, SOx and other trace chemical species, and carbon based soot particulates. In addition to CO_2 and NOx emissions,

formation of contrails and cirrus clouds (Figure 11) contribute significantly to radiative forcing (RF) which impacts the climate change. This last effect is unique to

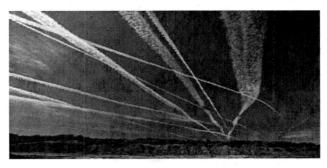

Figure 11: Contrails & Cirrus Clouds

aviation (in contrast to ground vehicles) because the majority of aircraft emissions are injected into the upper troposphere and lower stratosphere (typically 9-13 km in altitude). The impact of burning fossil fuels at 9-13 km altitude is approximately double of that due to burning the same fuels at ground level [12]. The present metric used to quantify the climate impact of aviation is radiative forcing (RF). Radiative forcing is a measure of change in earth's radiative balance associated with atmospheric changes. Positive forcing indicates a net warming tendency relative to pre-industrial times. Figures 12 and 13 show the IPCC (Intergovernmental Panel for Climate Change) estimated increase in total anthropogenic RF due to aviation related emissions (excluding that due to contrails and cirrus clouds) from 1992 to 2050 [13]. It should be noted that in Figures 12 and 13, RF scale is given in W/m^2. It is usually given in mW/m^2; then the numbers in Figures 12 and 13 should be multiplied by 1000 as shown. The horizontal line in Figures 12 and 13 is indicative of the current level of scientific understanding of the impact of each exhaust species.

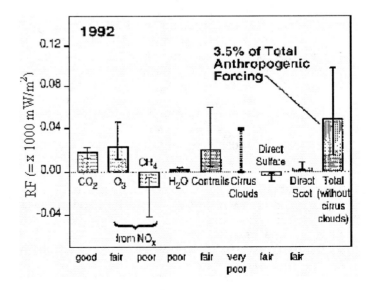

Figure 12: IPCC Estimated Radiative Forcing (RF) due to Emissions – 1992 [13]

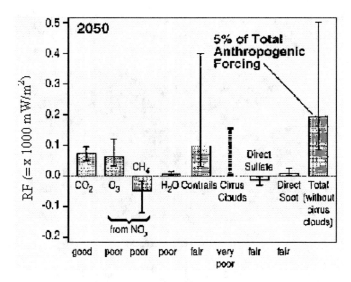

Figure 13: IPCC Estimated Radiative Forcing (RF) due to Emissions - 2050 [13]

It should be noted that the RF estimates for 2050 in Figure 13 are based on several assumptions about the growth in aviation, state of technology etc. which are most likely to change. Based on the RF estimates shown in Figures 12 and 13, aviation is expected to account for 0.05K of the 0.9K global mean surface temperature rise expected to occur between 1990 and 2050 [12]. However, RF is not a good metric for weighing the relative importance of short-lived and long-lived emissions. Most importantly, the range of uncertainty about the climate impact of contrails and cirrus cloud remains substantial. According to recent IPCC report, the best estimates for RF in 2005 from linear contrails were 10 (3-30)mW/m^2 and 30(10-80)mW/m^2 from total aviation induced cloudiness, the numbers in bracket give the range of the 2/3 confidence limit [14]. As noted in Reference [14], "the tradeoff estimate of the CO_2 RF in 2000 was 23.5mW/m^2. Despite the growth in CO_2 RF between 2000 and 2005, aviation induced cloudiness remains the greatest contributor to RF according to these estimates. Because of doubts of RF as a metric as well as data spread in cloudiness related RF, the relative contribution of the two (CO_2 and cloudiness) to climate change can not be ascertained with confidence at present time. However, the atmospheric conditions under which an aircraft will generate a persistent contrail – the Schmidt-Appleman criterion [15] – are well understood and can be predicted accurately for a particular aircraft.

Currently there is no technological fix to prevent contrail formation if the atmospheric conditions and engine exhaust characteristics satisfy the Schmidt-Appleman criterion. One assured way of reducing the persistent contrail formation is to reduce aircraft traffic through regions of supersaturated air in which the persistent contrail can form, by flying under, over or around these regions. However, this approach may not be acceptable commercially because of increase in fuel burn, disruption in airline schedule, added ATM workload, and

additional operating costs as well as increase in CO_2 and NOx emissions. Because contrail reduction involves an increase in CO_2 and NOx emissions, the best environmental solution is not the complete avoidance of contrails, but a balanced result that minimizes climate impact. This requires a better understanding of the relationship between the properties of the atmosphere (temperature, humidity etc.), the size of the aircraft, the quantity of its emissions (water and particulates), and extent of the persistent contrail and subsequent cirrus formation that results.

The adoption of synthetic kerosene produced by Fischer-Tropsch or some similar process offers the prospect of substantial reduction in sulfate and black carbon particulate emissions. This is likely to reduce the extent of contrail and cirrus formation, but the extent of reduction as well as to what extent it would reduce the fuel burn penalty of operational avoidance measures requires further research. Based on the current status, it appears that fuel additives do not offer a significant reduction in contrail formation. The contrail avoidance measures e.g. making modest changes in altitude can reduce contrail formation appreciably with a small penalty in additional fuel burn."

Increasing the cruise altitude and higher engine pressure ratio can reduce CO, HC, and CO_2 emissions as well as decrease the fuel burn (improve the fuel efficiency) and facilitate noise reduction. Since higher pressure ratio requires higher flame temperature, the NOx formation rate increases. On the other hand, decreasing the cruise altitude and reducing the engine overall pressure ratio can reduce the NOx but increase the CO_2 emissions. This should be an important consideration in the optimization of future aircraft and engine designs. Research is needed in understanding the impact of cruise altitude on climate. *In addition, there is a need for new optimized aircraft and engine designs that provide a compromise between minimizing the fuel burn and reducing the climate impact.* The lower NOx emissions can possibly be achieved by new combustor concepts such as flameless catalytic combustor and technological improvements in fuel/air mixers using alternative fuels (biofuels), aided by active combustion control. These concepts/technologies should make it possible to meet the N+1 and N+2 generation goals (Figure 5) of achieving the LTO NOx reductions by 60% and 75% respectively below the ICAO standard adapted at CAEP 6 (Committee on Aviation Environmental Protection). It should result in reducing the steepness of the trade-off between NOx and CO_2 emissions and should therefore also help in making a significant contribution to the aircraft performance goal by reducing the fuel burn by 33% and 40% for the N+1 and N+2 generation aircraft respectively.

Thus, there are three key drivers in emissions reductions as shown in Figure 14 [16]: (a) innovative engine technologies and aircraft designs, (b) the improvement in ATM and operations, and (c) the alternative fuels e.g.

biofuels. The three-prong approach can achieve the goals enunciated by ACARE and NASA by 2020 and beyond. These are discussed in next few sections.

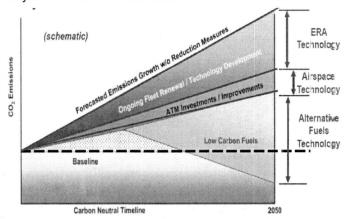

Figure 14: Key Drivers for Emissions Reductions [16]

INNOVATIVE ENGINE TECHNOLOGIES

In cruise condition, the amount of fuel burn varies in inverse proportion to propulsion efficiency and lift-to-drag ratio. Aircraft and engine manufacturers in U.S. and Europe along with several research organizations are developing new engine technologies aimed at improving the propulsion efficiency to reduce the fuel burn and also to simultaneously reduce NOx emissions and noise. The greatest gains in fuel burn reduction in the past sixty years (since the appearance of jet engine) have come from better engines. The earliest engines were turbojets in which all the air sucked in at the front is compressed, mixed with fuel and burned, providing thrust through a jet out the back (see Figure 9). Afterwards, more efficient turbofans were designed when it was realized that greater engine efficiency could be achieved by using some of the power of the jet to drive a fan that pushes some of the intake air through ducts around the core (see Figure 9). Other boosts in efficiency have come from better compressors and materials to let the core burn at higher pressure and temperature. As a result, according to International Airport Transport Association (IATA), new aircraft are 70% more fuel efficient than they were forty years ago. In 1998, passenger aircraft averaged 4.8 liters of fuel/100km/passenger; the newest aircraft – Airbus A380 and Boeing B787 use only three liters. Figure 15 shows the relative improvement in fuel efficiency of various aircraft engines since 1955 [17].

The current focus is on making turbofans even more efficient by leaving the fan in the open. Such a ductless "open rotor" design (essentially a high-tech propeller) would make larger fans possible; however one may need to address the noise problem and how to fit such engines on the airframe. In the short-to-medium-haul market, where most fuel is burned, the open rotor offers an appreciable reduction in fuel burn relative to a turbofan engine of comparable technology, but at the expense of some reduction in cruise Mach number. It is

worth noting here that in mid 1980's GE invested significant effort in advanced turbo-prop technology (ATP).The un-ducted fan (UDF) on a GE36 ultra high bypass (UHB) engine on MD-81 at Farnborough air show in 1988 (Figure 16 [18]) created enormous buzz in the air transportation industry. The author of this paper was at McDonnell Douglas during that period and played a small role in the airframe – engine integration study of MD81 with GE36 ATP. However, in spite of its potential for 30% savings in fuel consumption over existing turbofan engines with comparable performance at speeds up to Mach 0.8 and altitudes up to 30,000 ft, for a variety of technical and business reasons, the advanced turboprop concept never quite got-off the ground [19].

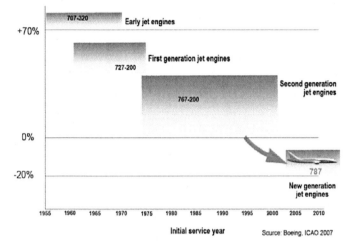

Figure 15: Relative Improvement in Fuel Efficiency of Various Aircraft Engines from 1955 to 2010 [17].

Figure 16: GE36 Turbo-Prop Demonstrator Engine on MD-81 Aircraft [18]

At present in Europe, under the auspices of NACRE (New Aircraft Concept Research Europe), Rolls-Royce and Airbus are making a joint study of the open rotor configurations (Figure 17), including wind-tunnel investigations of power plant installation effects. A key issue in future engine design is how to balance the conflicting aims of reducing fuel burn and NOx emissions (along with the other conflicting aims of reducing noise, weight, initial investment cost and maintenance cost). The results of these types of current and future projects should provide a sounder basis for making decisions between turbofan and open rotor engines for future aircraft. They should also take engine technology well towards its contribution to the goal of a 20% improvement in the installed engine fuel efficiency by 2020.

Figure 17: Open-Rotor Version of Pro-Active Green Aircraft in NACRE Study [14]

INNOVATIVE AIRCRAFT DESIGNS

As noted in Reference [14], "the classic swept-winged aircraft with a light alloy structure has been evolving for some sixty years and the scope for increasing its lift-to-drag ratio (L/D), if its boundary layers remain fully turbulent, is by now exceedingly limited. Nevertheless, it is well established that increasing L/D is one of the most powerful means of reducing fuel burn. The three ways of increasing L/D are (a) increase the wing span, (b) reduce vortex drag factor κ and (c) reduce the profile drag area. The vortex drag factor is a measure of the degree to which the span-wise lift distribution over the wing departs from the theoretical ideal. Current swept-wing aircraft are highly developed and there is little scope for further improvement. A flying wing may enable some additional small reduction in κ, however realistically; there is no real prospect of a significant reduction in fuel burn by altering span-wise loading distributions. Furthermore, increasing the wing span increases wing weight. Current long-range aircraft are optimized to minimize the fuel burn at current cruise Mach numbers. In a successful design the balance between the wing span and wing weight is close to optimum. However, the change to advanced composite materials for the wing structure should result in an optimized wing of greater span; both the B787 and Airbus A350 reflect this. If cruise Mach number is reduced, reducing wing sweep also enables the wing to be optimized at a greater span. The turbofan version of Pro-Active Green Aircraft (Figure 18) included in the NACRE study features a slightly forward swept wing optimized at a significantly higher than usual span. This aircraft is aimed at an appreciable increase in L/D at the expense of some reduction in cruise Mach number. The third option for increasing L/D is to reduce the profile

drag of the aircraft. This is seen as the option with the greatest mid-term and long-term potential. For large aircraft, the adoption of a blended wing-body (BWB) layout reduces profile drag by about 30%, providing an increase of around 15% in L/D (estimates of 15% - 20% have been published)."

Figure 18: Turbofan Version of Pro-Active Green Aircraft in NACRE Study [14]

The work on such configurations, both by Boeing (the X-48B, wind tunnel and flight tested at model scale by NASA [Figure 19]) and by Airbus within the NACRE project are proceeding. At present, the first applications of the Boeing BWB are envisaged to be in military roles or as a freighter, with 2030 suggested as the earliest entry to service date for a civil passenger aircraft.

Figure 19: Boeing/NASA X-48B BWB Technology Demonstrator Aircraft [20]

The other well known approach of reducing the profile drag is by the use of laminar flow control in one of its three forms - natural, hybrid or full. Natural laminar flow control was applied with great success in World War II on the P-51 Mustang fighter to give it an exceptional range. As a result there was significant effort devoted to the development of laminar flow airfoils after the end of World War II. In these airfoils, the reduction in friction drag was achieved by moving the transition farther back on the airfoil. In addition, the location of the maximum airfoil thickness was at about 60% of the chord which moved the shock system farther back and reduced the effects of boundary layer thickening and separation caused by it. However in spite of a large number of studies, the success in the laboratory in reducing the drag was never realized on medium size aircraft with swept wings. Therefore, its application has been restricted by a combination of size and wing sweep either to small aircraft with swept wings or medium-sized aircraft with zero or very little sweep. The Pro-Active Green Aircraft in the NACRE project (Figures 17 & 18) is designed to exploit natural laminar flow control and has slightly swept forward wings, to avoid contamination of the flow over the wing by the turbulent boundary layer on the fuselage.

"Hybrid laminar flow control employs suction over the forward upper surface of the wing to stabilize the boundary layer. This enables the drag reducing principles that underlie natural laminar flow control to be applied to larger, swept-winged aircraft up to typically the size of the A310. The use of suction to maintain laminar flow over the first half of an airfoil surface has been successfully demonstrated in flight on a B757 wing and an A320 fin. The aerodynamic principles are well understood but the engineering of efficient, reliable, lightweight suction systems requires further work. Thereafter, demonstration of the practicality of the system and assessment of the maintenance and other operational problems that it may encounter will require an extended period of operational validation. The application of suction to maintain laminar flow over the entire surface of a flying wing airliner was proposed by Handley Page in the early 1960s. The proposal was based on the substantial body of research into full laminar flow control, including flight demonstrations, over the preceding decade. Full laminar flow control may have potential to double L/D relative to current standards [14]."

Recently unveiled "Honda Jet" (Figure 20) has combined several innovative aircraft and engine design features namely a combination of over the wing (OTW) engine mount design, natural laminar flow wing (NLF), all composite fuselage, HF – 120 turbofan engine, which give it a 30-35% more fuel efficiency and higher cruise speed than conventional light business jets. This is the range of efficiency that can be achieved for the N+1 generation conventional tube and wing aircraft by 2015.

Figure 20: Honda Jet [21]

Saeed et al. [22] have recently conducted the conceptual design study of a Laminar Flying Wing (LFW) aircraft capable of carrying 120 passengers. They have estimated that, subject to the constraint of a low cruise Mach number of 0.58, LFC has the potential to reduce aircraft fuel-burn by just over 70%, to about 6 gram per passenger-km (PKM), with a trans-Atlantic range of 4125 nautical miles. Studies of this nature do show the promise of innovative aircraft designs to reduce the fuel burn.

Figure 5 shows the NASA goals of achieving a 33% and 40% reduction in fuel burn for N+1 and N+2 generation aircrafts respectively by using the advanced propulsion technologies, advanced materials and structures, and by improvements in aerodynamics and subsystems. Collier [23] from NASA Langley has provided a detailed outline as to how such savings in fuel burn can be achieved. He has estimated that for a N+1 generation conventional small twin aircraft (162 passengers and 2940nm range), 21% reduction in fuel burn can be achieved by using advanced propulsion technologies, advanced materials and structures, and by improvements in aerodynamics and subsystems. For an advanced small twin, additional 12.3% savings in fuel burn can be achieved by using hybrid laminar flow control as shown in Figure 21.

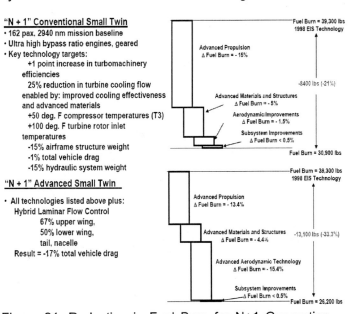

Figure 21: Reduction in Fuel Burn for N+1 Generation Aircraft Relative to Baseline B737/CFM56 Using Advanced Technologies [23]

For a N+2 generation aircraft (300 passengers and 7500 nm range) flying at cruise Mach of 0.85, 40% saving in fuel burn relative to baseline B777-200ER/GE90 can be achieved by a combination of hybrid wing-body configuration (with all composite fuselage), advanced engine and airframe technologies, embedded engines with BLI inlets and laminar flow as shown in Figure 22 [23]. For the baseline aircraft, the fuel burn at Mach 0.85 with 300 passengers for a 7500nm mission range is 237,000 lbs. The N+2 generation aircraft should require

141,100lbs of fuel. As discussed in next few sections, additional savings of 10% in fuel burn can be achieved by operational improvements.

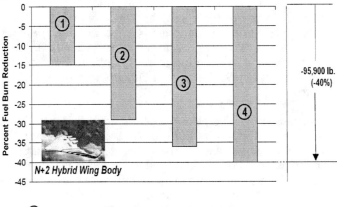

1 = Hybrid wing body configuration, including all composite fuselage
2 = 1 + advanced engine and airframe technologies (~2020 timeframe)
3 = 2 + embedded engines with BLI inlets
4 = 3 + laminar flow

Figure 22: Reduction in Fuel Burn for N+2 Generation Aircraft Relative to Baseline B777-200ER/GE96 Using Advanced Technologies [23]

OPERATIONAL IMPROVEMENTS/CHANGES

IMPROVEMENT IN AIR TRAFFIC MANAGEMENT (ATM) INFRASTRUCTURE

There are many improvements in operations that are being introduced, or will be introduced in the relatively near future that can reduce CO_2 emissions significantly. Foremost among these is the reduction of inefficiencies in ATM, which give rise to routes with dog-legs, stacking at busy airports, queuing for a departure slot with engines running, etc. U.S. Next Generation Air Transportation System (NextGen) architecture and the European air traffic control infrastructure modernization program, SESAR (Single European Sky ATM Research Program), are an ambitious and comprehensive attack on this problem. As described in the U.S. National Academy of Science (NAS) report [24], "NextGen is an example of active networking technology that updates itself with real time-shared information and tailors itself to the individual needs of all U.S. aircraft. NextGen's computerized air transportation network stresses adaptability by enabling aircraft to immediately adjust to ever-changing factors such as weather, traffic congestion, aircraft position via GPS, flight trajectory patterns and security issues. By 2025, all aircraft and airports in U.S. airspace will be connected to the NextGen network and will continually share information in real time to *improve efficiency, safety, and absorb the predicted increase in air transportation.*" Here it is worth noting that operational measures, which can apply to almost the entire world fleet, can have a greater impact, sooner, than the introduction of new aircraft and engine technologies, which can take perhaps 30 years to fully penetrate the world fleet.

AIR-TO-AIR REFUELING (AAR) WITH MEDIUM RANGE AIRCRAFT FOR LONG-HAUL TRAVEL

One particular operational measure that has been advocated is the use of medium-range aircraft, with intermediate stops, for long-haul travel. It has been estimated, using a simple parametric analysis, that undertaking a journey of 15,000km in three hops in an aircraft with design range of 5,000km would use 29% less fuel than doing the trip in a single flight in a 15,000km design. Hahn [25] and Creemers & Slingerland [26] have performed analyses to address this issue using sophisticated aircraft design synthesis methods. Hahn [25], analyzing the assessment for a 15,000km journey in one stage or three, predicted a fuel saving of 29%. Creemers & Slingerland [26], considering a B747-400 (range 13,334km) as the baseline long-range aircraft, designed an aircraft with the same fuselage and passenger capacity (420) but for half the design range (6,672km). This aircraft was predicted to do the long-haul journey in two hops with a 27% fuel saving and at a fuel cost of $70 per barrel, a DOC saving of 9%. Nangia [27] has shown that fuel burn savings of as much as 50% were achievable by using a 5,000km design for a 15,000km journey, since a medium range aircraft can carry a much higher share of their maximum payload as passengers. This difference — which appears essentially to be the difference between medium-range single and long-range twin-aisle aircraft — was not a feature of either the study of Hahn [25] or Creemers & Slingerland [26], which used the same fuselage for both long and medium range designs. This highlights the importance of cabin dimensions and layouts in considering future designs in which, both environmentally and commercially, seat-kilometers per gallon becomes an increasingly important objective. The full system assessment of this proposition, using optimized medium-range aircraft needs further investigation.

In order to avoid the intermediate refueling stops, air-to-air refueling (AAR) (Figure 23) has been suggested as a means of enabling medium-range designs to be used on long-haul operations. Nangia has now published a number of papers reporting his work on AAR, which indicate substantial fuel burn savings even after the fuel used by the tanker fleet is taken into account [27, 28].

Figure 23: Air-to-Air Refueling [27]

Nangia [28] has shown (Figure 24) that an aircraft with $L/D = 20$, would require 46,147 lbs, 161,269 lbs, and 263,073 lbs of fuel to cover a range of 3,000, 6,000 and 9,000 nautical miles (nm) respectively. With AAR, it will require 92,294 lbs and 138, 441 lbs of fuel for a range of 6,000 and 9,000 nm respectively indicating a savings of 43% and 47% in fuel burn relative to that required without AAR. Accounting for the fuel required by the air tanker – 9,000 lbs for one refueling for a range of 6,000nm and 18,000 lbs for two refueling for a range of 9,000nm, the net savings in fuel burn with AAR are 37% and 41% for a range of 6,000nm and 12,000 nm respectively. However it is paramount that with AAR, the absolute safety of the aircraft is assured.

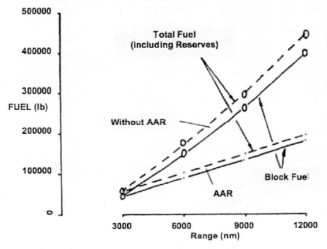

Figure 24: Savings in Fuel Burn with Air-to-Air Refueling (AAR) for Long Haul Flights [28]

CLOSE FORMATION FLYING (CFF)

The possibility of using CFF to reduce fuel burn or to extend range is well known. As stated by Nangia [28] "aircraft formations (Figure 25) occur for several reasons e.g. during displays or in AAR but they are not maintained for any significant length of time from the fuel efficiency perspective." The reason is that flying in formation will require extreme safety measures by use of sensors coupled automatically to control systems of individual aircrafts. Furthermore, flying a close formation through clouds or in gusty environment may not be practical. The obvious benefit of flying in formation is a more uniform downwash velocity field, which minimizes the energy transferred into it from propulsive energy consumption. Another benefit is the cancellation of vortices shed from the wing-tips of individual airplanes, except the two outermost ones. How effective this cancellation will be would depend upon the practicality of achievable spacing among the aircrafts. There would also be a substantial benefit in elimination of vortex contrails and cirrus clouds.

Recently, NASA conducted tests on two F/A-18 aircraft formations [29]. It was shown that the benefits of CFF occur at certain geometry relationships in the formation, namely the trailing aircraft should overlap the wake of the leading aircraft by 10-15% semi-span in this case. Jenkinson [30] suggested that the CFF of several large aircrafts is more efficient in comparison with flying a very large aircraft. The aircrafts could take-off from different airports and then fly in formation over large distances before peeling off for landing at required destinations. Bower at al. [31] have recently investigated a two aircraft echelon formation and a three aircraft formation of three different aircraft and analyzed the fuel burn. Their study determined the fuel savings and difference in flight times that result from applying CFF to missions of different stage lengths and different spacing between the cities of origin. For a two aircraft formation, the maximum fuel savings were 4% with a tip-to-tip gap between the aircraft equal to 10% of the span and 10% with a tip overlap equal to 10% of the span. For the three aircraft inverted-V formation, the maximum fuel savings were about 7% with tip-to-tip gaps equal to 10% of the span and about 16% with tip overlaps equal to 10% of the span.

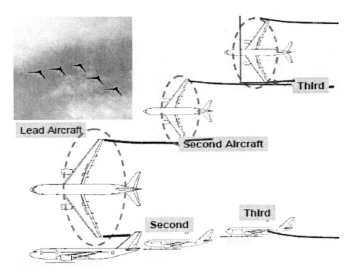

Figure 25: Three Different Aircraft Type in CFF [28]

Bower et al. [31] conducted a case study to examine the effect of formation flight on five FedEx flights from the Pacific Northwest to Memphis, TN. The purpose of this study was to quantify the fuel burn reduction achievable in a commercial setting without changing the flight schedule. With tip-to-tip gaps of about to 10% of the span it was shown that fuel savings of approximately 4% could be achieved for the set of five flights. With a tip-to-tip overlap of about 10% of the span the overall fuel savings were about 11.5% if the schedule was unchanged. This translated into saving of approximately 700,000 gallons of fuel per year for this set of five flights. Figure 26 shows the three types of aircrafts employed in the study – two Boeing B 727-200, two DC10-30 and one Airbus A300 – 600F.

It should be noted that in CFF, each aircraft will experience off-design forces and moments. It is important that these are adequately modeled and efficiently controlled. Simply using aileron may trim out the induced roll but at the expense of drag. But as Bower et al. [31] have shown, it is possible to realize savings in fuel burn by using the existing aircraft by suitably tailoring the formation.

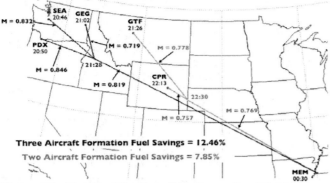

Figure 26: Five FedEx Aircraft in Formation Flight Enroute from Pacific Northwest to Memphis [31]

TAILORED ARRIVALS

Boeing [32] is working with several airports, airlines and other partners around the world in developing tools for "tailored arrivals" which can reduce fuel burn, lower the controller workload and allow for better scheduling and passenger connections (Figure 27). To optimize tailored arrivals, additional controller automation tools are needed. Boeing completed the trial of Speed and Route Advisor (SARA) with Dutch air traffic control agency (LVNL) and Eurocontrol in April/May 2009. SARA delivered traffic within 30 seconds of planned time on 80% of approaches at Schiphol airport in Netherlands compared to within 2 minutes on a baseline of 67%. At San Francisco airport, more than 1700 complete and partial tailored arrivals have been completed between December 2007 and June 2009 using the B777 and B747 aircraft. It has been found that tailored arrivals save an average of 950 kg of fuel and approximately $950 per approach. Complete tailored arrivals saved approximately 40% of the fuel used in arrivals. For one year period, four participating airlines saved more than 524,000 kg of fuel and reduced the carbon emissions by 1.6 million kg.

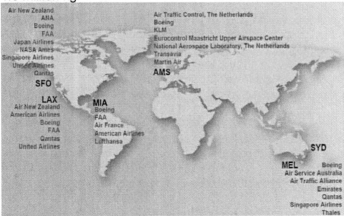

Figure 27: Airports and Partners Participating in the Concept of Tailored Arrivals [32]

SAVINGS IN FUEL BURN BY AIRCRAFT WEIGHT REDUCTION

It is well known that substantial savings in fuel burn can be achieved by reducing the ratio of the empty weight to payload of an aircraft. It can be accomplished by the development and use of lighter and stronger advanced composites, and by reducing the design range and cruise Mach number.

AIRCRAFT WEIGHT REDUCTION BY USE OF ADVANCED COMPOSITES

Reducing the weight of an aircraft is one of the most powerful means of reducing the fuel burn. Boeing and Airbus, as well as other Business and General Aviation aircraft manufacturers are investing in advanced composites which have the prospects of being lighter and stronger than the present carbon fiber composites (CFC). The replacement of structural aluminum alloy with carbon fiber composite is the most powerful weight reducing option currently available to the aircraft designer working towards a given payload-range requirement. The Boeing B787 and Airbus A350 have both taken this step, having wings and fuselage made with CFC. Most new designs are likely to take this path.

AIRCRAFT WEIGHT REDUCTION BY REDUCING THE DESIGN RANGE

Although the historic trend has been in the opposite direction, another powerful means of reducing the weight of an aircraft is to reduce its design range. The study by Hahn [25] has shown that by reducing the design range from 15,000km to 5,000km, with the fuselage and passenger accommodation fixed, it is possible to reduce the operational empty weight (OEW) by 29%. The study by Creemers & Slingerland [26] noted a 17% reduction in OEW by halving the design range from 13,334km to 6,672km. Nangia [27, 28] has also shown that, with the fuselage and number of passengers fixed, wing area increases rapidly to contain the fuel needed and to maintain C_L as the design range increases. Figure 28 shows the aircraft designs and maximum take-off weight MTOW for design range from 3,000 to 12,000 nm. From Nangia's study [28], it is clear that 3,000nm aircraft can provide substantial savings in fuel burn by having less weight and can be used for long range flight by using AAR. In past twenty years, each new aircraft type has achieved 10-15% gain in fuel efficiency. Additional achievements in fuel efficiency by improvements in airframe and engine design will take some time, however, several studies have shown that it is possible to reduce fuel burn significantly by instituting operational measures such as more efficient Air-Traffic Management (ATM), Air-to-Air Refueling (AAR), Close Formation Flying (CFF), Tailored Arrivals, and by reducing the ratio of empty weight to payload.

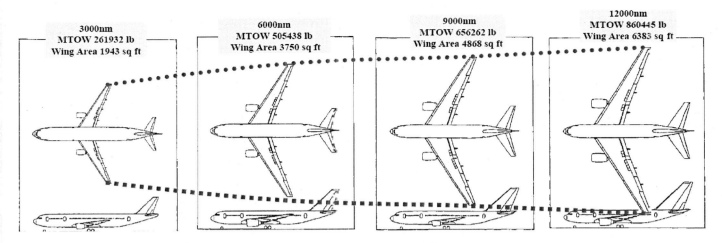

Figure 28: Aircraft Designs, with Fixed Fuselage, 250 Passengers and CL, for Different Ranges of Operation [27, 28]

ALTERNATIVE FUELS

All forms of powered ground and air transportation are experiencing the pressure of the need to mitigate greenhouse gas (GHG) emissions to arrest their impact on climate change. In addition the high price of fuel (oil reaching $149/barrel during summer of 2008) as well as the need for energy security are driving an urgent search for alternative fuels, in particular the biofuels. There is emphasis on both the improvements in energy efficiency and new alternative fuels. Aviation is particularly sensitive to these pressures since, for many years, no near term alternative to kerosene has been identified. Until recently, biofuels have not been considered cost competitive to kerosene. An important much desired characteristic of an alternative fuel is whether it can be used without any change to the aircraft or engines. The attractions of such a *drop-in fuel* are clear: it does not require the delivery of new aircraft but the environmental impact of all aircraft flying today can be significantly reduced. Non-drop-in fuels, such as hydrogen or methane hydrates, are unlikely to be used before 2050.

The key criteria in identifying that a new alternative fuel would be beneficial in reducing CO_2 emissions should be based on the life cycle analysis of CO_2; the life-cycle CO_2 generation must be less than that of kerosene. Many first generation biofuels have performed poorly against this criterion, though second generation biofuels appear to be far more promising. Furthermore, it is important that there are no adverse side-effects arising from production of the feedstock for biofuel generation, such as adverse impact on farming land, fresh-water supply, virgin rain-forests and peat-lands, food prices, etc. Algae and halophytes (salt-tolerant plants irrigated with sea/saline water) are emerging as potential sustainable feedstock solutions. The alternative fuels need to meet specific aviation requirements and essentially should have the key chemical characteristics of kerosene, that is they won't freeze at flying altitude and they would have a high enough energy content to power an aircraft's jet engine. In addition, the alternative fuel should have good high-temperature thermal stability

characteristics in the engine and good storage stability over time.

Interest in biofuels for civil aircraft has increased dramatically in recent years and the focus of the aviation industry on what is and what is not credible in this arena has sharpened. It is clear that a *'drop-in'* replacement for kerosene i.e. the synthetic kerosene appears to be the only realistic possibility in the foreseeable future. The potential of such bio-derived synthetic paraffinic kerosene (Bio-SPK) to reduce the net CO_2 emissions from aviation may well match or exceed that of advances in airframe and engine technologies, and perhaps may achieve reductions across the world fleet sooner than new technologies. In addition, since synthetic kerosene produces substantially less black carbon and sulphate aerosols than kerosene from oil wells, there is a possibility that its use will reduce contrail and cirrus formation as well.

Boeing, Airbus and the engine manufacturers believe that the present engine technology can operate on biofuels (tests are very promising) and that within 5 to 15 years, the aviation industry can convert to biofuels. On 19 June 2009, Billy Glover of Boeing made a presentation to the press at the Paris air show [32] describing the Boeing's "Sustainable Biofuels Research and Technology Program." Tables I and II show the comparisons of key fuel properties of currently used Jet A/Jet A-1 fuel with those with Bio-SPK fuel derived from three different feed-stocks (Jatropha, Jatropha/Algae, and Jatropha/Algae/Camelina) for neat fuel and blends respectively. All Bio-SPK blends met or exceeded the aviation jet fuel requirements. In this presentation, Boeing declared that they are preparing a comprehensive report on Bio-SPK fuels for submittal to ASTM International and expect an approval in 2010. Boeing is working across the industry on regional biofuel commercialization projects. There have already been a few experimental flights operated by several airlines using the biofuel blends and many more are planned in the near future.

Table I: Key Biofuel (Neat) and Jet/Jet A-1 Fuel Properties Comparison [32]

Property		Jet A/Jet A-1	ANZ Jatropha	CAL Jatropha/Algae	JAL Jatropha/Algae/Camelina
Freeze Point °C	Max	-40 Jet A -47 Jet A-1	-57.0	-54.5	-63.5
Thermal Stability JFTOT (2.5 hrs. at control temperature) Temperature °C	Min	260	340	340	300
Viscosity -20°C, mm²/s	Max	8.0	3.663	3.510	3.353
Contaminants Existent gum, mg/100mL	Max	7	<1	<1	<1
Metals ppm	Max	0.1 per metal	<0.1	<0.1	<0.1
Net Heat of Combustion MJ/kg	Min	42.8	44.3	44.2	44.2

ANZ = Air New Zealand, CAL = Continental Airline, JAL = Japan Airline

On 24 February 2008, Virgin Atlantic operated a B747-400 on a 20% biofuel/80% kerosene blend on a short flight between London-Heathrow and Amsterdam. This was the first time a commercial aircraft had flown on biofuel and it was the result of a joint initiative between Virgin Atlantic, Boeing and GE. On 30 December 2008, Air New Zealand (ANZ) conducted a two hour test flight of a B747-400 from Auckland airport with one-engine powered by 50-50 blend (B50) of biofuel (from Jatropha) and conventional Jet-A1 fuel. B50 fuel was found to be more efficient. ANZ has announced plans to use the B50 for 10% of its needs by 2013. The test flight was carried out in partnership with Boeing, Rolls-Royce and Honeywell's refining technology subsidiary UOP with support from Terasol Energy. On January 7th, Continental Airline (CAL) completed a 90-minute test flight using biofuel derived from algae and Jatropha. B737-800 flew from Houston with one engine operating on a 50-50 blend of biofuel and conventional fuel (B50) and the other using all conventional fuel for the purpose of comparison. The biofuel mix engine used 3,600 lbs of fuel compared to 3,700 lbs used by the conventional engine. On January 30, 2009, Japan Airline (JAL) became the fourth airline to use B50 blend of Jatropha (16%), algae (<1%) and Camelina (84%) on the third engine of a 747-300 in one-hour test flight. It was again reported that biofuel was more fuel efficient than 100% jet-A fuel.

It should be noted that in all the above demos, biofuel came from a sustainable feedstocks (see Tables I and II), sources that neither compete with staple food crops nor cause deforestation. It is worth mentioning that on 1 February 2008, Airbus A380 flew from Filton, U.K. to Toulouse, France with one of its Rolls-Royce engines powered by an alternative, synthetic gas-to-liquid (GTL) jet fuel. Airbus and Qatar Airways are now partners in a GTL consortium which also includes Shell International Petroleum to investigate the use of GTL neat/blend vis-à-vis conventional jet fuel. From an environmental standpoint, it is encouraging and very hopeful that both major manufacturers – Boeing and Airbus are positioning themselves to be at the forefront of alternative and bio-jet fuels. It is surmised that by 2050, with the use of synthetic kerosene derived from biomass, the world fleet CO_2 emissions per passenger-kilometer (PKM) could be lower at least by a factor of three, NOx emissions lower by a factor of 10 and contrail and contrail-induced cirrus formation lower by a factor of 5 to 15.

Table II: Key Biofuel (Blend) and Jet/Jet A-1 Fuel Properties Comparison [32]

Property		Jet A/Jet A-1	ANZ Jatropha	CAL Jatropha/Algae	JAL Jatropha/Algae/Camelina
Freeze Point °C	Max	-40 Jet A -47 Jet A-1	-62.5	-61.0	-55.5
Thermal Stability JFTOT (2.5 hours @control temperature)	Min	260	300	300	300
Viscosity -20°C mm2/s	Max	8.0	3.606	3.817	4.305
Contaminants Existent gum, mg/100mL	Max	7	1.0	<1	<1
Net Heat of Combustion MJ/kg	Min	42.8	43.6	43.7	43.5

ELECTRIC, SOLAR OR HYDROGEN POWERED GREEN AIRCRAFT

For many years, there have been several exploratory studies in academia and industry to build and fly aircraft using sources of energy other than Jet-kerosene or synthetic kerosene (biofuels). There have been several success stories in recent years. In March 2008, Boeing successfully conducted a test flight of a manned aircraft powered by PEM hydrogen fuel cells [33], shown in Figure 29. Since fuel cells convert hydrogen directly into electricity and heat without the products of combustions such as CO_2, they use a clean or green source of energy. Fuel cells propelled aircraft is also often called as "an all electric aircraft."

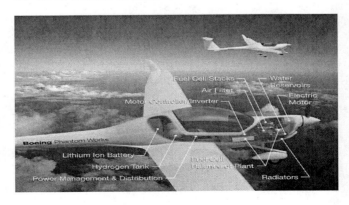

Figure 29: Boeing PEM Fuel Cell Powered Electric Aircraft [33]

Recently in June 2009, the prototype of a new solar-powered manned aircraft was unveiled in Switzerland by the company SOLARIMPULSE [34]. The airplane is designed to fly both day and night without the need for fuel. The aircraft has a wing span equal to that of a Boeing 747 but weighs only 1.7 tons. It is powered by 12,000 solar cells mounted on the wing to supply renewable solar energy to the four 10HP electric motors. During the day, the solar panels charge the plane's lithium polymer batteries, allowing it to fly at night.

Figure 30: Solar Power Aircraft HB-SIA from SOLARIMPULSE [34]

To be sure, the fuel-cell propelled electric aircraft and the solar energy driven aircraft are not likely to become feasible for mass air transportation. However, they can become viable for recreation and personal transportation, and possibly as business aircraft in not too distant future.

The idea of using liquid hydrogen as a propellant has been around for many decades, but is unlikely to become feasible for commercial aircraft, at least before 2050, because of many challenges that would have to be overcome. Figure 31 shows the artist's rendering of a hydrogen-powered version of A310 Airbus [35]. It is also called a "Cryoplane" because of the very visible cryogenic hydrogen tank located above the passengers. Cryogenic hydrogen is the only possibility for the airplane since the high pressure tanks would be too heavy. The physical properties of the liquid hydrogen determine the appearance of the Cryoplane. Liquid hydrogen occupies 4.2 times the volume of jet fuel for the same energy; therefore the tanks will have to be huge. Jet fuel weighs 2.9 times more than liquid H_2 for the same energy. The reduced weight partly compensates for the increased aerodynamic drag of the tanks. The Cryoplane would have less range and speed than A310. It will have higher empty weight. Furthermore, whatever energy source is used, 30% will be lost in hydrogen liquefaction. In addition, the cost, infrastructure and passenger acceptance issues would have to be addressed. The main advantage of using a hydrogen powered airplane is the reduced emissions as

shown in Figure 31 from Penner [36]. Since the use of H_2 does not produce any CO_2, it is dubbed as clean fuel.

Figure 31: Artist's rendering of a Hydrogen Powered Version of A310 Airbus [35]

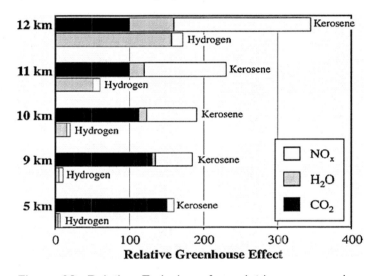

Figure 32: Relative Emissions from Jet-kerosene and Hydrogen at Various Altitudes [36]

MODELING ENVIRONMENTAL & ECONOMIC IMPACTS OF AVIATION

CAMBRIDGE UNIVERSITY AVIATION INTEGRATED MODELING PROJECT (AIM)

Institute for Aviation and the Environment at Cambridge University in U.K. has developed one of the most comprehensive projects – called the Aviation Integrated Modeling (AIM) project to develop a policy assessment capability to enable comprehensive analyses of aviation, environment and economic interactions at local and global levels. It contains a set of inter-linked modules of the key elements which include models of aircraft/engine technologies, air transport demand, airport activity and airspace operations, all coupled to global climate, local environment and economic impact blocks. A major benefit of AIM architecture is the ability to model data flow and feedback between the modules allowing for the policy assessment to be conducted by imposing policy effects on upstream modules and determining the implications through down stream modules to the output metrics, which can then be compared to the baseline case [37].

These modules include: (a) an *Aircraft Technology and Cost Module* to simulate aircraft fuel use, emissions production and ownership/operating costs for various airframe/engine technology evolution scenarios which are likely to have an effect during the period of the forecast; (b) an *Air Transport Demand Module* to predict passenger and freight demand into the future between origin-destination pairs within the global air transportation network; (c) an *Airport Activity Module* to investigate the air traffic growth as a function of passenger and freight growth, to calculate delays and future airline response to them, and to model ground and low altitude operations and congestion to determine LTO emissions as a function of growth in air traffic operations within the vicinity of the airport; (d) an *Aircraft Movement Module* to simulate airborne trajectories between city-pairs, accounting for airspace inefficiencies and delays for given Air Traffic Control (ATC) scenarios and to identify the locations of emissions release from aircraft in flight; (e) a *Global Climate Module* to investigate global environmental impact of aircraft movements in terms of multiple emissions species and contrails; (f) a *Local Air Quality and Noise Module* to investigate local environmental impacts from dispersion of critical air pollutants and noise from landing and take-off (LTO) operations; and (g) a *Regional Economics Module* to investigate positive and negative economic impacts of aviation in various parts of the world, including the increase in direct and indirect employment opportunities in the region. The schematic of the AIM general architecture is shown in Figure 33 [37].

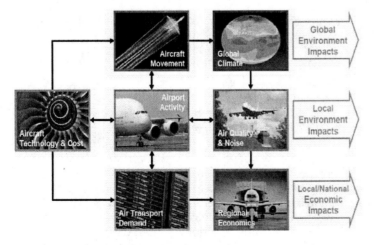

Figure 33: AIM Architecture [37]

The details of the seven modules and interaction among them are not given here but can be found in many papers listed on the website of the Institute for Aviation and the Environment of Cambridge University in U.K (http://www.iae.damtp.cam.ac.uk/innovation.html). Here

we briefly describe the power of the AIM architecture by reproducing some results from Reynolds et al. [37].

Employing the AIM architecture, Reynolds et al. [37] have performed a case study of the U.S. transportation system, which provides a forecast of air transport passenger demand between 50 major airports in U.S. from 2000 to 2030. The flights between these 50 airports represent over 40% of U.S. scheduled domestic departures in 2000 and nearly 20% the world's scheduled flights. Reynolds et al. [37] conducted simulations under three scenarios: 1. Unconstrained/No Feedback (air transport passenger demands and resulting operations were assumed to grow unconstrained), 2. Feedback of Delay Effects (a simplified airline response to delay is modeled by assuming that the 50% of the cost incurred by the airlines due to delays are passed directly to passengers in the form of higher fares), and 3. Feedback of Delay Effects Plus Per-Km Tax Policy (This is same as scenario 2 , but with a per-Km tax applied to tickets from 2020 onwards with the objective of reducing the Revenue Passenger Km (RPKM) demand in 2020 to 2000 levels, so that the resulting delays and emissions can be directly compared). *Reynolds et al. [37] state that these three scenarios, their associated forecasts and environmental impact results are for illustrative purposes only to show the capabilities of AIM; they do not represent realistic evolutions of the U.S. air transportation system.* The main focus of the scenarios is on interactions between the Air Transport Demand and the Airport Activity Modules. However, one can calculate the en route and local emissions utilizing the capabilities of other modules in AIM integrated structure as given in [37]. Details of the data and assumptions used in the simulation are not presented here. The reader is referred to the paper by Reynolds et al. [37].

Forecasts from 2000 to 2030 for annual demand in terms of Revenue Passenger-Km (RPKM) from the Air Transport Demand Module; and total system aircraft operations, system average arrival delay and local NOx emissions at Chicago O'Hare (ORD) from the Airport Activity Module for the above three scenarios are presented in Figures 34 – 37 from Reynolds et al. [37]. The demand forecasts in Figure 34 include those from Airbus (for U.S market), and Boeing, ICAO and AERO-MS for the North American (NA) market for the purpose of comparison. Since they apply to different route groups and time periods, the start year total RPKM value in each case has been normalized to the historical value for the 50 airports extracted from U.S department of transportation T100 data.

Figure 34 shows that for scenario 1, the demand growth measured by increase in RKPM will be 3.5 times the 2000 level by 2030. This is higher than the published estimates as expected given the unconstrained nature of the scenario 1. In scenario 2, the relatively modest feedback of 50% of the increased operating cost to the passenger has a significant effect, particularly over longer time frames. Demand forecast shows a 20% reduction (Figure 34), annual systems operations show a 15% reduction (Figure 35) and average arrival delays show a 50% reduction (Figure 36). Under scenario 3, Figures 34-36 show the effects of distance-based tax; in order to reduce the RPKM demand to 2000 levels in 2020, a 7.7 cents/km charge is required, equating to an additional $300 on a ticket from New York to Los Angeles. Figure 37 shows the annual local emissions at Chicago O'Hare (ORD); all scenarios show an initial gradual increase in emissions which can be explained in conjunction with Figures 34-36 accounting for the increase in RPKM, aircraft operations and arrival delays. The sharp decrease in emissions in scenario 3 in 2020 is due to the reduced operations caused by the introduction of distance-tax policy. The Local Air Quality and Noise Module of AIM architecture can provide results for local air quality at ORD e.g. the annual average NOx concentration at ORD as well as en route CO_2 emissions and global radiative forcing. These results demonstrate that significant insights about environmental and economic impact of aviation can be gained by AIM architecture. It should be noted that many improvements and enhancements to AIM architecture are currently under development at Cambridge.

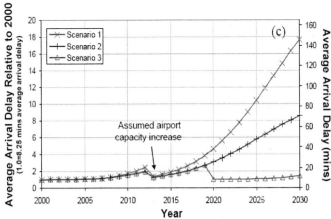

Figure 36: System Average Arrival Delays at O'Hare [37]

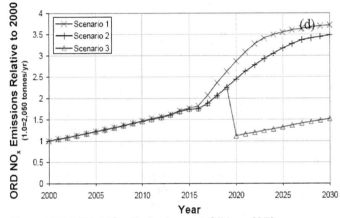

Figure 37: LTO NOx Emissions at O'Hare [37]

SUSTAINABLE AIRPORTS

The airports and associated ground infrastructure constitute an integral part of Green Aviation. To address the issues of energy and environmental sustainability, the Clean Airport Partnership (CAP) was established in U.S. in 1998 [38] and is the only not-for-profit corporation in the U.S devoted exclusively to improving environmental quality and energy efficiency at airports. CAP believes "that efficient airport operations and sound environmental management must go hand in hand. This approach can reduce costs and uncertainty of environmental compliance; facilitate growth, while setting a visible leadership example for communities and the nation." The airport expansion and the development of new airports should include both the environmental costs and life-cycle costs. Sustainable growth of airports requires that they be developed as inter-modal transport hubs as part of an integrated public transport network. The ground infrastructure development should include low emission service vehicles, LEEDS certified green buildings with low energy requirements, and recyclable water usage. There should be effective land use planning of the area around the airports (including securing land for future development) with active investments into the surrounding communities. Airport expansion must also consider the issue of noise and its impact on the surrounding communities, and should be

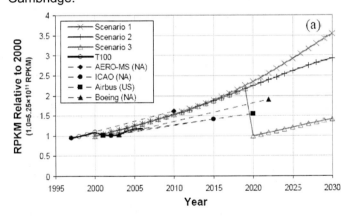

Figure 34: Forecast of System Revenue Passenger-Km (RPKM) Growth at O'Hare [37]

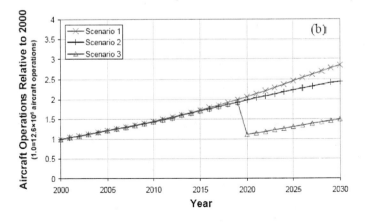

Figure 35: Forecast of Total System Aircraft Operations at O'Hare [37]

involved in its mitigation by engaging in the flight path design. The air quality near the airports should be monitored and measures for its continuous improvement should be put in place. In addition, there should be regulatory requirements to set risk limits.

OPPORTUNITIES & FUTURE PROSPECTS

It is clear that the expected three fold increase in air travel in next twenty years offers enormous challenge to all the stakeholders – airplane manufacturers, airlines, airport ground infrastructure planners and developers, policy makers and consumers to address the urgent issues of energy and environmental sustainability. The emission and noise mitigation goals enunciated by ACARE and NASA can be met by technological innovations in aircraft and engine designs, by use of advanced composites and biofuels, and by improvements in aircraft operations. Some of the changes in operations can be easily and immediately put into effect, such as tailored arrivals and perhaps AAR. Some innovations in aircraft and engine design, use of advanced composites, use of biofuels, and overhauling of the ATM system may take time but are achievable by concerted and coordinated effort of government, industry and academia. They may require significant investment in R&D. It is now recognized by the industry (airlines and manufacturers) as well the relevant government agencies and the policy makers that there is urgent need for action to meet the challenges of climate change; aviation is becoming an important part of it. It is worth noting that in July 2008 in Italy, G8 countries (U.S, Canada, Russia, U.K., France, Italy, Germany and Japan) called for a global emission reduction target of "at least 50%" by 2050, which is in line with goal established by IATA members at their June 2009 Annual General Meeting in Kuala Lumpur, Malaysia. IATA further committed to carbon-neutral traffic growth by 2020. These challenges provide opportunities for breakthrough innovations in all aspects of air transportation.

ACKNOWLEDGMENTS

The author is grateful to Dr. Tom Reynolds of the Institute of Aviation and Environment at Cambridge University to allow the use of material used in the section on "Modeling Environmental & Economic Impacts of Aviation." All the material in this section has been taken from Reference [37]. The author is also grateful to Dr. Raj Nangia for helpful discussions and allowing the use of material from several of his papers [27, 28]. Reference [14] has also provided a significant amount of material for several sections. The author is thankful to Dr. Richard Wahls of NASA Langley for his permission to use the material from NASA Langley presentations on "Environmentally Responsible Aviation." The author would like to thank Professor Raimo J. Hakkinen for reading the manuscript and for making many helpful suggestions that have improved the paper. Finally, it should be noted that the material used in this review paper has been used from a variety of sources listed in the references; any omission is completely unintentional.

REFERENCES

1. Schafer, A., Heywood, J.B., Jacoby, H.D., and Waitz, I.A., *Transportation in a Climate Constraint World,* MIT Press, Cambridge, MA, 2009.
2. http;//www.boeing.com/randy/archives/2006/07/in_the_year_202.html
3. http://www.acare4europe.com/
4. NRC Meeting of Experts on NASA's Plans for System-Level Research in Environmental Mitigation, National Harbor, MD, 14 May 2009; Presentation by R.A. Wahls; http://www.aeronautics.nasa.gov/calendar/20090514.htm
5. Mankins, J.C.,"Technology Readiness Levels," http://www.hq.nasa.gov/office/codeq/trl
6. Aerospace International, *The Green Issue,* Aerosociety, U.K., March 2009.
7. Smith, M.J.T., *Aircraft Noise,* Cambridge University Press, Cambridge, U.K., 1989.
8. Erickson, J.D., "Environmental Acceptability" Office of Environment and Energy, Presented to FAA, 2000.
9. http://silentaircraft.org/
10. Reynolds, T.G., "Environmental Challenges for Aviation – An Overview," Presented to Low Cost Air Transport Summit, London, 11-12 June 2008.
11. www.iea.org
12. Lee, J.J., Lukachko, S.P., Waitz, I. A., and Schafer, A., "Historical & Future Trends in Aircraft Performance, Cost, and Emissions," Annu. Rev. Energy Environ, Vol. 26, pp. 167-200, 2001.
13. Penner, J.E., *Aviation and the Global Atmosphere,* Cambridge University Press, Cambridge, U.K., pp. 76-79, 1999.
14. Royal Aeronautical Society Annual Report, "Air travel - Greener by Design Annual Report 2007-2008," April 2008 (http://www.greenerbydesign.org.uk/).
15. Schumann, U., "On Conditions for Contrail from Aircraft Exhaust," Meteor. Zeitsch, Vol. 5, pp. 3-22, 1996.
16. NRC Meeting of Experts on NASA's Plans for System-Level Research in Environmental Mitigation, National Harbor, MD, 14 May 2009; Presentation by A. Strazisar; http://www.aeronautics.nasa.gov/calendar/20090514.htm
17. www.boeing.com
18. www.b-domke.de/AviationImages/Propfan/0815
19. www.flightglobal.com/articles/2007/06/12/214520
20. http://www.dfrc.nasa.gov/Gallery/Photo/X-48B/HTML/ED08-0092-13.html
21. http://hondajet.honda.com/
22. Saeed, T.I, Graham, W.R., Babinsky, H., Eastwood, J.P., Hall, C.A., Jarrett, J.P., Lone, M.M. and Seffen, K.A., "Conceptual Design of a Laminar Flying Wing Aircraft," AIAA 2009-3616, 27th AIAA Applied

Aerodynamics Conference, San Antonio, TX, 22-25 June 2009.

23, Collier, F.S., NASA Langley, "Progress in Environmental Aeronautics," Presentation at Aviation & Environment – A Primer for North American Stakeholders Meeting; http://www.airlines.org/NR/rdonlyres/A78FA93B-986C-4D95-BA87-B4DD961CC369/0/11collier.pdf

24. National Academy of Science (NAS) Report, "Assessing the Research and Development plan for the Next Generation Air Transportation System: Summary of a Workshop," (http://www.nap.edu/catalog/12447.html), 2008.

25. Hahn. A.S., "Staging Airliner Service," AIAA 2007-7759, 7th AIAA ATIO Conference, Belfast, 18-20 Sept. 2007.

26. Creemers, W.L.H. and Slingerland, R., "Impact of Intermediate Stops on Long-Range Jet-Transport Design," AIAA 2007-7849, 7th AIAA ATIO Conference, Belfast, 18-20 Sept. 2007.

27. Nangia, R.K., "Air to Air Refueling in Civil Aviation," Paper #9, Royal Aeronautical Soc. "Greener by Design" Conference, London, 7 October 2008.

28. Nangia, R.K., "Way Forward to a Step Jump for Highly Efficient & Greener Civil Aviation – An Opportunity for the Present and a Vision for the Future," Personal Publication RKN-SP-2008-120, September 2008.

29. Wagner, E., Jacques, D., Blake, W., and Pachter, M., "Flight Test Results for Close Formation Flight for Fuel Savings," AIAA 2002-4490, AIAA Atmospheric Flight Mech. Conf., Monterey, CA, 5-8 August 2002.

30. Jenkinson, L.R., Caves, R.E, and Rhodes, D.R., "A Preliminary Investigation into the Application of Formation Flying to Civil Operation," AIAA 1995-3898, 1995.

31. Bower, G.C., Flanzer, T.C. and Kroo, I.M., "Formation Geometries and Route Optimization for Commercial Formation Flight," AIAA 2009-3615, 27th AIAA Applied Aerodynamics Conference, San Antonio, TX, 22-25 June 2009.

32. Boeing Presentation at Paris Air Show by Billy Glover, June 2009 (http://www.boeing.com/paris2009/media/presentation/june17/glover_enviro_briefing/).

33. www.boeing.com

34. www.solarimpulse.com

35. http://www.planetforlife.com/h2/h2vehicle.html

36. Penner, J.E., Aviation and the Global Atmosphere, Cambridge University Press, Cambridge, U.K., p. 257, 1999.

37. Reynolds, T.G., Barrett, S., Dray, L.M., Evans, A.D., Kohler, M.O., Morales, M.V., Schafer, A., Wadud, Z., Britter, R., Hallam, H., and Hunsley, R., " Modeling Environmental & Economic Impacts of Aviation: Introducing the Aviation Integrated Modeling Project," AIAA 2007-7751; 7th AIAA Aviation Technology, Integration and Operations Conference, Belfast, 18-20 Sept. 2007.

38. http://www.cleanairports.com

CONTACT

Ramesh Agarwal
Washington University in St. Louis
Phone: 314-935-6091
Email: rka@wustl.edu

NEW TECHNOLOGIES

In-Vehicle Networking Technology for 2010 and Beyond	2010-01-0687 Published 04/12/2010

Christopher A. Lupini
Delphi Corporation

ABSTRACT

This paper is an overview of the current state (calendar year 2010) of in-vehicle multiplexing and what pertinent technologies are emerging. Usage and trends of in-vehicle networking protocols will be presented and categorized. The past few years have seen a large growth in the number and type of communication buses used in automobiles, trucks, construction equipment, and military, among others. Development continues even into boating and recreation vehicles. Areas for discussion will include SAE Class A, B, C, Diagnostics, SafetyBus, Mobile Media, Wireless, and X-by-Wire. All existing mainstream vehicular multiplex protocols (approximately 40) are categorized using the SAE convention as well as categories previously proposed by this author. Top contenders will be pointed out along with a discussion of the protocol in the best position to become the industry standard in each category.

INTRODUCTION

CURRENT MULTIPLEX CATEGORIES

The multiplexing of automotive electrical data onto communication buses dates back to the late 1970s. It was originally hoped that a single bus protocol could handle the needs of any vehicle. Gradually that expanded to the SAE categorization of Class A, B, and C and the realization that up to three protocols and/or networks may be necessary.

By 1995 the need for multiple buses per vehicle was becoming apparent [1]. The cost tradeoff, especially, was studied - do you put everything on one bus or split it up into several buses? Which is more economical? Which is more efficient?

This paper continues to discuss the idea that multiple in-vehicle networks will be necessary - mainly on high-end

vehicles in the coming years. These categories include (besides the existing SAE classes) diagnostics, airbag, mobile media, X-by-Wire, and wireless. Each area needs its own protocol and one or more networks running that protocol. Sometimes this is for safety reasons, such as with airbags or X-by-Wire. Regardless of vehicle function partitioning, we now have distinct classes of signals that will communicate over their own network, or networks (i.e. multiple sub-buses for "smart connector/actuator") [2].

Although not discussed at length in this paper, there is a distinction between protocol and network. Conceivably one might have the same protocol running on several networks - say CAN for both a body bus and a powertrain bus. So even though there could be eight or more networks, there may actually be fewer protocols used. Also, not all protocols are complete - meaning they specify attributes of all seven layers of the OSI model [3]. Some are only physical layers (i.e. GM UART, J1708). Some are only higher layers (i.e. TTP).

BACKGROUND

Much as changed in the area of in-vehicle multiplexing since the late 1970s. Incredible progress has been made, and much work remains to be done. Serial data communication continues to be the "glue that holds everything together" for automotive electronics but cost has been and remains the primary hurdle. Quite a few late model vehicles now have three, four, or more serial data networks present.

CURRENT STATUS OF THE IN-VEHICLE CATEGORIES

CLASS A

Usage is for low-end, non-emission diagnostic, general-purpose communication. Bit rate is generally less than 10 Kb/s and must support event-driven message transmission. Cost is generally about "x" adder per node. These days a very rough estimate of $0.50 to $1 may be used for the value of "x". This cost includes any silicon involved (i.e. microprocessor module or transceiver, etc.), software, connector pin(s), service, etc. The "cost" data discussed in this paper is for estimate purposes and is only to be used to compare with other categories.

There have been great strides in this area and LIN has become the defacto standard. Proprietary protocols continue to disappear. Figure 1 illustrates a typical example of LIN reducing the wire count into a door from dozens of wires to a minimum of three (LIN, power, ground).

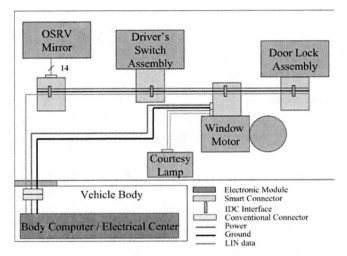

Figure 1. Multiplexed Car Door Example

Some examples of Class A protocols are listed in Table 1a.

<table 1a here>

Most of these Class A protocols are UARTs. UART is very simple and economical to implement. Most microcontrollers have the necessary SCI module built-in, or it can be implemented without a microprocessor. The transceiver is smaller and cheaper than those of other protocols. The transceiver IC may be a custom chip combining multi-protocol capability with regulators, drivers, etc. Right now the leading candidate for a Class A world standard is LIN [4].

Table 1b compares some of the major attributes of the Class A protocols from Table 1a.

<table 1b here>

CLASS B

The usage here is for the vast majority of non-diagnostic, non-critical communication. Speed is between 10 Kb/s and approximately 125 Kb/s. Must support event-driven and some periodic message transmission plus sleep/wakeup. Cost is around 2x per node. Protocols used for Class B networks are listed in Table 2a.

<table 2a here>

The world standard in this area is still CAN. In particular, ISO 11898-2 or ISO 11898-3 at around 80-125 Kb/s for car applications and J1939 at 250 Kb/s for Truck & Bus applications. Both of these use the same digital circuitry and transceiver in many cases. J1850 continues its usage and has actually started a steep decline in volume. The "usual" passenger 500 Kb/s CAN networks are more Class C than Class B.

The ISO 11898-3 (old 11519-2) "fault-tolerant" low speed 2-wire CAN interface [5] found a niche in some car applications, but is still a small percentage of implementations. This CAN physical layer is slower and costs more than an ISO 11898-2 interface, but the bus fault detection capability is enticing. Table 2b compares some of the major attributes of the Class B protocols from Table 2a.

<table 2b here>

CLASS C

Usage is for faster, higher bandwidth systems such as engine timing, fuel delivery, etc. Bit rate is between 125 Kb/s and 1 Mb/s. Must support real-time periodic parameter transmission (perhaps in the few milliseconds range). Unshielded twisted pair is the medium of choice instead of shielded twisted pair or fiber optics. Cost is about 3x to 4x per node, unless STP or fiber optics is involved - which is typically necessary above 500 Kb/s. Some example protocols used are listed in Table 3a.

<table 3a here>

J1939 is commonly used for Class B and Class C applications for truck & bus, construction, agriculture, marine, and other industries. Most passenger car applications run ISO 11898-2 at 500 Kb/s for their Class C network. The big difference from CAN in Class B applications is the type of nodes that are connected. Total CAN usage, according to CAN in Automation (CiA) is in the hundreds of millions of nodes worldwide.

Table 3b compares some of the major attributes of the Class C protocols from Table 3a. GMLAN is not shown due to its confidentiality to GM.

<table 3b here>

EMISSIONS DIAGNOSTICS

Usage is to satisfy OBD-II, OBD-III, E-OBD, Tier-2, etc. so must be a "legally acceptable" protocol. Protocols used today are listed in Table 4a. There is overlap with some of the other categories.

<table 4 a here>

Since this data link is only needed between the engine controller and the off-board connector, a simple approach is sufficient. Most automakers and truck makers are using KW2000 (ISO 14230) and this has been an emissions diagnostic standard. In the U.S., high-speed CAN was phased-in as the "OBD-III" emissions test interface beginning in MY04 (Model Year 2004). This has been the only legally acceptable protocol since MY07. SAE J2480 was an initiative to develop a CAN emissions diagnostic interface, but it was found to overlap with ISO 15765 so it has been abandoned. General information on these protocols is shown in Table 4b.

<table 4b here>

MOBILE MEDIA

Usage is for "PC-on-wheels" applications. At least two different networks and protocols may be necessary. These sub-categories will be referred to as low speed and high speed. Beginning with this paper wireless is now in its own category. The necessary bit rate for mobile media applications is between 1 Mb/s and 100 Mb/s+.

Example usage is illustrated by the Figure 2.

<figure 2 here>

Low Speed Mobile Media - Usage is for telematics, diagnostics, and general information passing. Cost is around 3x per node. IDB-C, a token-passing form of CAN at 250 Kb/s, has fallen out of favor. Most OEMs already have a mid-speed bus based on CAN to handle low-end telematics communication functions. There has been recent interest in a lower cost "high-end" network that can handle digital audio streams, but not necessarily video. Toward that end, the D2B and MOST developers are working on copper-based solutions. Depending on the EMC performance, 10 Mb/s, 25 Mb/s, or even 50 Mb/s bit rates are being used. Table 5a lists some possible low-speed mobile media protocols.

<table 5a here>

High Speed Mobile Media - Usage is for real-time audio and video streaming. Cost is around 15x to 25x, mainly due to fiber optics. Fiber optics will be necessary due to the high speed required to pass real-time video streams from multiple sources to multiple outputs. Sometimes has been compatible with industry-standard systems such as Connected Car PC or AutoPC. D2B has seen the first usage (Mercedes 1999 S-class) but MOST appears to be the top contender at this time. One of the better choices in this area is Firewire via an effort led by the IEEE 1394 Automotive working group. Ethernet is beginning to be discussed as one of the candidates. Table 5b lists some possible high-speed mobile media protocols. Table 5c is a summary of details on these methods.

<table 5b here>

<table 5c here>

Wireless

Usage is quite pervasive, but a single standard has not emerged. Will be necessary (initially) for cell phones and palm PCs (PDAs). Eventual use may include cameras, pagers, etc. Cost target is around 5x per node. Figure 3 is just an idea of the type of products that could be connected via RF.

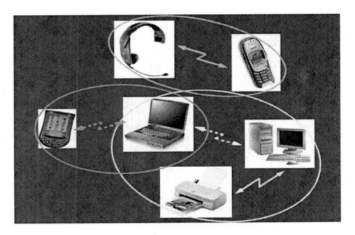

Figure 3. Wireless Devices in a Vehicle

Much of the advertisement attention has been with Bluetooth. However, 802.11 also has its proponents, so many groups and suppliers are studying co-location so that products containing either standard can exist near each other. Ultrawideband (UWB) is still seen as a possibility for future in-car communication. Approved by the U. S. Federal Communications Commission (FCC) in February 2002, it is essentially "white noise" communication. Using precise clocking, tiny amounts of information are transported across a very wide range of frequencies at very low power (perhaps 1/10000 that of a cell phone). Compared with spread spectrum which uses a small range of frequencies one at a time, UWB uses a wide range of frequencies all at once. Leading wireless protocols are listed in Table 6.

<table 6 here>

SAFETYBUS

Usage is for airbag systems. There may be two, or more, buses such as one for firing and one for sensing. Must support at least 64 nodes consisting of squibs, accelerometers, occupant sensors, seatbelt pretensioners, etc. Cost is (hoped to be) 1x to 2x per node. The USCAR "SafetyBus" committee was attempting to standardize on a suitable protocol, but degraded into separate, independent, camps. The two main ones are Safe-by-Wire and BST. Byteflight is in production in at least one BMW vehicle. Table 7a is the list of current airbag network protocols.

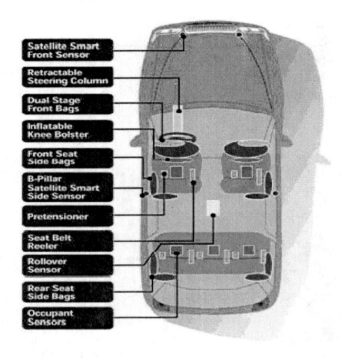

<table 7a here>

Many issues here involving packaging constraints, existing mechanical envelop, legalities, etc. The winning protocol may well be a hybrid of several existing proposals. In fact, BST and Safe-by-Wire are actually hybrids of earlier protocols. For now there is no clear industry direction. Table 7b compares some of these protocols.

<table 7b here>

DRIVE-BY-WIRE

Usage is for brake-by-wire, throttle-by-wire, steer-by-wire, etc. applications. Bit rate is between 1 Mb/s and 10 Mb/s. Fiber optics will be necessary due to the increased speed. The utmost in reliability, performance, and real-time capability is required. Cost is around 15x+ per node.

Some possible candidate protocols are given in Table 8a.

<table 8a here>

TTP had the momentum during the 1990s, but CAN was usually implemented to avoid the costs of time-triggered protocols). FlexRay continues to win support and went into production with BMW for 2007 model year. A major issue is how much fault tolerance is really required. Any scheme will require dual bus interfaces, dual microprocessors, bus watchdogs, timers, etc. Cost is a big problem. The level of fault-tolerance needed requires a lot of silicon and software which, of course, is expensive. The consortium TTAgroup (www.ttagroup.org) was trying to standardize on a protocol. Table 8b is a comparison of these protocols' details.

<table 8b here>

ISSUES

More networks brings more cost. Obviously the number of ECUs per vehicle can't increase forever. An interesting way to cut down on the number of ECUs is to bundle functions into something called "domain controllers". This is a concept that has been talked about for awhile - sometimes called "regional computing" or generic "electric and electronic controllers (EECs). More progress needs to be made in vehicle electronic architectures. The industry has quite of experience in gateways, but not in routers, backplanes, or backbones.

Another question is what networks will be needed to support hybrid and full electric vehicles? Many of the same protocols will suffice, but connecting dozens of battery packs or cells together, for example, is a new challenge.

SUMMARY/CONCLUSIONS

Multiple buses per vehicle have been a reality for some time. Primarily this has been because of re-use, and carryover of existing networks and protocols. However, in the near future new functions such as smart connectors, drive-by-wire, and mobile multi media enhancements have been forcing the need for additional protocols and networks and the means to route data between them. Despite the best intentions of OEMs, this will continue. There is no "one bus fits-all". Instead, there are different buses for different things.

REFERENCES

1. SAE 950293 - "Aspects and Issues of Multiple Vehicle Networks" Emaus

2. SAE 2001-01-0072 "LIN Bus and its Potential for Use in Distributed Multiplex Systems" Ewbank, Lupini, Perisho, DeNuto, Kleja

3. ISO 7498 - Data Processing Systems, Open Systems Interconnection Standard Reference Mode.

4. www.lin-subbus.org

5. "A CAN-Based Architecture for Highly Reliable Communication Systems" Hilmer, Kochs, Dittmar. Proceedings from 5th ICC 1999

6. "A Double CAN Architecture for Fault-Tolerant Control Systems" Ferriol, F. Navio, Pons J. Navio.. Proenza, Julia. Proceedings from 5th ICC 1999

7. "Automakers Aim to Simplify Electronic Architectures" Murray Charles, Electronic Design News July 2009

CONTACT INFORMATION

Christopher A. Lupini
christopher.a.lupini@delphi.com
www.delphi.com
765-451-3207

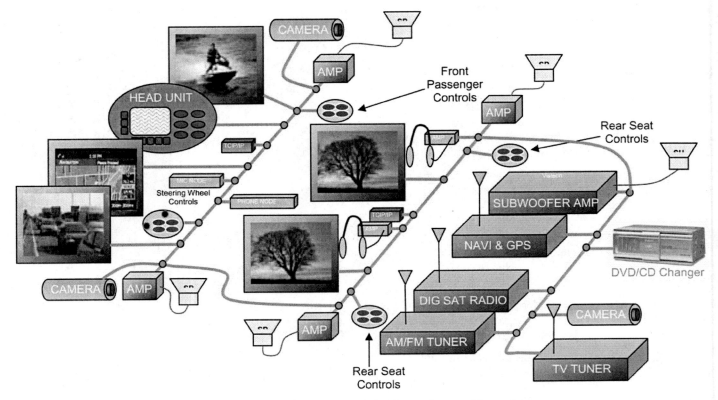

Figure 2. Possible Mobile Media Entertainment Network

Table 1a. Some Class A Protocols

NAME:	USER:	USAGE:	MODEL YEARS:	COMMENTS:
UART	GM	Many	1985 - 2005+	Being phased out
Sinebus	GM	Audio	2000+	Radio steering wheel controls
E&C	GM	Audio/HVAC	1987 - 2002+	Being phased out
I²C	Renault	HVAC	2000+	Used little
J1708/J1587/J1922	T&B	General	1985 - 2002+	Being phased out
CCD	Chrysler	HVAC, audio, etc.	1985 - 2002+	Being phased out
ACP	Ford	Audio	1985 - 2002+	
BEAN	Toyota	Body	1995+	
UBP	Ford	Rear backup	2000+	
LIN	many OEMs	Smart Connector	2003+	LIN Consortium developing

FEATURE	UART(ALDL)	SINEBUS	E & C	I²C	SAE J1708	ACP	BEAN	LIN
	BUS NAME							
AFFILIATION	GM	DELCO	GM	PHILIPS	TMC - ATA	FORD	TOYOTA	Motorola
APPLICATION	GENERAL & DIAGNOSTICS	AUDIO	GENERAL		CONTROL & DIAGNOSTICS	AUDIO CONTROL	BODY CONTROL & DIAGNOSTICS	SMART SENSORS
MEDIA	SINGLE WIRE	SINGLE WIRE	SINGLE WIRE	TWISTED PAIR	TWISTED PAIR	TWISTED PAIR	SINGLE WIRE	SINGLE WIRE
BIT ENCODING	NRZ	SAM	PWM	AM	NRZ	NRZ	NRZ	NRZ
MEDIA ACCESS	MASTER/ SLAVE	MASTER/ SLAVE	CONTENTION		MASTER/ SLAVE	MASTER/ SLAVE	CONTENTION	MASTER/ SLAVE
ERROR DETECTION	8-bit CS	NONE	PARITY	ACK bit	8-bit CS	8-bit CS	8-bit CRC	8-bit CS
HEADER LENGTH	16 BITS	2 BITS	11 - 12 BITS		16 BITS	12 - 24 BITS	25 BITS	2 BITS/BYTE
DATA LENGTH	0 - 85 BYTES	10 - 18 bits	1 - 8 BITS			6 - 12 BYTES	1 - 11 BYTES	8 BYTES
OVERHEAD	Variable	75 %	Variable	45 %	Variable	25 %	28 %	2 BYTES
IN-MESSAGE RESPONSE	NO	NO	NO		NO	NO	NO	NO
BIT RATE	8192 b/s	66.6 KHz – 200 KHz	1000 b/s	1 - 100 Kb/s	9600	9600 b/s	10 Kb/s	20 Kb/s
MAXIMUM BUS LENGTH	Not Specified	10 METERS	20 METERS	Not Specified	Not Specified	40 METERS	Not Specified	40 METERS
MAXIMUM NODES	10		10			20	20	16
µ NEEDED?	YES	NO	YES		YES	YES	YES	NO
SLEEP/WAKEUP	NO	NO	NO		NO	NO	NO	
H/W AVAIL?	YES	NO	YES		YES	YES	YES (?)	NO
COST	LOW	LOW	LOW		MEDIUM	LOW	LOW	LOW

Table 2a. Some Class B Protocols

NAME:	USER:	USAGE:	MODEL YEARS:	COMMENTS:
GMLAN (SWC)	GM	Many	2002+	GM only user; J2411
GMLAN (mid)	GM	Infotainment	2002+	ISO 11898-2 CAN
ISO 11898	Europe	Many	1992+	Various speeds – 47.6 Kb/s to 500 Kb/s in use
J2284	GM, Ford, Chry.	Many	2001+	125, 250 Kb/s; based on ISO 11898
Fault-tol CAN	Europe	Many	2000+	ISO 11898-3 CAN
Class 2	GM	Many	Until 2002+	J1850; being phased out
PCI	Chrysler	Many	Until 2002+	J1850
SCP	Ford	Many	Until 2002+	J1850
J1939	T&B	Many	1994+	Replacing J1708/1587/1922

Table 2b. Comparison of Class B Protocols

FEATURE	SINGLE-WIRE CAN (SWC)	CAN 2.0 ISO 11898-1,2,3 ISO 11992 J2284	J1850 ISO 11519-4			SAE J 1939
AFFILIATION	SAE/ISO	BOSCH/SAE/ISO	GM	FORD	CHRYSLER	TMC - ATA
APPLICATION	DIAGNOSTICS	CONTROL & DIAGNOSTICS	GENERAL & DIAGNOSTICS	GENERAL & DIAGNOSTICS	GENERAL & DIAGNOSTICS	CONTROL & DIAGNOSTICS
TRANSMISSION MEDIA	SINGLE WIRE	TWISTED PAIR	SINGLE WIRE	TWISTED PAIR	SINGLE WIRE	TWISTED PAIR
BIT ENCODING	NRZ-5 MSb first	NRZ-5 MSb first	VPW MSb first	PWM MSb first	VPW MSb first	NRZ-5 MSb first
MEDIA ACCESS	CONTENTION	CONTENTION	CONTENTION	CONTENTION	CONTENTION	CONTENTION
ERROR DETECTION	CRC	CRC	CRC	CRC	CRC	CRC
HEADER LENGTH	11 BITS	11 or 29 BITS	32 BITS	32 BITS	8 BITS	29 BITS
DATA FIELD LENGTH	0-8 BYTES	0-8 BYTES	0-8 BYTES	0-8 BYTES	0-10 BYTE	8 BYTES
MESSAGE OVERHEAD	9.9 %	9.9 % - 22 %	33.3 %	33.3 %	8.3 %	9.9 % - 22 %
IN-MESSAGE RESPONSE	NO	NO	Optional Normally NO	Optional Normally YES	Optional Normally YES	NO
BIT RATE	33.33 Kb/s 83.33 Kb/s	10 Kb/s to 1 Mb/s	10.4 K b/s	41.6 K b/s	10.4 K b/s	250 Kb/s
MAXIMUM BUS LENGTH	30 METERS	Not Specified 40 (Typical)	35 METERS (5 Meters for scan tool)	35 METERS (5 Meters for scan tool)	35 METERS (5 Meters for scan tool)	40 METERS
MAXIMUM NODES	16	Not Specified 32 (Typical)	32	32	32	30 FOR STP 10 FOR UTP
µ NEEDED?	YES	YES	YES	YES	YES	YES
SLEEP/WAKEUP	YES	NO	YES	NO	NO	NO
H/W AVAIL?	NO	YES	YES	YES	YES	YES
COST	LOW	MEDIUM	LOW	LOW	LOW	MEDIUM

Table 3a. Some Class C Protocol

NAME:	USER:	USAGE:	MODEL YEARS:	COMMENTS:
GMLAN (high)	GM	All	2002+	500 Kb/s; J2284
HSCAN	Ford	Various	2004+	500 Kb/s; J2284
HSCAN	Chrysler	Various	2004+	500 Kb/s; J2284
ISO 11898	Europe	Most	1992+	Various speeds of CAN
J1939	T&B	Most	1994+	250 Kb/s CAN

Table 3b. Comparison of Class C Protocols

FEATURE	BUS NAME		
	CAN 2.0 ISO 11898 ISO 11519-2 ISO 11992 J2284 J1939	SAE J1939	Intellibus
AFFILIATION	BOSCH/SAE/ISO	TMC - ATA	Boeing/SAE
APPLICATION	CONTROL & DIAGNOSTICS	CONTROL & DIAGNOSTICS	CONTROL & DIAGNOSTICS
TRANSMISSION MEDIA	TWISTED PAIR	TWISTED PAIR	TWISTED PAIR
BIT ENCODING	NRZ-5 MSb first	NRZ-5 MSb first	Manchester Bi-phase
MEDIA ACCESS	CONTENTION	CONTENTION	Master/Slave
ERROR DETECTION	CRC	CRC	CRC, Parity
HEADER LENGTH	11 or 29 BITS	29 BITS	16 - 48 Bits
DATA FIELD LENGTH	0-8 BYTES 11 or 29-bit ID	MOST ARE 8 BYTES 29-bit ID	0 - 32 Bytes
MESSAGE OVERHEAD	9.9 % - 22 %	9.9 % - 22 %	28% - 75%
IN-MESSAGE RESPONSE	NO	NO	Optional
BIT RATE	10 Kb/s to 1 Mb/s	250 Kb/s	12.5 Mb/s
MAXIMUM BUS LENGTH	Not Specified 40 (Typical)	40 METERS	30 METERS
MAXIMUM NODES	Not Specified 32 (Typical)	30 W/ SHIELDED TWISTED PAIR 10 W/ UNSHIELDED TP	64
µ NEEDED?	YES	YES	NO
SLEEP/WAKEUP	NO	NO	YES
H/W AVAIL?	YES	YES	FPGA
COST	MEDIUM	MEDIUM	MEDIUM

Table 4a. Some Emission Diagnostics Protocols

NAME:	USER:	USAGE	MODEL YEARS:	COMMENTS:
ISO 15765-4	Europe	E-OBD	2000+	E-OBD CAN
J 1850	GM, Ford, Chry.	OBD-II	1994+	Not accepted in Europe
ISO 9141-2	Europe	OBD-II, general	1994+	Old OBD-II UART
ISO 14230-4	Many	OBD-II, OBD-III	2000+	Keyword 2000

Table 4b. Comparison of Emission Diagnostics Protocols

FEATURE	BUS NAME					
	ISO 15765	J1850 ISO 11519-4			ISO/DIS 9141 ISO/DIS 9141-2	KEYWORD XX
AFFILIATION	ISO	GM	FORD	CHRYSLER	WORLD	Various
APPLICATION	EMISSIONS DIAGNOSTICS	GENERAL & DIAGNOSTICS	GENERAL & DIAGNOSTICS	GENERAL & DIAGNOSTICS	DIAGNOSTICS ONLY	DIAGNOSTICS
TRANSMISSION MEDIA	TWISTED PAIR	SINGLE WIRE	TWISTED PAIR	SINGLE WIRE	SINGLE WIRE	1-WIRE
BIT ENCODING	NRZ	VPW MSb first	PWM MSb first	VPW MSb first	NRZ (strt, 7D, P, stop) LSb first	NRZ
MEDIA ACCESS	CONTENTION	CONTENTION	CONTENTION	CONTENTION	TESTER/SLAVE	MASTER/ SLAVE
ERROR DETECTION	CRC	CRC	CRC	CRC	PARITY (odd)	x-bit CS
HEADER LENGTH	11 and 29-BITS	32 BITS	32 BITS	8 BITS	Not Specified	16 BITS
DATA FIELD LENGTH	8 BYTES	0-8 BYTES	0-8 BYTES	0-10 BYTE	Not Specified	0 - 85 BYTES
MESSAGE OVERHEAD	About 50%	33.3 %	33.3 %	8.3 %	Variable	Variable
IN-MESSAGE RESPONSE	NO	Optional Normally NO	Optional Normally YES	Optional Normally YES	NO	NO
BIT RATE	250 or 500 Kb/s	10.4 K b/s	41.6 K b/s	10.4 K b/s	<10.4 Kb/s	5 b/s - 10.4 Kb/s
MAXIMUM BUS LENGTH	40 METERS	35 METERS (5 Meters for scan tool)	35 METERS (5 Meters for scan tool)	35 METERS (5 Meters for scan tool)	Limited by total impedance to ground	Not Specified
MAXIMUM NODES	32	32	32	32	Limited by total impedance to ground	10
µ NEEDED?	YES	YES	YES	YES	YES	YES
SLEEP/ WAKEUP	YES	YES	NO	NO	NO	NO
H/W AVAIL?	YES	YES	YES	YES	YES	YES
COST	LOW	LOW	LOW	LOW	LOW	LOW

Table 5a. Some Low-Speed Mobile Media Bus Protocols

NAME:	USER:	MODEL YEARS:	COMMENTS:
IDB-C	none long term	2002+	250 Kb/s CAN; www.idbforum.org
D2B SmartwireX	unknown	2005+	www.candc.co.uk/candc_company
MOST over copper	unknown	2009+	www.mostcooperation.com

Table 5b. Current High-Speed Mobile Media Bus Protocols

NAME:	USER:	YEAR:	COMMENTS:
D2B	Mercedes, Jaguar	1999+	www.candc.co.uk/candc_company
MOST	BMW, GM, Daimler, Ford, VW, Toyota	2000+	www.mostcooperation.com
Firewire	Chrysler, Ford, Toyota	2007+	www.1394ta.org
USB	Clarion, various autos	1998+	www.autopc.com
IntelliBus	Boeing (no passenger autos)	2004+	www.intellibusnetworks.com
Ethernet	tbd	2014+	Referenced as "Ethernet AV" etc.

Table 5c. Comparison of Mobile Media Protocols

FEATURE	IDB-C	Intellibus	MOST	SmartWireX	MML	USB	IEEE 1394
AFFILIATION	SAE	Boeing/SAE	Oasis	C&C	DELCO	Commercial	IEEE
APPLICATION	Aftermarket Entertainment	CONTROL & DIAGNOSTICS	Stream Data & Control	STREAM DATA & CONTROL	STREAM DATA & CONTROL	PC DEVICES	PC DEVICES
TRANSMISSION MEDIA	2-Wire	TWISTED PAIR	Optical	TWISTED PAIR	OPTICAL FIBER	SHIELDED TWISTED PAIR	SHIELDED TWISTED PAIR
BIT ENCODING	NRZ	Manchester Bi-phase	BiPhase	PWM	NRZ	NRZ	NRZ
MEDIA ACCESS	Token-slot	Master/Slave	Master/Slave	Master/Slave	Master/Slave	Contention	Contention
ERROR DETECTION	15-bit CRC	CRC, Parity	CRC	Parity	CORRECTING (optional)	CRC	CRC
HEADER LENGTH	11 BITS	16 - 48 Bits			1 BYTE		
DATA LENGTH	8 BYTES	0 - 32 Bytes			1 - 200+ BYTES		
MESSAGE OVERHEAD	~ 32 BITS	28% - 75%			5 - 10 %	25 %	25 - 30 %
IN-MESSAGE ACK.	1 ACK BIT	Optional	No	No	No		
BIT RATE	250 Kb/s	30 Mb/s	25 Mb/s	tbd kb/s	110 Mb/s	12 Mb/s	98 - 393 Mb/s
MAXIMUM BUS LENGTH	TBD	30 METERS	TBD	150 METERS	10 METERS		72 METERS
MAXIMUM NODES	16	64	24	50	16	127	16
μ NEEDED?	YES	NO	YES	YES	YES	YES	YES
SLEEP/WAKEUP	YES	YES	YES	YES	YES	NO	NO
H/W AVAIL?	NO	FPGA	YES	YES	NO	YES	YES
COST	LOW	LOW	HIGH	HIGH	HIGH	MEDIUM	MEDIUM

Table 6. Current Wireless Mobile Media Bus Protocols

NAME:	USER:	MODEL YEARS:	COMMENTS:
Bluetooth	tbd	2005+	**www.bluetooth.com**
IEEE 802.1	tbd	tbd	**www.ieee802.org/11**
UWB	tbd	tbd	**www.uwb.org**

Table 7a. Current SafetyBus Protocols

NAME:	USER:	YEARS:	COMMENTS:
Safe-by-Wire	tbd	2002+	Delphi-TRW-Philips-Autoliv-SDI
BST	tbd	2002+	Bosch-Siemens-Temic
DSI	tbd	2002+	Motorola/AMP
Byteflight	BMW	2002+	"ISIS", SI
PSI-5	many	2006+	www.psi5.org

Table 7b. Comparison of SafetyBus Protocols

FEATURE	BUS NAME			
	BST	SafeByWire	DSI	Byteflight
AFFILIATION	Bosch-Siemens-Temic	Delphi-Philips-TRW-Autoliv-SDI	Motorola	BMW
APPLICATION	Airbag	Airbag	Airbag	Airbag
TRANSMISSION MEDIA	2-WIRE	2-WIRE	2-WIRE	2-WIRE or 3-WIRE or optical
BIT ENCODING	Manchester Biphase	3-level voltage	3-level voltage	
MEDIA ACCESS	MASTER/ SLAVE	MASTER/ SLAVE	MASTER/ SLAVE	MASTER/ SLAVE
ERROR DETECTION	Odd Parity and/or CRC	8-bit CRC	4-bit CRC	16-bit CRC
HEADER LENGTH	Various	Various	Various	Various
DATA FIELD LENGTH	1 byte	1 byte	1 - 2 bytes	1 byte
MESSAGE OVERHEAD				
IN-MESSAGE ACK.	NO	NO	NO	NO
BIT RATE	31.25 Kb/s, 125 Kb/s, 250 Kb/s	150 Kb/s	5 Kb/s - 150 Kb/s	10 Mb/s
MAXIMUM BUS LENGTH	TBD	25 - 40 m	TBD	TBD
MAXIMUM NODES	12 squibs, 62 slaves	64	16	
µ NEEDED?	NO	NO	NO	NO
SLEEP/WAKEUP	NO	NO	NO	NO
H/W AVAIL?	YES	YES	YES	YES
COST	LOW	LOW	LOW	LOW

Table 8a. Current Drive-by-Wire Protocols

NAME:	USER:	YEARS:	COMMENTS:
TTP/C	BMW, Audi	2004+	www.tttech.com
TTCAN	tbd	tbd	www.can-cia.de
FlexRay	BMW, Audi, others	2004+	www.flexray-group.com

Table 8b. Comparison of Drive-by-Wire Protocols

	BUS NAME			
	TTP		**FlexRay**	**TTCAN**
AFFILIATION	U-VIENNA	Boeing/SAE	Motorola	CiA
APPLICATION	Safety Control	Safety Control	Safety Control	Safety Control
TRANSMISSION MEDIA	Not Specified	Twisted pair	Twisted pair or fiber	Twisted pair
BIT ENCODING	Not Specified	Manchester Bi-phase	NRZ	NRZ-5
MEDIA ACCESS	Isochronous	Master/Slave	Time or Priority	Time or Contention
ERROR DETECTION	16-bit CRC	CRC, Parity	24-bit CRC	15-bit CRC
HEADER LENGTH	1 Byte	16 - 48 Bits	40 Bits	11 - 29 Bits
DATA FIELD LENGTH	16 Bytes	0 - 32 Bytes	0 – 246 Bytes	0 - 8 Bytes
MESSAGE OVERHEAD	18.75 %	28% - 75%	3% - 100%	9.9 - 22%
IN-MESSAGE ACK.	YES	Optional	NO	YES
BIT RATE	Not Specified	12.5 Mb/s	10 Mb/s	1 - 2 Mb/s
MAXIMUM BUS LENGTH	Not Specified	30 meters	Not Specified	40 meters
MAXIMUM NODES	Not Specified	64	Not Specified	32
μ NEEDED?	YES	NO	YES	YES
SLEEP/WAKEUP	NO	YES	YES	YES
H/W AVAIL?	NO	FPGA	NO	YES
COST	HIGH	LOW	MEDIUM	LOW

Development of Injector for the Direct Injection Homogeneous Market using Design for Six Sigma	2010-01-0594 Published 04/12/2010

Edwin A. Rivera, Noreen Mastro, James Zizelman, John Kirwan and Robert Ooyama
Delphi Powertrain Systems

ABSTRACT

Gasoline direct injection (GDi) engines have become popular due to their inherent potential for reduction of exhaust emissions and fuel consumption to meet increasingly stringent environmental standards. These engines require high-pressure fuel injection in order to improve the fuel atomization process and accelerate mixture preparation. The injector is a critical part of this system. The injector technology needed to satisfy the market demands is constantly changing.

This paper focuses on how the Design for Six Sigma innovation methodology was successfully used to develop a new injector for the homogeneous direct injection market. The project begins with the work to understand the market needs and market drivers then decomposes those needs into functional requirements and concepts. The concepts are evaluated and the best concept is selected. The project ends with the optimization of the critical functions including fuel flow control and fuel spray control.

INTRODUCTION

Figure 1 indicates the rollout of US and European light-duty emission standards and fuel consumption targets over the next several years. Dramatic fuel consumption / CO_2 reductions are necessary, both near-term and long-term, while tailpipe emission standards are becoming increasingly more stringent. Gasoline direct injection engines are a key technology for improved fuel economy and engine performance. GDi is already available in a number of vehicles and GDi powertrains are being developed and implemented in production vehicles with increased frequency (see, for example, [1,2,3,4,5,6,7,8,9,10]). As a result, the fraction of GDi vehicles worldwide is projected to markedly increase in the coming years (see Figure 2).

GDi offers a number of fundamental advantages that enable improved engine performance compared to traditional port fuel injection. In-cylinder injection offers advantages during engine warm-up. GDi improves fuel control compared to Port Fuel injection (PFI) in a cold engine when fuel vaporization characteristics are compromised in the intake port. GDi also enables a split injection strategy during engine warm-up. Split injection can provide a locally rich mixture near the spark plug. This improves combustion robustness to enable greater spark retard for catalyst heating while the globally leaner mixture provides reduced HC emissions compared to PFI [1 - 2].

Additionally, GDi offers key features to enable improved low end torque with turbocharged engines. Direct injection decouples air delivery from fuel delivery. Because GDi delivers fuel after the exhaust valve closes, improved volumetric efficiency through aggressive scavenging is possible without short-circuiting unburned fuel into the exhaust port. In-cylinder fuel vaporization with GDi also cools the intake charge to reduce knock propensity and allow greater boosting at lower engine speeds. With increased low end torque, homogeneous GDi engines can be downsized and downspeeded for substantial fuel economy benefits for a significant reduction in fuel consumption compared to a PFI engine baseline. Lean, stratified GDi combustion provides even further benefits for reduced fuel consumption such that the CO_2 reduction benefit of a boosted, downsized stratified GDi vehicle approaches that of a turbo-diesel. However, injector requirements are more stringent for stratified combustion to ensure that a combustible fuel air charge is properly prepared in space and time for ignition by the spark plug [11]. This leads to increased cost for stratified GDi fuel systems. Additionally, homogeneous engines are attractive because they typically run at stoichiometric air-fuel ratio, thus avoiding the issues and increased cost of lean NOx aftertreatment facing stratified GDi engines.

This paper describes Delphi's development of the Multec multi-hole homogeneous gasoline direct injector. Design for Six Sigma (DFSS) - IDDOV was used as the primary product development method for this innovation project. First is a general overview of homogeneous GDi with a description of the key injector requirements. This is followed by a description of the steps in the concept development process. These steps include:

• Identify and initiate the project

• Define - understanding market needs and market drivers, decomposition of those needs into functional requirements

• Develop - concept creation, concept evaluation and concept selection

• Optimize - key product functions while reducing sensitivity to noise variation

• Verify and validate

The discussion ends with an overview of the optimization process for critical functions including fuel flow control and fuel spray control.

<figure 1 here>

<figure 2 here>

HOMOGENEOUS GASOLINE DIRECT INJECTION OVERVIEW

GDi imposes substantially greater requirements on fuel delivery compared to PFI. In a PFI engine the fuel is traditionally injected onto the back of a closed intake valve. During standard, warmed-up PFI engine operation, heat from the intake valve rapidly vaporizes the fuel in the port before the fuel and air are simultaneously inducted into the cylinder during the intake stroke. By contrast, fuel vaporization and mixing for GDi engines must occur rapidly in the cylinder. This process is largely influenced by fuel spray characteristics such as droplet size and spray penetration. These characteristics are achieved through careful injection system design with injectors operating at moderate fuel pressures currently extending up to 200 bar.

Figure 3 shows the major components for Delphi homogeneous GDi fuel systems comprising Multec inwardly-opening, multi-hole GDi injectors, a fuel rail and an engine-driven high pressure fuel pump. Key injector requirements for homogeneous GDi injection are the capability to operate at fuel pressures up to 200 bar, good linearity over a wide flow range to ensure precise delivery over the full engine map, low injector noise, and spray generation that provides good vaporization and mixing without wetting in-cylinder surfaces and without after injection caused by valve bounce. Injection is typically during the intake stroke to improve vaporization and mixing.

Two different injector lengths are shown in the figure. Depending on the engine application, the injectors may be in either side-mount (as shown in Figure 4) or central-mount configurations. The longer injector shown in the figure is required for some engines with central-mount injection. Side-mount injectors are frequently easier to package in an engine, but the off-axis mounting location makes uniform mixture preparation more challenging, necessitates significant skew in the spray pattern and increases concerns for impingement of the spray on the cylinder wall or piston top that causes increased smoke emissions. Central-mount injectors have a more symmetric location that improves mixing and generally reduces the potential for impingement. However in-cylinder access through the head often is prohibited due to packaging conflicts with the valvetrain components and spark plug. Regardless of the mounting location, multi-hole injectors produce distinct spray streams from each hole as shown in Figure 5. Characteristics of these streams are specific to a given engine to conform to spray targeting needs, and can differ substantially between applications (see Figure 6). Designing the injector utilizes both experimental and modeling tools to simultaneously optimize the parameters required for spray formation appropriate to the specific engine [12].

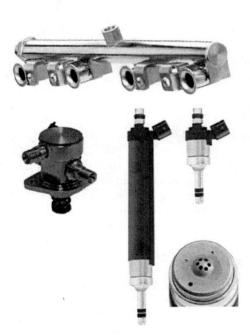

Figure 3. Homogeneous GDi fuel system components.

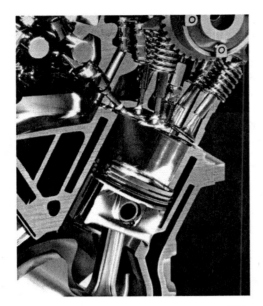

Figure 4. Side-mounted homogenous GDi.

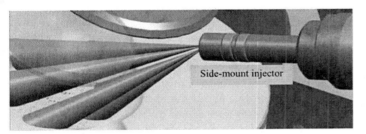

Side-mount injector

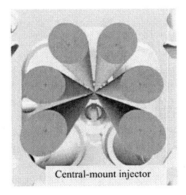

Central-mount injector

Figure 5. Schematic of spray streams for homogeneous GDi injectors.

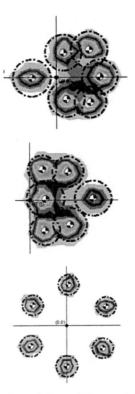

Figure 6. Plan view of three different injector targeting patterns.

DEFINING PROJECT REQUIREMENTS

<figure 7 here>

OVERVIEW OF THE QUALITY FUNCTION DEPLOYMENT PROCESS

In order to connect the voice of the customer and market needs directly to the product Functional Requirements the Quality Function Deployment (QFD) tools were used. QFD is a planning tool used to translate customer needs and expectations into the appropriate design actions. [13] This is accomplished by incorporating the "Voice of the Customer" (VOC) into all phases of the product development. QFD is especially useful for managing large numbers of requirements. The primary objective is to achieve high customer satisfaction. There are four tables (houses) typically used.

• House 1 - is used to translate customer and market needs and expectations into measures that are "critical to satisfaction" (CTS)

• House 2 - is used to translate CTS's to Functional Requirements (FRs)

• House 3 - is used to translate Functional Requirements into product Design Parameters (DPs)

• House 4 - is used to translate Design Parameters into Process Parameters

Figure 7 shows House 1 of the QFD tool, sometimes referred to as the House of Quality, used to begin the Direct Injection EMS project. This EMS level house served as the foundation of our GDi injector development project. Beginning with this EMS project allowed the team to consider what the EMS system requires from a direct injector. This combined with the knowledge gained about the direct injector market from discussions with potential customers allowed the team to form a complete set of needs and requirements.

The QFD House-1 used for the Multec GDi project is shown in Figure 8. It shows the complete list of customer needs for a Homogeneous GDi injector. This simple list is the result of an extensive search internal and external to Delphi. In order to build this House the team used the following steps.

1. Determine the voice of the customer through interviews and definition of market needs. This included visiting vehicle OEMs throughout the world and using Delphi's own EMS project as one of the customers. Initially, special focus was given to European vehicle OEMs and specific OEMs which would supply vehicles to North America. Once the feedback from these OEM's was summarized the team focused on obtaining feedback from the other growing markets. In addition to these interviews the team used a Value Proposition to define the market needs. The Value Proposition is an active list of product characteristics which engine manufacturers view as competitive differentiators.

2. Complete the customer assessment on the right side of the table. This allowed the team to document the customer view from various potential customers at the start of the project. By evaluating the feedback specifically for each major customer need we were able to better prioritize the product development work.

3. Develop CTS measures. This is where the relationship between the market needs and CTSs was developed. Three symbols are used. The solid circle indicates a strong relationship between the market need and the CTS measure. A strong relationship indicates high confidence in being able to measure performance to the specific need or requirement. Symbols for moderate and weak relationships are also used. The focus should be on the strong relationships. Measures were defined that could directly assess each customer need to ensure the team that the performance to customer needs could be quantified.

4. Define targets to CTS measures along the bottom of the table. This area describes project targets developed by the team to define project success.

5. Define correlations between CTS measures; sometimes called the roof of House 1. Some are positive and some are negative. These were later used in conjunction with the Axiomatic Design Matrix in order to identify barriers to functional optimization.

<figure 8 here>

The next major step in the Design for Six Sigma methodology is to translate the CTS's from House 1 to House 2. Once this was complete the development of the FRs (Functional Requirements) was pursued. Defining the FRs proved a significant challenge. It was important for the team to spend time ensuring that the Functional Requirements identified in the QFD House 2 were of the appropriate hierarchy. For example, it would have been a mistake to assess "Actuate Valve" and "Open Valve" on House 2 and consider measures for each at the same functional level. It would drive redundant design parameters and appear in House 3 (the next step) to be coupled (conflicting) functions when in reality one is a subset of the other. From past experience, if the Functional Requirements identified in House 2 are not at the same hierarchy, the development of concepts and the assessment of coupling becomes confused and overly difficult. The team decided to build a function model specifically for the purpose of identifying the correct functional hierarchy. (See Figure 9).

The function model tool is helpful in defining the appropriate relationship between higher and lower level functions by encouraging the user to ask "How and Why". By asking "How" a function is accomplished we build the model to the right. By asking "Why" a function is needed, we check our work to the left. The function model is useful once it has been created, but the team found that the process of creating this function model was most useful because it allowed the team to agree on the fundamental relationship of these high level functions.

<figure 9 here>

Once this function model was completed and the hierarchy was defined the relationship matrix on House 2 was completed. In Figure 10 we see House 2 which defines the relationships between the CTS's and the Functional Requirements. The reader will notice that some CTS's do not have a defined relationship to any of the defined Functional Requirements (i.e., the entire row of the matrix for some CTS's is blank). These CTS's are outside the scope of this paper. In the present analysis, we will focus only on the CTS's in House 2 that are related to injector performance.

<figure 10 here>

As taught by the American Supplier Institute, the steps involving definition of Functional Requirements, Concept Generation and definition of design parameters (DP's) is best accomplished using a method called "zigzagging" [14]. The team initiated the zigzag by defining the higher level FRs, then defining the DP's which pertained to those FRs. The next step was to decompose the FRs and DPs defining lower levels. This zigzag process encourages innovation at the higher functional levels first, and then helps to guide the innovation at lower functional levels. The Axiomatic Design process, developed by Dr. Nam Suh, was used at this time to complete the coupling analysis. (See Figure 11) During this process the team identified several contradictions (or conflicts) listed below.

1. Opening vs. Closing Response

2. Opening vs. Pintle Stroke

3. Pintle Stroke vs. Static Flow

4. Pintle Stroke vs. Audible Noise

5. Seal Diameter vs. Static Flow

6. Opening vs. Seal Diameter

7. Opening vs. Audible Noise

8. Closing vs. Audible Noise

9. Opening vs. Valve Bounce

10. Closing vs. Valve Bounce

11. Opening vs. Durability

12. Closing vs. Durability

13. Seal Diameter vs. Tip Leakage

14. Seal Diameter vs. Spray Quality

<figure 11 here>

The cells in Figure 11 containing an "x" define a direct relationship. The following question was used to determine coupled relationships. "As DP_n is used to optimize FR_n is the ability to optimize other FRs negatively impacted?" If the answer is yes, then an "x" is placed in the box to indicate a direct relationship. The red boxes indicate coupled relationships. This means that more than one design parameter significantly affects more than one function.

In order to develop concepts which could be used to overcome these coupled relationships (also known as contradictions) the team used the TRIZ methodology to define new concepts. Here is a summary of the potential ideas which could be leveraged to improve the concepts. Each idea is followed by the TRIZ principals in parenthesis ( ).

• FR1 Actuate Valve: "Want and Don't Want a Long Stroke"

  ○ Slide Hammer Armature / Pintle (Segmentation)

  ○ Pilot Valve (Segmentation, Preliminary Counteraction)

  ○ Slide Valve (Curvature Increase, Equipotentiality)

  ○ Rotational Valve (Curvature Increase, Equipotentiality)

  ○ Pressure Balanced Valve (Equipotentiality)

  ○ 3-Way Valve / Master-Slave (Hydraulics)

  ○ Compliant Pintle

• FR2 Control Spray Pattern: "Want and Don't Want to Generate Spray at Restriction"

  ○ Seat Geometry Generates Spray Shape / Atomization (Segmentation)

  ○ Shape Change Valve (Parameter Change)

  ○ Swirler (Curvature Increase)

  ○ Deflector (Segmentation)

• FR3 Control Static Flow: "Want and Don't Want to Meter Flow at Spray Generator"

  ○ Moving Seat (Other Way Around) (Do It In Reverse)

  ○ Outward Opening (Segmentation)

  ○ Seat Geometry Generates Spray Shape / Atomization (Segmentation)

  ○ Shape Change Valve (Parameter Change)

  ○ Different pressure drop ratio balance

CONCEPT SELECTION

Upon completion of concept generation, the concept selection development process began. Several of the concepts required development before the selection process could begin. The team used the Pugh Concept Selection tools to aid in the selection of the best concepts. One notable winning concept was the one listed first in the idea list above "Slide Hammer Armature / Pintle (Segmentation)". Some performance advantages of this design are discussed later in this paper under the topic "Minimizing Pintle Bounce".

After selecting concepts for each key functional requirement, the task of developing and optimizing them followed.

DESIGN DEVELOPMENT AND OPTIMIZATION

Taguchi Methods® for robust optimization [15] provides the best strategy for developing designs that are insensitive to sources of variation, at the lowest possible cost. Taguchi's method for parameter design reverses the traditional approach of hitting the target first and reducing variability last. Instead, Taguchi requires that robustness be improved first, and the target adjusted last. This method, called Taguchi 2- Step Parameter Design, provides a more efficient process to the engineer.

Optimization of Meter Fuel

We achieve the high level function "meter fuel" by actuating the valve, controlling static flow, and controlling the spray pattern, as shown in the injector function model, Figure 9. We optimized these functional requirements independently.

Figure 12 shows the ideal and typical valve displacement profiles that occur in response to an input control signal. The "typical" profile identifies the types of deviations that can occur during valve actuation, including delay, fly time (amount of time the valve is in motion) and bounce. These deviations lead to variation in the amount of fuel delivered for a given pulse.

<figure 12 here>

Control Static Flow

Delivery of a consistent amount of fuel when the injector valve is fully open reduces overall variation in fuel metering. A robust optimization study to control static (full open) flow was performed using software simulation.

The scope of this optimization is summarized in the Parameter Diagram (Figure 14), which shows the design variables (control factors), noise (variation), signal and output. Each of the elements in the Parameter Diagram (P-diagram) was chosen after defining an ideal performance function. The ideal function response (Figure 15) represents perfect performance. Using target static flow as the signal, we would expect the actual flow (response) to equal the target if everything were perfect. Therefore, the ideal function in this case is a one-to-one correspondence of target to actual static flow, represented by a straight line with a slope of one.

The control factors studied consisted of design features relating to the metering valve (sketch of metering valve shown in Figure 13).

Noise factors, defined as variables not controllable, or too expensive to control in the design, included the manufacturing tolerances for geometry and flow features. Levels of these noise factors were compounded, or stacked up, to either increase or decrease the flow from its expected value. Compounding noise reduces the number of runs needed in the study. Figure 16 shows the control and noise factors, and number of levels for each.

An L18 orthogonal array was chosen for the design space because it accommodates three-level control factors, and evenly distributes the interaction effects between control factors. This reduces the potential for confounding interaction effects with main effects. Figure 17 shows an L18 array. Static flows were computed for both compounded noises, N1 and N2, at each of the 18 treatment combinations for three static flow signal levels. Signal-to-Noise ratio (S/N) and slope (Beta) were computed from the static flow data at each treatment combination. S/N is a measure of how well the design performs in the presence of noise, i.e. variation. Beta

is a measure of how well the design has achieved the ideal target.

These values of S/N and Beta were compiled into response tables that isolate the main effects of each control factor on the actual static flow (response). Figure 18 graphs the response table results. In this case, B3, D3 and E3 clearly provide the maximum S/N for each of those control factors. Factor C has a flat effect on S/N, so any level can be chosen without affecting robustness to variation, and would thus make a good candidate for adjusting the target.

The second step of the two step optimization is to tune the design parameters to achieve the target. In this case, a slope of one corresponds to a perfect targeting of the static flow. By selecting factors B3 and D3 to maximize S/N, we are moving further from our target flow, however the greater improvement in S/N at these factor levels was deemed more beneficial than the small improvement in targeting. In addition, factor C has minimal effect on S/N within the range studied, and can be adjusted as necessary (i.e. select C3) to improve targeting.

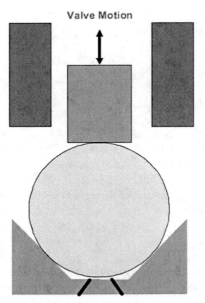

Figure 13. Metering Valve Sketch for Control Static Flow Optimization

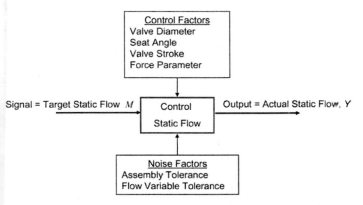

Figure 14. Parameter Diagram for Control Static Flow
Optimization

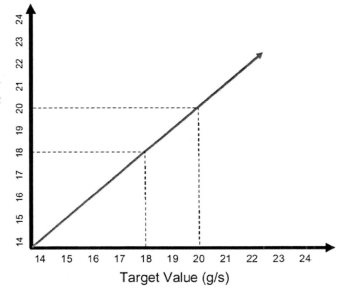

Figure 15. Ideal Function for Static Flow Optimization

	Description	Level 1	Level 2	Level 3
A	Dummy 1	x	x	
B	Valve Diameter	B1	B2	B3
C	Seat Angle	C1	C2	C3
D	Valve Stroke	D1	D2	D3
E	Force Parameter	E1	E2	E3
F	Dummy 2	x	x	x
G	Dummy 3	x	x	x
H	Dummy 4	x	x	x

Noise Factors	Assembly Tolerance	Flow Variable Tolerance
N1 = Lean	LL	LL
N2 = Rich	UL	UL

Figure 16. Control and Noise Factor Levels

	A	B	C	D	E	F	G	H
1.	1	1	1	1	1	1	1	1
2.	1	1	2	2	2	2	2	2
3.	1	1	3	3	3	3	3	3
4.	1	2	1	1	2	2	3	3
5.	1	2	2	2	3	3	1	1
6.	1	2	3	3	1	1	2	2
7.	1	3	1	2	1	3	2	3
8.	1	3	2	3	2	1	3	1
9.	1	3	3	1	3	2	1	2
10.	2	1	1	3	3	2	2	1
11.	2	1	2	1	1	3	3	2
12.	2	1	3	2	2	1	1	3
13.	2	2	1	2	3	1	3	2
14.	2	2	2	3	1	2	1	3
15.	2	2	3	1	2	3	2	1
16.	2	3	1	3	2	3	1	2
17.	2	3	2	1	3	1	2	3
18.	2	3	3	2	1	2	3	1

Figure 17. L18 Array

<figure 18 here>

Actuate Valve

The next functional requirement to be optimized was "actuate valve". This function covers the dynamic aspect of fuel metering. Referring again to Figure 9, delay and fly time contribute to variation in dynamic actuation in both the opening and closing directions. The opening and closing events combined make up the function named "Actuate Valve". In the opening direction, less fuel is delivered than expected due to both delay and fly time, while in the closing direction, excess fuel is delivered. There were several potential candidates for the ideal function in this case, however we recognized from modeling that the majority of fueling error occurred due to the excess fuel delivered during the closing event. Additionally, more variation occurs in the closing response than in the opening response due to expected product and usage variation. (See Figure 19 and Figure 20 for modeling results of opening and closing response variation). We therefore constructed an experiment to minimize excess fuel delivered during closing.

215

Figure 21 shows the ideal function. The target for excess fuel is zero (Slope, Beta = 0). Following the Taguchi method, our first step was to reduce variation of excess closing fuel, and the second step to minimize the amount of excess fuel.

We used simulation of the closing event to perform the experiment. In each run, we maintained a constant closing force margin in order to evaluate each treatment combination on an equal basis. We adjusted the closing force margin to the appropriate level by changing the hold current value just before executing the command to close.

The P-diagram, shown in Figure 22, summarizes the scope of the optimization study. Again, we used a compound noise strategy where we grouped noise factors that cause excess fuel to increase or decrease. We chose an L18 orthogonal array for the design space because it provides for many control factors, and as in the previous optimization, it minimizes confounding of main and interaction effects, and also allows for three-level control factors. Figure 23 shows the response graphs with initial and optimum (maximized S/N) levels indicated.

All but one of the control factor responses exhibits trends that reduce variation while also reducing mean excess fuel. This consistency simplifies the selection of levels. The exception, Factor B, requires a compromise in its selection leading to a tradeoff between variation and targeting.

Figure 24 shows the average excess closing fuel at each static flow for the initial and optimum designs. The excess fuel variation is reduced from the initial levels, and the target values decreased, as desired. Resulting Opening Response and Closing Response times are compared to other injector designs in Figure 25.

<figure 19 here>

<figure 20 here>

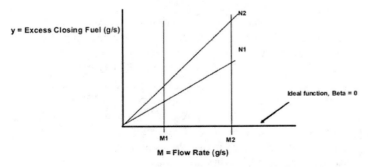

Figure 21. Ideal Function for Actuate Valve Optimization

<figure 22 here>

<figure 23 here>

<figure 24 here>

<figure 25 here>

Control Spray Pattern

The last functional requirement optimized, "control spray pattern", includes fuel spray elements such as fuel mass distribution, atomization and penetration. All are critical to proper engine operation. In addition, design features that control spray pattern can also affect static flow. Figure 26 shows the P-diagram for the study to optimize spray, which defines the scope of the experiment. Because we have known target values for plume location, fuel mass distribution, fuel penetration and static flow, we chose the Nominal-the-Best (Figure 27) ideal function for each of these parameters. This study was done experimentally, and the proprietary results will not be shown here.

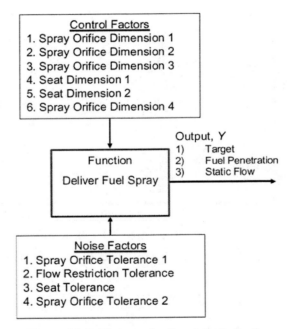

Figure 26. P-Diagram for Spray Optimization

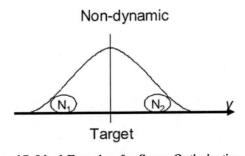

Figure 27. Ideal Function for Spray Optimization Study

Minimizing Bounce

The TRIZ principal of Segmentation led the team to a decoupled armature and valve group concept. This design enabled the "coupled" functions identified through Dr. Nam Suh's Axiomatic Design process to be "decoupled" both physically and functionally (i.e. Armature vs. Valve Group and Opening Response vs. Closing Response)

In simple terms, Opening Response (OR) is a function of magnetic force; Closing Response (CR) is a function of closing spring force. On opening, the initial travel of the armature is independent of the valve group. This slide hammer effect reduces OR. On closing, the closing force acts primarily on the valve group instead of the armature and valve group assembly. The mass of the valve group is much less than the total assembly which reduces CR and audible noise. Bounce was then minimized using Taguchi Methods for Robust Design, all enabled by the decoupled design.

In order to develop an effective design to minimize bounce during valve closing, it was necessary to devise a way to reliably measure how much bounce occurs in hardware. This would allow us to make a quantitative comparison between designs. The methodology developed uses digital imaging, and derives a scalar value that represents the amount of fuel present after the injector valve closes. This scalar value is called bounce flux. A smaller value represents less fuel present, and thus less bounce. Conversely, a larger value represents more fuel present, and thus more bounce. Figure 28 shows the bounce flux measured for the Delphi injector, and others. Vehicle tests showed the affect of bounce on emissions. Figure 29 shows cumulative HC emissions when using injectors with large bounce flux compared to those with small bounce flux. There is a clear improvement when bounce is minimized. This vehicle test comparison also establishes bounce flux thresholds for good and bad emissions performance.

<figure 28 here>

<figure 29 here>

Audible Noise

One of the benefits of optimizing the desired function of a product is that there is less energy available to contribute to undesired functions. One such undesired function is audible noise. Figure 30 compares the total sound pressure level of several injectors. The results are a summary from a Robust Assessment which includes some of the most popular injectors on the market today. The Delphi injectors show lower sound pressure levels than the other injectors. The Delphi injector is the quietest direct injection injector the team has measured to date. The energy delivered to Delphi injectors is used primarily for the desired functions (Actuate Valve, Control Static Flow and Control Spray Pattern) so less

energy is available to produce unwanted outputs like audible noise and vibration.

<figure 30 here>

Fuel Flow Control

One of the more important results in the optimization of an injector is the fuel flow control performance. Figure 31 shows two graphs. The small graph is a linear flow range comparison of five injector designs. The large graph is an expansion of the low pulse width range comparing the performance of the five designs below 1.2 milliseconds. One typical measure of assessing low pulse width linearity is to note the pulse width at which an injector exceeds 5% deviation from linear flow. Injectors with low deviation from linear flow are desirable for enabling fuel control at low pulse widths and enabling strategies that use multiple injections per cylinder event. The results of the optimized Delphi injector show that the design performs well down to pulse widths below 0.5 milliseconds. One additional benefit of this optimization was the resulting low variation in shot to shot (injection to injection) variation. Figure 32 shows the results of the comparison of these injectors at various pulse widths. The optimized Delphi injector performs well and does not exceed the chosen 3% limit above 0.3 milliseconds. Low shot to shot variation is also desirable for low flow control.

<figure 31 here>

<figure 32 here>

SUMMARY / CONCLUSION

The Design for Six Sigma innovation methodology was used to develop a competitive injector for the homogeneous gasoline direct injection market. DFSS provided the best opportunity to identify and optimize the critical functions required to meet or exceed customer expectations. The key DFSS tools used for this development were the House of Quality, Function Modeling, Axiomatic Design, TRIZ, Pugh Concept Selection Process and Taguchi methods for robust design. In addition to delivering this first injector, the process inherently leaves a comprehensive base of knowledge that can drive future incremental and generational improvements. The process of documenting and analyzing customer needs and desires and then translating to clearly defined performance functions and requirements is applicable or adaptable to future development.

The results show that the development and optimization was successful. One key feature used from the concept generation step was the decoupled armature. This feature resulted in low bounce flux leading to reduced vehicle tailpipe hydrocarbon emissions. An additional benefit was reduced audible noise as a result of minimized energy loss during injector actuation. The optimization of the Actuate Valve and Static Flow functions resulted in short response times leading to low pulse width flow variation and low shot to shot flow variation.

In order to define the Functional Requirements, a function model was used to manage the complexities of an injector. The function model is now used as a tool for further design work, especially as an aid in the analysis of failure modes and effects, and defining the validation plan.

The TRIZ methodology was used during the concept generation stage. This methodology can be difficult to use without extensive training. An expert was consulted in order to produce the usable benefit within the project timing.

Optimization focused on three separate functions: Actuate Valve, Control Static Flow and Control Spray Pattern Optimizing the entire product function at once is typically recommended so all interactions are included. In this case, separation of the functions was required to allow a manageable optimization plan.

The optimization of the "Actuate Valve" function resulted in more of the energy used to open and close the injector efficiently; leaving less energy to produce the undesired functions. The desired functions, Opening and Closing Response, Low Pulse Width Linearity and Shot to Shot Repeatability are all benchmark performance. Less unwanted energy then minimized the undesired functions, audible noise and unwanted fueling (valve bounce).

DFSS was a key strategy for the development of this world class injector. Moreover, this DFSS project has laid a foundation that will allow this injector to meet the ever growing demands of the developing GDi market.

REFERENCES

1. Lutterman, C. and Mährle, W., "BMW High Precision Fuel Injection in Conjunction with Twin-Turbo Technology: a Combination for Maximum Dynamic and High Fuel Efficiency," SAE Technical Paper 2007-01-1560, 2007.

2. Davis, R.S., Mandrusiak, G.D., and Landenfeld, T., "Development of the Combustion System for General Motors' 3.6L DOHC 4V V6 Engine with Direct Injection," SAE Technical Paper 2008-01-0132, 2008.

3. Königstein, A., Larsson, P.-I., Grebe, U. D., and Wu, K.-J., "Differentiated Analysis of Downsizing Concepts," presented at 29th Vienna Motor Symposium, Austria, 2008.

4. Yi, J., Wooldridge, S., Coulson, G., Hilditch, J. et al., "Development and Optimization of the Ford 3.5L V6 EcoBoost Combustion System," *SAE Int. J. Engines* 2(1): 1388-1407, 2009.

5. Klauer, N., Klüting, M., Steinparzer, F., and Unger, H., "Turbocharging and Variable Valve Trains - Fuel Reducing Technologies for Worldwide Use," presented at 30th Vienna Motor Symposium, Austria, 2009.

6. Boccadoro, Y., Tranchant, O., Pionnier, R., and Engelhardt, H., "The New RENAULT TCe 130 1.4 l Turbocharged Gasoline Engine," presented at 30th Vienna Motor Symposium, Austria, 2009.

7. Lee, H. S., "Hyundai-Kia's Powertrain Strategy for Green and Sustainable Mobility," presented at 30th Vienna Motor Symposium, Austria, 2009.

8. Böhme, J., Müller, H., Ganz, M., and Marques, M., "The New Five Cylinder 2.5l TFSI Engine for the Audi TT RS," presented at 30th Vienna Motor Symposium, Austria, 2009.

9. Lückert, P. Kreitmann, F., Merdes, N., Weller, R., Rehberger, A., Bruchner, K., Schwedler, K., Ottenbacher, H., and Keller, T., "The New 1.8-Litre 4-Cylinder Petrol Engine with Direct Injection and Turbocharging for All Passenger Cars with Standard Drivetrains from Mercedes-Benz," presented at 30th Vienna Motor Symposium, Austria, 2009.

10. Hadler, J., Szengel, R., Middendorf, H., Kuphal, A., Siebert, W., and Hentschel, L., "Minimum Consumption - Maximum Force: TSI Technology in the New 1.2l Engine from Volkswagen," presented at 30th Vienna Motor Symposium, Austria, 2009.

11. Husted, H.L., Piock, W., and Ramsay, G., "Fuel Efficiency Improvements from Lean, Stratified Combustion with a Solenoid Injector," SAE Technical Paper 2009-01-1485, 2009.

12. Das, S., Chang, S.-I., and Kirwan, J., "Spray Pattern Recognition for Multi-Hole Gasoline Direct Injectors using CFD Modeling," SAE Technical Paper 2009-01-1488, 2009.

13. Yang, K., and Basem EH., "Desgn for Six Sigma, A Roadmap for Product Development," McGraw-Hill, 2003.

14. Suh, N., "Axiomatic Design, Advances and Applications", Oxford University Press, 2001

15. Taguchi, G., Chowdhury, S., and Wu, Y. "Taguchi's Quality Engineering Handbook", John Wiley & Sons, 2005.

CONTACT INFORMATION

Edwin A. Rivera
DFSS Black Belt
Edwin.A.Rivera@Delphi.com

Noreen L. Mastro
DFSS Green Belt
Noreen.L.Mastro@Delphi.com

ACKNOWLEDGMENTS

The authors are indebted to the team members who were dedicated to the development of a world class injector. Thanks to the following project team members for contributing their expertise and talents to the success of this project: Brent Wahba, Charles Braun, Kevin Allen, Ronald Krefta and Kevin Keegan. Thanks also to the DFSS coaching expertise provided by Craig D. Smith DFSS MBB, Alan Wu, American Supplier Institute (ASI) and Pearse Johnston ASI. Also thanks to Jay Sofianek for his sponsorship throughout most of the project.

DEFINITIONS/ABBREVIATIONS

GDi

Gasoline Direct Injection

PFI

Port Fuel Injection

DFSS

Design for Six Sigma

QFD

Quality Function Deployment

EMS

Engine Management System

CTS

Critical to Satisfaction measures

OEM

Original Equipment Manufacturer

FR

Functional Requirement

DP

Design Parameter

S/N

Signal to Noise Ratio

UL

Upper Limit

LL

Lower Limit

ECE

Economic Commission for Europe

EUDC

Extra-Urban Drive Cycle

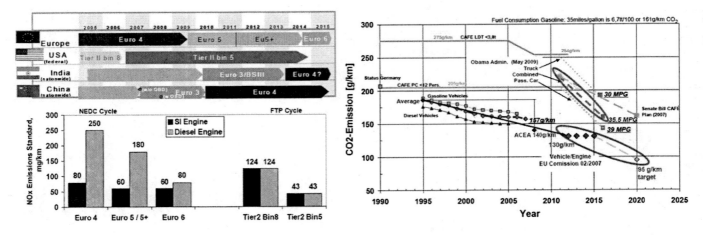

Figure 1. Rollout of light duty regulated emissions and fuel consumption/CO₂ targets.

Powertrain Architecture Evolution

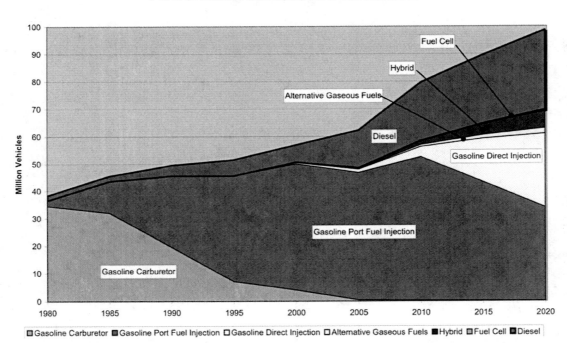

Figure 2. Projected global market penetration for major powertrain architectures. Forecast developed by Delphi using market share data from IHS Global Insight.

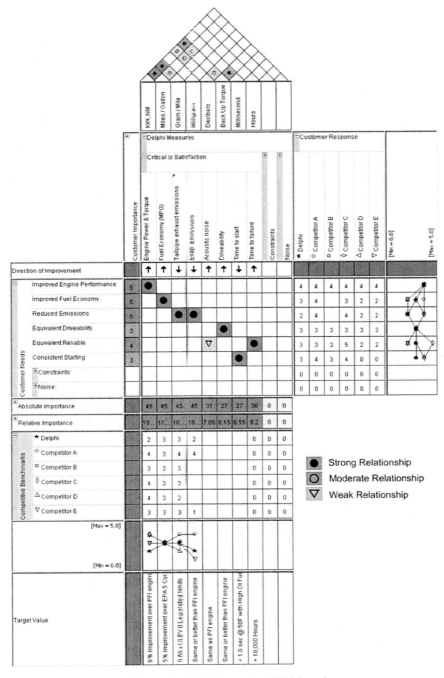

Figure 7. QFD House 1 - EMS Level

221

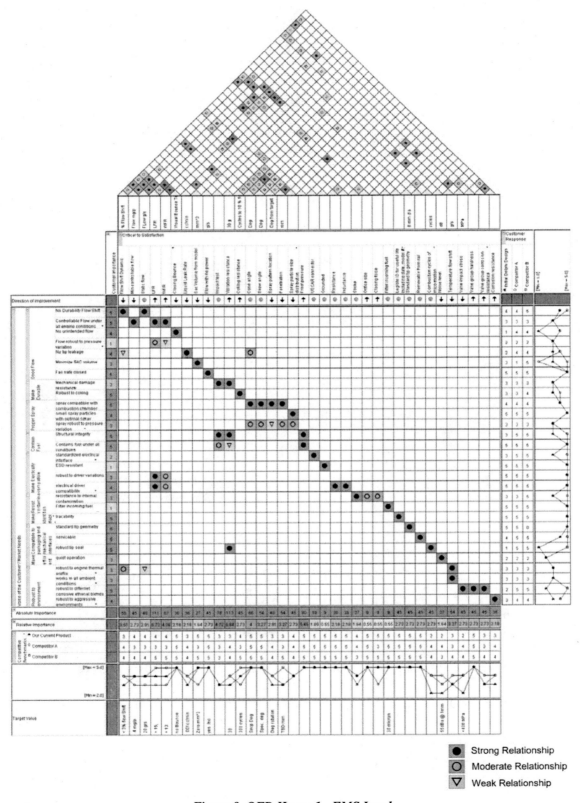

Figure 8. QFD House 1 - EMS Level

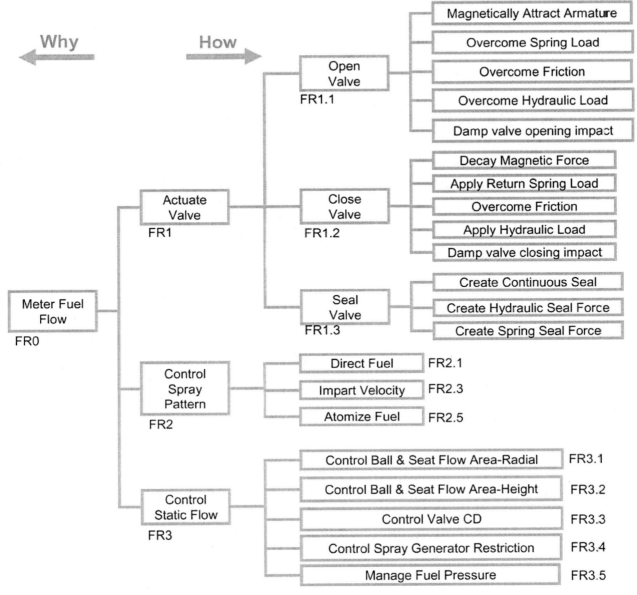

Figure 9. Injector Function Model - Top Level

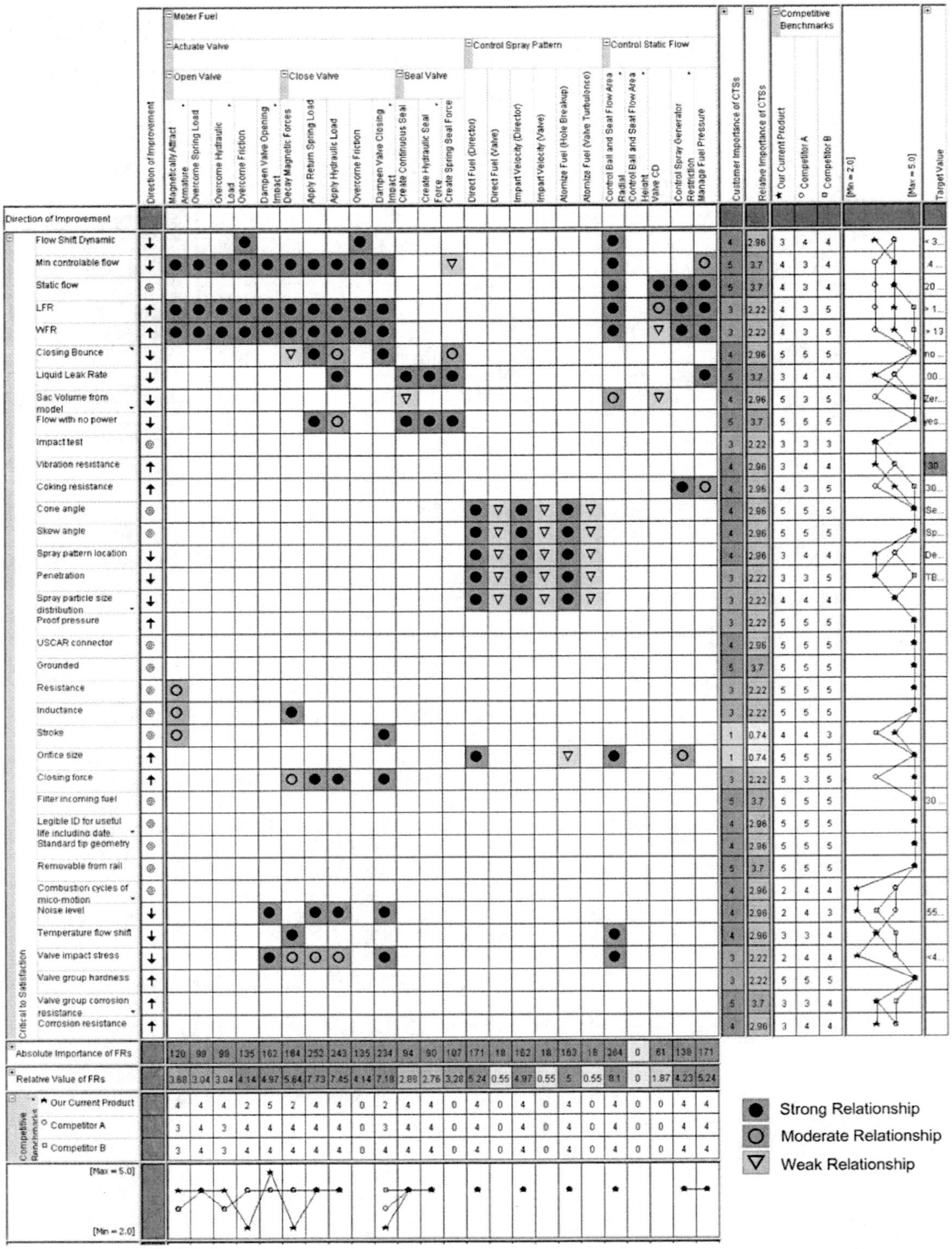

Figure 10. QFD House 2 - Translate CTS Measures to Functional Requirements

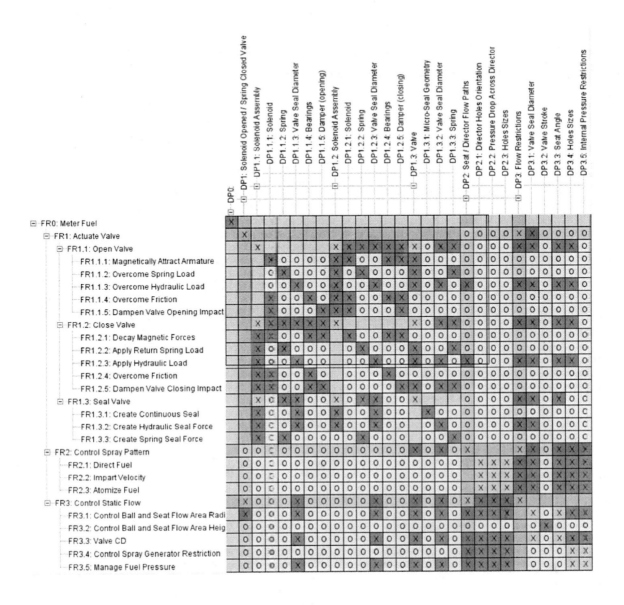

O Uncoupled Relationship (no significant relationship)

X Decoupled Relationship (optimization possible is optimization order is considered)

X Coupled Relationship (optimization difficult due to coupled interactions)

Figure 11. Axiomatic Design Matrix

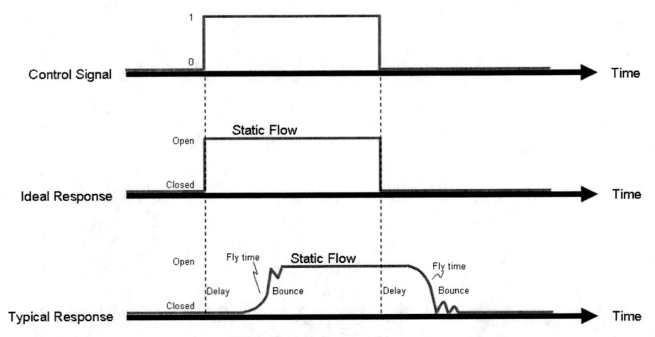

Figure 12. Valve Response Diagram

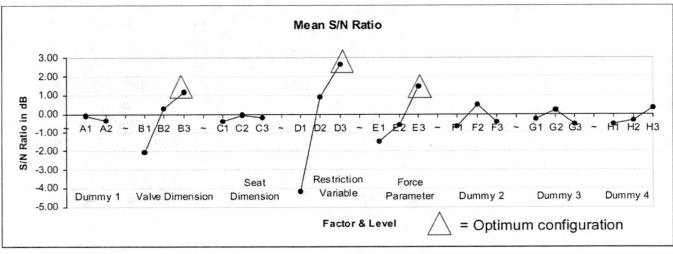

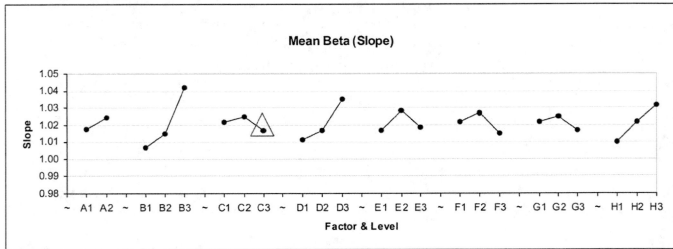

Figure 18. S/N and Slope results for Control Static Flow Optimization. B3, D3 and E3 chosen for best robustness to variation (highest S/N); C3 chosen for targeting.

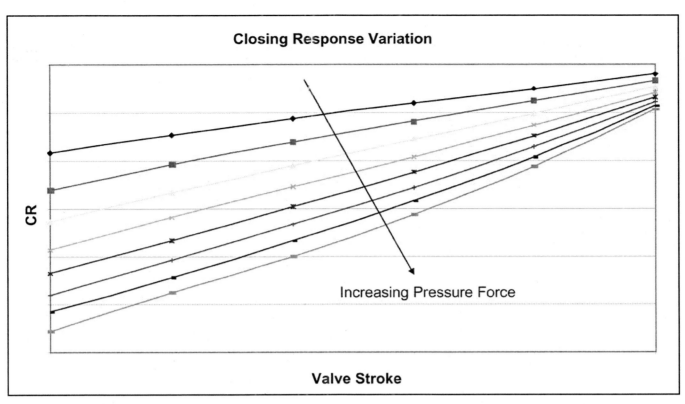

Figure 19. Modeling results of Closing Response Variation due to restriction and pressure force.

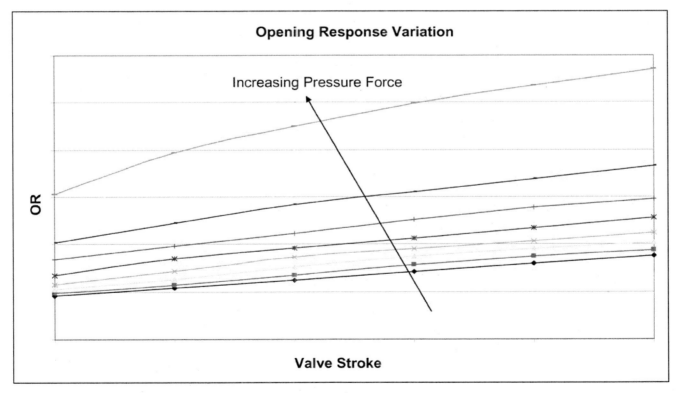

Figure 20. Modeling results of Opening Response Variation due to restriction and pressure force.

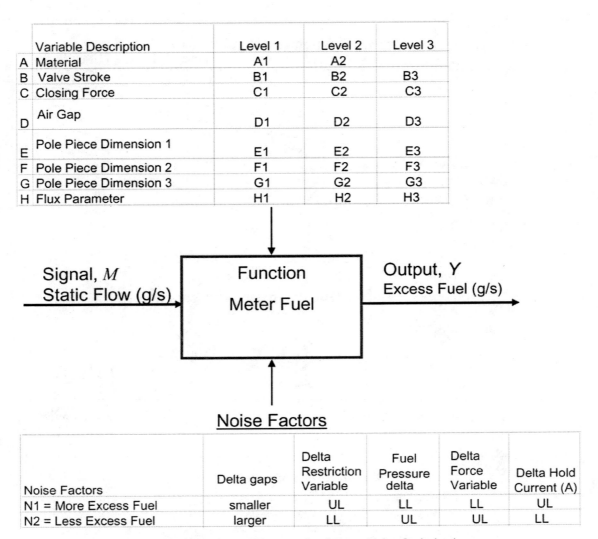

Variable Description	Level 1	Level 2	Level 3
A Material	A1	A2	
B Valve Stroke	B1	B2	B3
C Closing Force	C1	C2	C3
D Air Gap	D1	D2	D3
E Pole Piece Dimension 1	E1	E2	E3
F Pole Piece Dimension 2	F1	F2	F3
G Pole Piece Dimension 3	G1	G2	G3
H Flux Parameter	H1	H2	H3

Signal, M
Static Flow (g/s)

Function

Meter Fuel

Output, Y
Excess Fuel (g/s)

Noise Factors

Noise Factors	Delta gaps	Delta Restriction Variable	Fuel Pressure delta	Delta Force Variable	Delta Hold Current (A)
N1 = More Excess Fuel	smaller	UL	LL	LL	UL
N2 = Less Excess Fuel	larger	LL	UL	UL	LL

Figure 22. Parameter Diagram for Actuate Valve Optimization

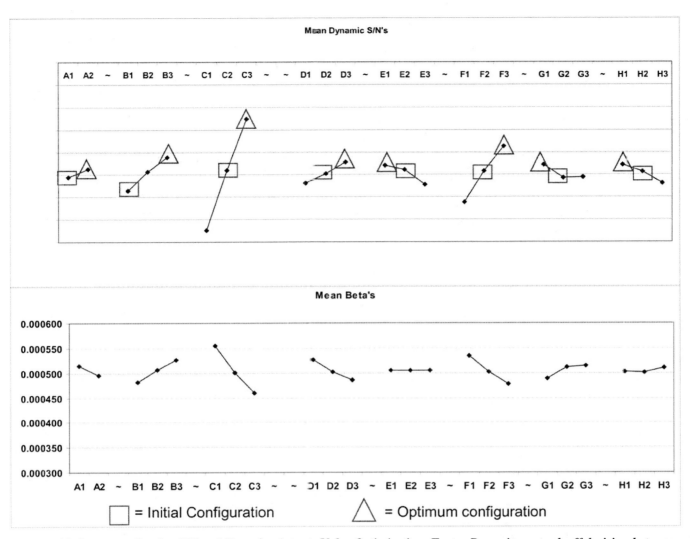

Figure 23. Response Graphs, S/N and Beta, for Actuate Valve Optimization. Factor B requires a tradeoff decision between reducing variation and targeting to minimize excess fuel.

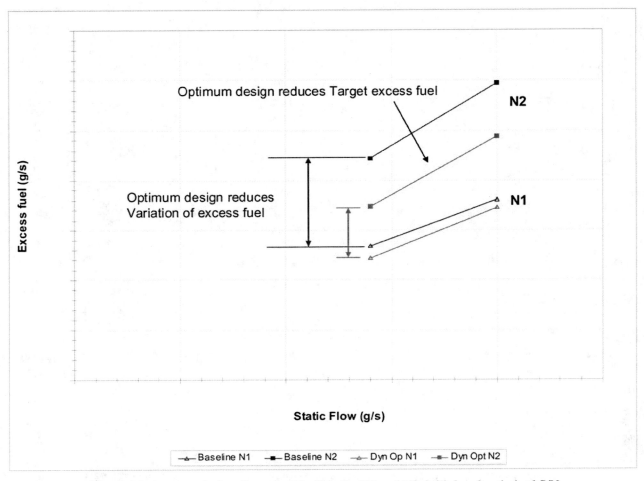

Figure 24. Average closing flow vs. static flow for N1 and N2, initial and optimized S/N

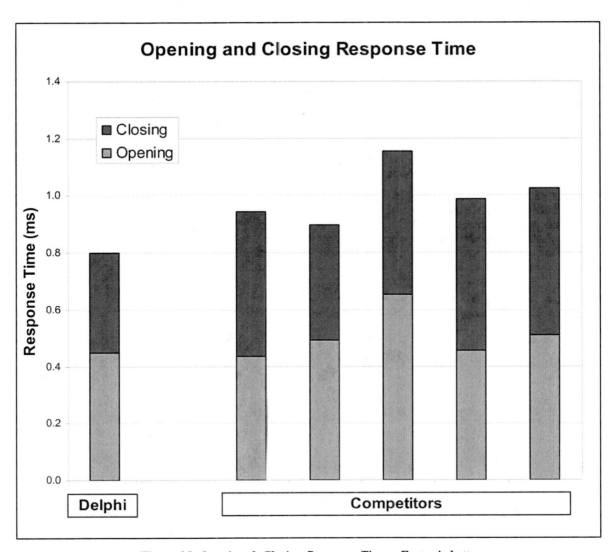

Figure 25. Opening & Closing Response Times. Faster is better

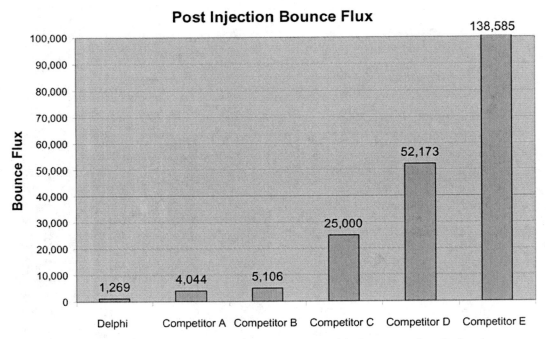

Figure 28. Bounce flux represents the amount of fuel present after closing

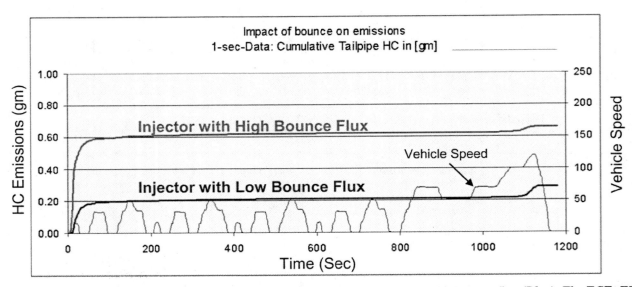

Figure 29. Note significant HC improvement from 25,000 bounce flux (Red) to <5,000 Bounce flux (Blue). The ECE+EUDC drive cycle was used for this test.

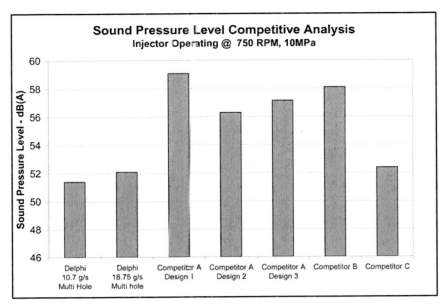

Figure 30. Sound Pressure Level Comparison

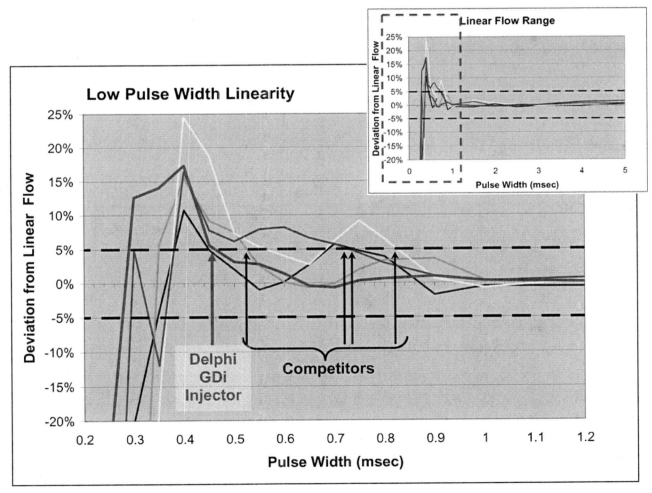

Figure 31. Low Pulse Width Linearity. The point at which a flow curve exceeds 5% deviation from linear flow defines the low end of the linear range. All of the injectors compared had similar static flow measurements.

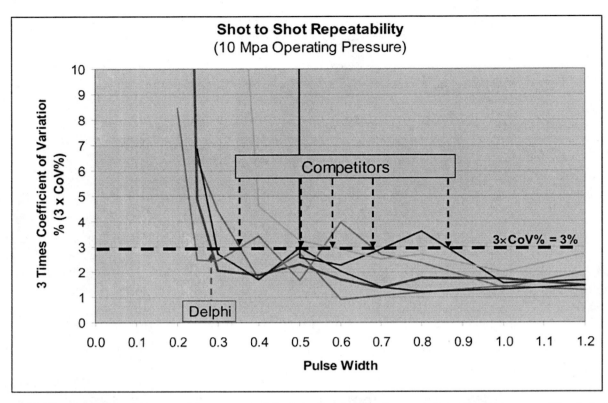

Figure 32. Shot to shot performance. (Assessment of the injection to injection variation of various competitive injectors.)

Boosted HCCI for High Power without Engine Knock and with Ultra-Low NOx Emissions - using Conventional Gasoline	2010-01-1086 Published 04/12/2010

John E. Dec and Yi Yang
Sandia National Laboratories

ABSTRACT

The potential of boosted HCCI for achieving high loads has been investigated for intake pressures (P_{in}) from 100 kPa (naturally aspirated) to 325 kPa absolute. Experiments were conducted in a single-cylinder HCCI research engine (0.98 liters) equipped with a compression-ratio 14 piston at 1200 rpm. The intake charge was fully premixed well upstream of the intake, and the fuel was a research-grade (R+M)/2 = 87-octane gasoline with a composition typical of commercial gasolines.

Beginning with P_{in} = 100 kPa, the intake pressure was systematically increased in steps of 20 - 40 kPa, and for each P_{in}, the fueling was incrementally increased up to the knock/stability limit, beyond which slight changes in combustion conditions can lead to strong knocking or misfire. A combination of reduced intake temperature and cooled EGR was used to compensate for the pressure-induced enhancement of autoignition and to provide sufficient combustion-phasing retard to control knock. The maximum attainable load increased progressively with boost from a gross indicated mean effective pressure ($IMEP_g$) of about 5 bar for naturally aspirated conditions up to 16.34 bar for P_{in} = 325 kPa. For this high-load point, combustion and indicated thermal efficiencies were 99% and 47%, respectively, and NOx emissions were < 0.1 g/kg-fuel. Maximum pressure-rise rates were kept sufficiently low to prevent knock, and the COV of the $IMEP_g$ was < 1.5%. Central to achieving these results was the ability to retard combustion phasing (CA50) as late as 19° after TDC with good stability under boosted conditions. Detailed examination of the heat release rates

shows that this substantial CA50 retard was possible because intake boosting significantly enhances the early autoignition reactions, keeping the charge temperature rising toward the hot-ignition point despite the high rate of expansion at these late crank angles. Overall, the investigation showed that well-controlled boosted HCCI has a strong potential for achieving power levels close to those of turbo-charged diesel engines.

INTRODUCTION

Homogeneous charge compression ignition (HCCI) engines can provide both high efficiency and low NOx and particulate emissions, but their limited power output remains a significant barrier to widespread implementation. The load limitation typically occurs because the maximum cylinder-pressure rise rate (PRR) increases with fueling rate, eventually causing engine knock. To mitigate these high PRRs with increased fueling, the combustion phasing can be retarded, but the amount of allowable retard is limited by excessive cycle-to-cycle variation in the power output, as measured by the gross indicated mean effective pressure ($IMEP_g$),[1] and eventually misfire [1,2]. As the fueling is increased, it can become increasingly difficult to maintain sufficiently retarded combustion to prevent run-away knock [3] without having poor combustion stability that can drift into misfire [1,2,4]. Although rapid feedback, closed-loop control systems can help extend this knock/stability limit [1], the maximum load for naturally aspirated HCCI is still low relative to traditional spark-ignition or turbo-charged diesel engines.

Additionally, depending on the fuel-type and operating conditions, the fueling can reach the point that the

---

[1]$IMEP_g$ refers to the IMEP over the compression and expansion strokes only, while $IMEP_{net}$ refers to the IMEP over the entire cycle, including the gas-exchange strokes.

combustion is no longer sufficiently dilute to maintain low combustion temperatures, and NOx emissions begin to rise [1,3]. For these conditions, the value of using HCCI could be diminished since NOx aftertreatment would be required.

Intake-pressure boosting is widely used to increase the power output of internal combustion engines, and it offers a means for increasing power while still maintaining the dilute conditions required for low NOx. However, the application of boost to HCCI has been limited because it can be difficult to control knocking. The knocking propensity of HCCI increases with boost for two reasons. First, the greater charge mass with boost results in a greater pressure rise with combustion, which increases the PRR for the same combustion phasing. Second, boost enhances the autoignition process, resulting in an advancement of the combustion phasing, which would increase the PRR even without the greater pressure rise with boost. Because of these difficulties, most previous attempts to achieve high loads with boosted HCCI used special fuels and/or reduced compression ratios.

Early work on boosted HCCI was presented by Christensen et al. [5] who showed that boost could significantly enhance the IMEP, but the combustion phasing became advanced causing knocking. However, they noted that ethanol and natural gas showed less advancement with boost than iso-octane. This behavior allowed them to reach an $IMEP_{net}$ = 14 bar using natural gas compared to only about 9.7 bar for iso-octane, at an absolute intake pressure (P_{in}) = 3 bar. Later, Christensen and Johansson [6] extended the high load to $IMEP_g$ = 16 bar using natural gas with an iso-octane pilot, combined with exhaust gas recirculation (EGR) and a variable compression ratio (P_{in} = 2.5 bar). In a later work from the same research group, Olsson et al. [7] achieved a brake mean effective pressure (BMEP) of 16 bar for P_{in} = 3 bar absolute. In this work, dual fueling with variable amounts of ethanol and n-heptane was used to control combustion phasing, with the fuel mixture being mostly ethanol for high loads. Ethanol was selected as the less-reactive fuel for this work because iso-octane and gasoline were found to ignite too early at high load. In all these works, the charge was well premixed using port injection, compression ratios ranged from 17 to 19, and NOx emissions were well below the US 2010 standard of 0.27 g/kWh.

Using a somewhat different approach based on early direct injection (DI) with diesel fuel injectors, Bessonette et al. [8] examined the high load limits of HCCI-like combustion for an engine with a compression ratio of 12. Several different fuel blends were investigated, and the study found that a fuel between diesel and gasoline worked best. A BMEP of 16 bar was achieved with both a low-octane gasoline and a low-cetane diesel fuel. NOx emissions were generally low, but exceeded the US 2010 standard at some conditions, probably due to incomplete charge mixing.

Although the above-mentioned studies did not use conventional liquid fuels, conventional gasoline was used by Kalghatgi et al. [9] in an investigation of partially premixed low-temperature combustion that has some similarities to HCCI. In this work, gasoline was directly injected into a diesel engine with about 75% of the fuel injected 11° before top dead center (bTDC) and the remainder injected early (150° bTDC). Due to the relatively long ignition delay with gasoline, substantial premixing occurred, resulting in NOx and smoke levels that were far lower than for traditional diesel operation. A high load point of IMEP = 15.95 bar (gross or net IMEP is not specified) was achieved for P_{in} = 2 bar absolute, but with smoke levels of FSN = 0.067 and NOx emissions of 0.58 g/kWh, the latter being more than twice the US 2010 limit. In summary, these previous works achieved power levels on the order of 16 bar IMEP, but issues such as non-conventional fuels, dual fuels, high PRRs, or high NOx emissions limit applicability.

The objective of the current work is to determine the potential of boosted HCCI for producing high power levels using conventional gasoline without engine knock. Maintaining NOx emissions below the US 2010 standard and high thermal efficiency were also priorities. For this study, the intake charge was fully premixed, which eliminated the potential of local regions with high combustion temperatures that might produce NOx. A compression ratio of 14 was selected as being sufficiently high for good thermal efficiency, but low enough to mitigate friction losses. With this compression ratio, autoignition of gasoline for naturally aspirated conditions required heating the intake air to temperatures of 130°C or higher (in practical applications, retained hot residuals could be substituted for this intake heat). However, the required intake temperature dropped rapidly with boost due to the pressure-induced enhancement of autoignition. Controlling engine knock under highly boosted conditions required substantial combustion-phasing retard. Fortunately, under boosted conditions, good combustion stability could be maintained with much greater combustion-phasing retard than was possible for naturally aspirated conditions. This allowed high fueling rates to be maintained for boosted operation, as will be discussed in detail later. To achieve the required retard despite the pressure-induced enhancement of autoignition, a combination of reduced intake temperature and cooled EGR was used.

The next section describes the experimental facility, fuel specifications, data acquisition and analysis techniques, and the knock-limit criterion. Following this, the results are presented in two parts. First, engine performance data for various levels of intake boost are presented and discussed. The second part presents an analysis and discussion of the cylinder temperature and heat release rates (HRR) that explain why boosted operation worked well for the conditions

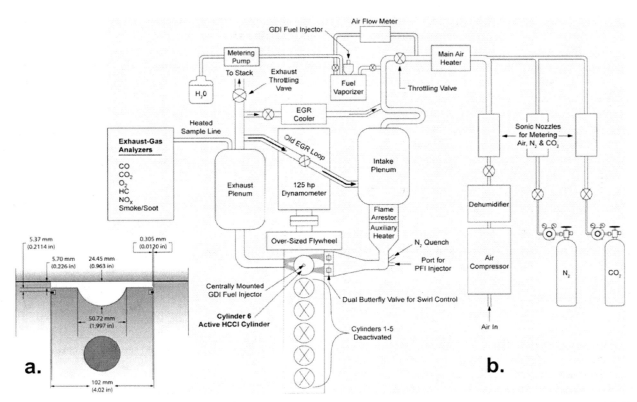

Figure 1. Schematic of the HCCI engine facility (b), and combustion chamber geometry of the CR = 14 piston, at TDC (a).

studied. Finally, in the last section, the study is summarized and conclusions are drawn.

EXPERIMENTAL SETUP & DATA ACQUISITION

ENGINE FACILITY

The HCCI research engine used for this study was derived from a Cummins B-series six-cylinder diesel engine, which is a typical medium-duty diesel engine with a displacement of 0.98 liters/cylinder. Figure 1 shows a schematic of the engine, which has been converted for single-cylinder operation by deactivating cylinders 1-5. The engine specifications and operating conditions are listed in Table 1. The configuration of the engine and facility is nearly the same as for previous studies (e.g. Refs. [2,3,4,10,11]).

<figure 1 here>

TABLE 1. Engine Specifications and Operating Conditions

Displacement (single-cylinder)	0.981 liters
Bore	102 mm
Stroke	120 mm
Connecting Rod Length	192 mm
Geometric Compression Ratio	14:1
No. of Valves	4
IVO	0° CA*
IVC	202° CA*
EVO	482° CA*
EVC	8° CA*
Swirl Ratio	0.9
Fueling system	Fully Premixed
Engine Speed	1200 rpm
Intake Temperature	45 – 146°C
Intake Pressure (abs.)	100 – 325 kPa
Coolant Temperature	100°C

*0° CA is taken to be TDC intake. The value-event timings correspond to 0.1 mm lift.

The active HCCI cylinder is fitted with a compression-ratio (CR) = 14 custom piston as shown in Fig. 1a. This piston provides an open combustion chamber with a large squish clearance and a quasi-hemispherical bowl. Compared to the broad shallow bowl used in most previous studies, *e.g.* [4,12], this bowl is deeper and has a smaller diameter, thus allowing a broad flat squish region. This change in bowl geometry was made to facilitate safer operation for concurrent testing of a variable valve actuation system (not reported here). Comparison tests showed minimal differences in performance and emissions compared to the CR = 14 shallow-bowl piston (see Ref. [12] for a schematic). A more complete discussion of the comparison between pistons may be found in Ref. [3]. Both of these custom-designed CR = 14 pistons provide a small topland ring crevice, amounting to only 2.1% of the top-dead-center (TDC) volume, including the volume behind the top piston ring.

As shown in Fig. 1b and discussed in Ref. [10], the engine facility is equipped with two EGR loops and the capability of supplying simulated EGR (combustion products) by adding N_2, CO_2 and H_2O. For the current study, all EGR was supplied using the cooled EGR loop. The EGR cooler (a gas-to-water heat exchanger) is necessary for maintaining the desired intake temperature as EGR levels are increased. In addition, this loop introduces the EGR well upstream of the intake plenum, so the EGR travels with the intake air through a series of bends before reaching the intake plenum, to insure that the intake charge is well mixed. With this configuration, the exhaust pressure must be greater than the intake pressure for EGR to flow into the intake. The required back pressure was achieved by throttling the exhaust flow using the valve shown in Fig. 1b.

The intake air was supplied by an air compressor and precisely metered by a sonic nozzle as shown in Fig. 1b. For operation without EGR, the air flow was adjusted to achieve the desired intake pressure, as measured by a pressure transducer on the intake runner. For the current study, the intake pressure was varied from 100 kPa (simulating naturally aspirated conditions) to 325 kPa. All pressures given are absolute. For operation with EGR, the air flow was reduced from the amount required to achieve the desired intake pressure with air alone, and the valve on the EGR line was opened. The exhaust back-pressure throttle valve was then adjusted to produce enough EGR flow to reach the desired intake pressure. This typically resulted in the exhaust pressure being about 2 kPa greater than the intake pressure. For consistency, the back pressure was maintained at 2 kPa above the intake pressure, even when EGR was not used. The EGR fraction was varied by adjusting the amount of supplied air, and then adjusting the exhaust throttle to maintain the desired intake pressure.

For this study, the intake charge was fully premixed using the fueling system shown at the top of the schematic in Fig. 1b. This fueling system consists of a gasoline-type direct injector mounted in an electrically heated fuel vaporizing chamber and appropriate plumbing to ensure thorough premixing with the air and EGR upstream of the intake plenum. A positive displacement fuel flow meter was used to determine the amount of fuel supplied. The fuel was a research-grade gasoline supplied by Chevron-Phillips Chemical Co., and its specifications are listed in Table 2.

TABLE 2. *Chevron-Phillips Research-Grade Gasoline**

Antiknock Index (R+M)/2	87.0
RON	90.8
MON	83.2
Specific gravity	0.746
API gravity	58.2
Carbon, wt%	85.59
Hydrogen, wt%	13.88
Oxygen, wt%	0.00
A/F Stoichiometric	14.79
Lower Heating Value, gas-phase fuel (MJ/kg)	43.17
LHV for stoichiometric charge (MJ/kg)	2.734

Hydrocarbon Type, vol%	
Aromatics	22.7
Olefins	4.2
Saturates	73.1

Distillation, °C	
5%	53
10%	59
30%	75
50%	93
70%	115
90%	145
95%	157

*Based on analysis provided by Chevron-Phillips

Prior to starting the experiments, the engine was fully preheated to 100°C by means of electrical heaters on the "cooling" water and lubricating-oil circulation systems. In addition, the intake tank and plumbing were preheated to 50 - 60°C to avoid condensation of the fuel or water from the EGR gases. An auxiliary heater mounted close to the engine provided precise control of the intake temperature to maintain the desired combustion phasing. Intake temperatures ranged from 45° - 146°C. All data were taken at an engine speed of 1200 rpm.

DATA ACQUISITION

Cylinder pressure measurements were made with a transducer (AVL QC33C) mounted in the cylinder head approximately 42 mm off center. The pressure transducer signals were digitized and recorded at ¼° CA increments for one hundred consecutive cycles. The cylinder-pressure transducer was pegged to the intake pressure near bottom dead center (BDC) where the cylinder pressure reading was virtually constant for several degrees. Intake temperatures were monitored using thermocouples mounted in the two intake runners close to the cylinder head. Firedeck temperatures were monitored with a thermocouple embedded in the cylinder head so that its junction was about 44 mm off the cylinder center and 2.5 mm beneath the surface. Surface temperatures were estimated by extrapolating the thermocouple reading to the surface, using the thickness of the firedeck and assuming that its back surface was at the 100°C cooling-water temperature [13]. For all data presented, 0° crank angle (CA) is defined as TDC intake (so TDC compression is at 360°). This eliminates the need to use negative crank angles or combined bTDC, aTDC notation.

The crank angle of the 50% burn point (CA50) was used to monitor the combustion phasing. CA50 was determined from the cumulative apparent heat-release rate (AHRR), computed from the cylinder-pressure data (after applying a 2.5 kHz low-pass filter [14]). Computations were performed for each individual cycle, disregarding heat transfer and assuming a constant ratio of specific heats [15]. The average of 100 consecutive individual-cycle CA50 values was then used to monitor CA50 and for the CA50 values reported. The reported PRRs and ringing intensities are computed from the same low-pass-filtered pressure data. For each cycle, the maximum PRR was analyzed separately with a linear fit over a moving ±0.5° CA window. Similar to CA50, these individual-cycle values were then averaged over the 100-cycle data set.

A second method of computing the HRR was used for detailed HRR-curve analysis presented in the second part of the results. Here, the heat release was computed in a more refined way from the ensemble-averaged pressure trace (with the 2.5 kHz low-pass filter applied), using the Woschni correlation for heat transfer [15]. Using the ensemble-averaged pressure trace has benefits from the standpoint of reduced noise on the heat-release traces. On the other hand, it can lead to overestimated burn durations if the cycle-to-cycle variations are large. However, for the condition where this was applied, the phasing was fairly stable with the standard deviation of CA50 over 100 fired cycles averaging about 1.2° CA. Moreover, the results of this detailed HRR analysis are mainly used for comparisons of the early part of the heat release, leading up to hot ignition. Since these early autoignition reactions persist for 15° - 25° CA, the relatively small cycle-to-cycle variation in CA50 will have little effect.

Although these detailed HRR curves appear quite smooth when plotted on a scale showing the entire combustion event, small quasi-periodic fluctuations were evident when viewed on the highly amplified scale used for comparisons of the early heat release. This is believed to be an artifact of the

HRR being based on the derivative of the pressure trace (which amplifies noise in the pressure data) and the low-pass filtering process [14]. To mitigate these fluctuations, the HRR data (in Figs. 12b, 13, 14, and 15) were filtered using a piecewise smoothing function in which the value of each point was adjusted based on a weighted average of its original value and the values of its neighbors. For this filtering, the original value is given the greatest weight, and the weighting of the values of the neighboring points is reduced linearly with distance over a specified number of neighbors. The number of neighbors specified determines the amount of smoothing. Care was taken to compare the filtered and unfiltered HRR curves and to adjust the number of neighbors to insure that the filtered curves faithfully reproduced the trends in the unfiltered HRR data. Finally, it should be noted that although the filtering facilitates comparisons between the curves, all observations discussed regarding the comparisons of the HRR between conditions are evident without the filtering, *i.e.* the differences discussed are greater than the noise removed with this filtering.

Charge temperatures during the closed part of the cycle (*i.e.* compression/expansion) are computed using the ideal-gas law in combination with the measured pressure (ensemble-averaged over 100 cycles), the known cylinder volume, and the trapped mass. The trapped mass equals the sum of the supplied air, fuel, EGR, and residuals, the latter being computed by applying the ideal gas law to the clearance volume at TDC-valve-overlap, using a pressure determined from cycle-simulation modeling (using Ricardo's WAVE program) and an estimated residual temperature based on exhaust blowdown to this pressure. The average molecular weight used for the calculation of the charge temperature during the compression stroke corresponds to that of the trapped gases - including fresh air, EGR, retained residuals, and fuel. During the combustion event, and in proportion to the mass fraction burned, the molecular weight in the calculations is gradually changed to that of the measured exhaust composition.[2] This exhaust molecular weight is then used for the remainder of the expansion stroke and for computing the trapped mass.

Exhaust emissions data were also acquired, with the sample being drawn from the exhaust plenum using a heated sample line (see Fig. 1b). CO, CO_2, HC, NOx, and O_2 levels were measured using standard exhaust-gas analysis equipment. In addition, a second CO_2 meter monitored the intake gases just prior to induction into the engine. This allowed the EGR fraction of the intake gases to be computed from the ratio of the intake and exhaust CO_2 concentrations.

---

[2] Dissociation of the combustion products is neglected since it is expected to be small due to the relatively low peak-combustion temperatures.

RESULTS AND DISCUSSION - ENGINE PERFORMANCE

The goal of the engine performance testing was to determine the maximum load that can be achieved with boosted HCCI while maintaining acceptable knock and NOx emissions. To accomplish this, the intake pressure was systematically increased, and for each P_{in}, fueling was increased up to the knock/stability limit [1,2,3]. Because increased pressure enhances autoignition, a greater degree of combustion phasing "adjustment" was required as the boost level was increased. A combination of intake-temperature reduction and cooled-EGR addition was used to achieve the necessary combustion-phasing retard.

KNOCK/STABILITY LIMIT

As mentioned in the Introduction, for each P_{in}, increasing the fueling increases the PRR, eventually causing the engine to knock. On the other hand, retarding the combustion phasing reduces the PRR [2] since it slows the HRR by amplifying the benefit of the naturally occurring thermal stratification [16]. Additionally, the larger combustion-chamber volume for retarded combustion reduces the total pressure rise, further reducing the PRR for the same HRR [3]. This reduced PRR allows the fueling to be increased without knock, and the process of incrementally increased combustion retard followed by increased fueling can be continued up to the stability limit, as demonstrated in Ref. [2].

The allowable combustion retard is limited by poor combustion stability as manifested by excessive cycle-to-cycle variation in the $IMEP_g$. This occurs because the expansion rate increases with increased timing retard, producing a higher rate of charge cooling. This cooling counteracts the temperature rise produced by the early autoignition reactions, decreasing the rate of temperature rise toward the point of hot ignition [4]. As retard is increased, the temperature-rise rate decreases, eventually reaching the point that small random variations in the compressed-gas temperature cause significant variations in hot-ignition timing. This causes large variations in the $IMEP_g$ and can lead to run-away knock or decay to misfire [2,3,4].[3]

As a result, the maximum fueling rate is limited by the amount of combustion-phasing retard that can be applied with acceptable combustion stability. In practice, it is difficult to precisely determine the maximum load point since operating parameters such as the intake temperature must be adjusted precisely and in small increments to prevent run-away knock or misfire when operating near the knock/stability limit. For the current work, the fuel was incrementally increased in small steps, amounting to a 2 - 3% increase in $IMEP_g$, as the maximum fueling rate was approached, and the step size was often reduced if operation became unstable. Therefore, the highest loads reported for the various intake pressures are expected to be very close to the true knock/stability limit, attainable without fast closed-loop feedback controls [1].

Knock-Limit Criterion

The acceptable knock limit for HCCI engines is often defined in terms of a maximum allowable PRR ($dP/d\theta$, where θ is a variable representing °CA). However, this does not correctly reflect the potential for knock under boosted conditions where the cylinder pressure changes significantly. In this work, the correlation for ringing intensity developed by Eng [17] is used as a measure of engine knock.

$$\text{Ringing Intensity} \approx \frac{1}{2\gamma} \cdot \frac{\left(0.05 \cdot \left(\frac{dP}{dt}\right)_{max}\right)^2}{P_{max}} \cdot \sqrt{\gamma R T_{max}}$$

(1)

Where $(dP/dt)_{max}$, P_{max}, and T_{max} are the maximum values of PRR (in real time), pressure, and temperature, respectively, γ is the ratio of specific heats (c_p/c_v), and R is the gas constant. The ringing is a measure of the acoustic energy of the resonating pressure wave that creates the sharp sound commonly known as engine knock. Based on the onset of an audible knocking sound and the appearance of obvious ripples on the pressure trace, a ringing criterion of 5 MW/m^2 was selected as the ringing limit. This corresponds to about 8 bar/°CA at 1200 rpm, naturally aspirated. It should be noted that as P_{in} is increased above 100 kPa, $dP/d\theta$ can exceed 8 bar/°CA since the corresponding increased value of P_{max} in the denominator of Eq. 1 reduces the ringing intensity for a given PRR. At all boost levels tested, perceived engine knock correlated well with the ringing intensity rising above 5 MW/m^2, giving confidence in this correlation.

ACHIEVING MAXIMUM LOAD WITH BOOST

No EGR Addition

As a starting point for determining the maximum load with intake boost, P_{in} was set to 100 kPa, simulating naturally aspirated conditions. For this P_{in}, intake heating was required to obtain autoignition, even though no EGR was used. The fueling was initially set somewhat arbitrarily to a moderately

---

[3]For example, if one cycle ignites a couple of degrees too early, it will knock which increases wall heat transfer. Then, the hotter walls cause the following cycle to be even more advanced, resulting in even harder knock and greater wall heat transfer, and so on. Once started, this run-away knock process is very rapid [3,10], and can lead to engine damage. Similarly, a cycle that is a bit too retarded will not burn as completely, leading to lower residual and wall temperatures, causing the next cycle to be more retarded and burn even less completely, and so on until full misfire.

high load for HCCI, corresponding to an equivalence ratio (ɸ) = 0.38. For this fueling, CA50 was adjusted to 368° CA by setting the intake temperature (T_{in}) to 146°C, to provide stable combustion with low ringing. The fueling was then incrementally increased while CA50 was incrementally retarded, by reducing T_{in}, to maintain stable combustion with acceptable ringing. This process was repeated until reaching the knock/stability limit, which occurred at ɸ = 0.48, $IMEP_g$ = 506 kPa, CA50 = 373° CA, and T_{in} = 131°C. These data are shown in Fig. 2 as the series of points with increasing $IMEP_g$ for P_{in} = 100 kPa.

Figure 2 also shows all the other data points acquired without EGR. For intake pressures of 130 and 160 kPa, the procedure was the same as outlined above for P_{in} = 100 kPa, with three exceptions. First, a lower T_{in} was required to compensate for the enhancement of autoignition by the higher intake pressures. Second, more CA50 retard was required for the same ɸ to mitigate ringing since the pressure rise with combustion increases with boost due to the greater charge mass. The combination of these two effects resulted in intake temperatures ranging from 117 - 105°C for P_{in} = 130 kPa, and from 94 - 74°C for P_{in} = 160 kPa, over the range of fueling rates ($IMEP_g$) shown. Third, the maximum attainable ɸ was a little lower than the 0.48 value obtained for P_{in} = 100 kPa.

Further increasing P_{in} to 180 kPa required T_{in} = 60°C for the lowest fueling rate tested (ɸ = 0.38). Increasing the load above this value without EGR addition would require temperatures below 60°C to retard the combustion phasing sufficiently to prevent knock. Although the engine facility allows T_{in} to be reduced as low as about 30°C, T_{in} = 60°C was selected as the lower limit for this engine-performance study for the following reasons. First, 60°C is a representative out-of-intercooler temperature for production boosted engines. Second, maintaining T_{in} = 60°C prevents condensation of water from EGR (as presented in the next section) since temperatures are above the dew point of stoichiometric engine exhaust. Third, preliminary testing showed that reducing T_{in} much below 60°C results in significant low-temperature heat release (LTHR) under boosted conditions (an example will be presented later in Fig. 14). Since transitioning into or out of conditions with LTHR can sometimes require large adjustments to maintain combustion-phasing [18], it was deemed beneficial to avoid LTHR.

Increasing the boost above 180 kPa, with T_{in} at the 60°C limit and without EGR, requires a reduction in ɸ below the

baseline value of 0.38 in order to maintain ringing ≤ 5 MW/m². As a result, the maximum $IMEP_g$ actually decreases for boost levels greater than 180 kPa, as shown in Fig. 2. Thus, the highest load attainable without EGR addition is $IMEP_g$ = 8.8 bar at P_{in} = 180 kPa.

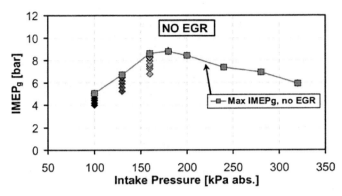

Figure 2. Maximum $IMEP_g$ without EGR addition at the knock/stability limit ($100 \leq P_{in} \leq 160$ kPa) or the knock limit ($180 \leq P_{in} \leq 320$ kPa).

With EGR Addition

For $P_{in} \geq 180$ kPa, increasing the load above the values shown in Fig. 2 requires further CA50 retard to maintain acceptable ringing. This was accomplished by adding cooled EGR to the intake mixture [12].[4] With this additional means of combustion retard, the fueling was incrementally increased above the values in Fig. 2, while the ringing was controlled by incrementally increasing the amount of EGR (similar to the procedure outlined above in which T_{in} was reduced to retard combustion). This process was continued until reaching the knock/stability limit, as shown in Fig. 3. T_{in} was held constant at 60°C for all data points with EGR addition. It should also be noted that EGR does not have a significant direct effect on PRR, rather, combustion retard produced by the EGR is the main cause for the reduced PRR/ringing [11].

As can be seen in Fig. 3, the use of cooled EGR allowed the load to be increased significantly. Unlike the no-EGR case, the maximum $IMEP_g$ rises monotonically with P_{in} up to the highest pressures tested, P_{in} = 325 kPa, for which $IMEP_g$ = 16.34 bar. The maximum-load line with EGR (Fig. 3) shows that the $IMEP_g$ increases almost linearly with boost up to P_{in} = 200 kPa. However, above P_{in} = 200 kPa, the slope of the maximum-load line becomes less steep. There are two main mechanisms responsible for this reduction in the slope.[5] Initially, 200 kPa ≤ P_{in} ≤ 240 kPa, this occurs because CA50 could not be retarded beyond the amount used for P_{in} = 200

---

[4]Substituting EGR for some of the excess air retards HCCI autoignition for two main reasons: 1) compressed-gas temperatures are lower due to the higher specific heat (lower γ) of the combustion-product gases, and 2) oxygen concentrations are reduced, slowing the autoignition reactions [12].

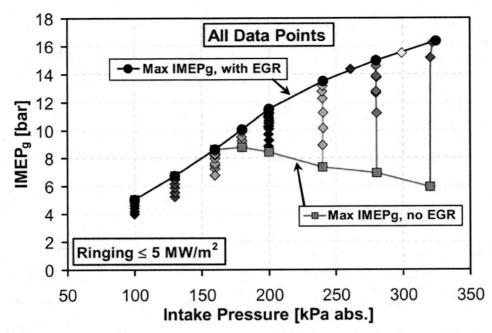

Figure 3. Maximum IMEP$_g$ at the knock/stability limit for various intake pressures, both with EGR and without.

kPa with acceptable stability (discussed in more detail later in the paper), which limits the allowable fueling rate. For higher intake pressures, P$_{in}$ ≥ 260 kPa, the maximum load is no longer limited by the ability to retard CA50 because fueling is limited by the oxygen content of the charge. This occurs because the amount of EGR required to control combustion phasing has increased to the point that almost all the excess air has been replaced with combustion products, *i.e.* the mixture is virtually stoichiometric, but dilute with EGR. Because the charge contains proportionally less reactants (fuel and air), less CA50 retard is required. Furthermore, as P$_{in}$ increases from 260 to 325 kPa, progressively more EGR is necessary to overcome the autoignition enhancement with boost, resulting in further reductions in the mass fraction of reactants in the charge, *i.e.* the charge/fuel (C/F) ratio increases. This further reduces the slope of the maximum load line.

<figure 3 here>

Figure 4 shows the percentage of EGR used for the maximum load points discussed above. For lean operation, EGR contains both excess air and combustion products, but it is only the combustion products that affect combustion phasing. Accordingly, Fig. 4 also shows the percentage of combustion products in the intake gas with EGR addition. This is designated as the fraction of complete stoichiometric products (CSP) because the composition of the combustion

products was computed as if the fuel burned to completion in a stoichiometric mixture. With the onset of EGR addition at P$_{in}$ = 180 kPa, the percentage of CSP is less than the EGR. However, as the amount of EGR increases with increasing P$_{in}$, the two lines converge, and have almost the same value for P$_{in}$ ≥ 260 kPa where the mixture is almost stoichiometric. As can be seen, at the highest load, the EGR and CSP percentages reach ~60%. Note that to insure good combustion efficiency, the mixture for the last three data points shown was kept slightly lean of stoichiometric, with about 0.5% O$_2$ remaining in the exhaust.

Finally, although the slope of maximum-load line in Fig. 3 is decreasing at the higher boost levels tested, it still has a significant upward slope at P$_{in}$ = 325 kPa. Therefore, even higher power (IMEP$_g$) could be obtained with further increases in boost. Currently, the highest load point shown is limited by the 170 bar rating of the cylinder head. Although at 149 bar, the 100-cycle-average peak-pressure for the maximum load point with P$_{in}$ = 325 kPa is somewhat below this limit (see Fig. 5), and individual cycles do not exceed 158 bar peak pressure, it is difficult to stabilize the engine at this operating point without excursions to conditions that produce peak pressures over 170 bar. With better control systems or an engine rated for higher pressures, these data indicate that higher maximum loads for HCCI engines should be readily attainable.

---

[5] In addition to the mechanisms discussed, the slope of the maximum IMEP$_g$ line may be reduced relative to its value at lower P$_{in}$ because T$_{in}$ is no longer being reduced with increasing P$_{in}$ ≥ 180 kPa, so changes in T$_{in}$ are no longer contributing to the increase in the intake-charge density with P$_{in}$. Also, the increasing concentrations of EGR reduce thermal efficiency [11].

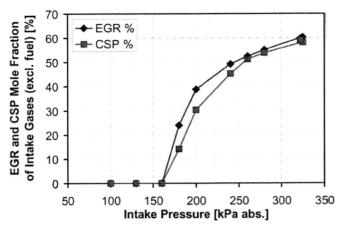

Figure 4. Intake-gas mole fractions of the total EGR and the combustion product portion (CSP) of the EGR for the maximum load (IMEP$_g$) points shown in Fig 3.

Potential for Turbo-Charging

As mentioned above, the exhaust back pressure was set to P$_{in}$ + 2 kPa for all data presented. In practice, however, it is desirable to use a turbo-charger to supply the intake boost. To investigate the potential for turbo-charging, additional tests were conducted at P$_{in}$ = 200 kPa for IMEP$_g$ = 10.71 bar, which is just a little less that the maximum-load, 200 kPa point shown in Fig. 3. For this condition, the back pressure was increased to 240 and 260 kPa, which correspond to pressures required to drive turbo-chargers with overall efficiencies (combined turbine and compressor) of 40 and 36%, respectively. The exhaust temperature was approximately 450°C. The increased exhaust pressure had almost no effect on the IMEP$_g$, reducing it from 10.71 to 10.69 bar. However, the IMEP$_{net}$ fell from 10.52 bar to 10.09 and 9.91 bar for exhaust pressures of 240 and 260 kPa, respectively, due to the increased pumping work. This moderate (~5%) reduction in IMEP$_{net}$, suggests that turbo-charging could be a viable option for providing the intake boost.

MAXIMUM-LOAD POINTS

To better understand the engine performance and emissions at the maximum-IMEP$_g$ points shown in Fig. 3, Figs. 5,6,7,8,9, 10 present data for several engine parameters.

Cylinder Pressure

Cylinder-pressure traces corresponding to these maximum-IMEP$_g$ points are shown in Fig. 5. As can be seen, the pressure at TDC, which is representative of a motored pressure, increases from 30 to 106 bar as P$_{in}$ increases from 100 to 325 kPa, and peak pressures increase from 47 to 149 bar. For the highest boost condition, two data points are shown, the highest load point for which P$_{in}$ = 325 kPa and

IMEP$_g$ = 16.34 bar, and the second-highest load point for which P$_{in}$ = 322 kPa and IMEP$_g$ = 16.22 bar. Although this second-highest load is only slightly less than the highest, the peak pressure is significantly lower, only 132 bar, due to the more retarded timing. Also, despite the greater retard, combustion stability is still very good. This indicates that retarding the timing offers the potential of allowing a further increase in boost to achieve even higher loads without exceeding the maximum cylinder head pressure, albeit with a small decrease in thermal efficiency due to the greater retard. Finally, it is also evident from the pressure traces that combustion-phasing becomes more retarded as P$_{in}$ increases from 100 to 200 kPa, and that the combustion phasing becomes more advanced again as P$_{in}$ is increased from 240 to 325 kPa, as will be discussed with respect to Fig. 10 below.

<figure 5 here>

Thermal and Combustion Efficiencies

Figure 6 presents the indicated-thermal and combustion efficiencies for the maximum-IMEP$_g$ points. The indicated thermal efficiency is defined as the (gross-indicated-work)/ LHV-fuel. As can be seen, there is a general trend of increasing thermal efficiency with increased boost, with values generally being about 43 - 44%. The highest load point reaches 47%.

Combustion efficiencies also increase with the boost pressure, and they are quite high compared to typical moderate-load HCCI combustion, ranging from 97 to 99% for P$_{in}$ = 100 to 325 kPa, respectively. The high combustion efficiency at P$_{in}$ = 100 kPa is mainly the result of high combustion temperatures due to the high fueling rate (ϕ = 0.48, C/F = 31), a relatively high intake temperature (131°C), and high wall temperatures (firedeck temperature = 139°C). As P$_{in}$ increases, the combustion efficiency increases, for two reasons: 1) the wall temperatures progressively increase, reaching 164°C at P$_{in}$ = 325 kPa, which helps maintain reactions close to the wall, and 2) high levels of EGR reduce the net flow of gases through the engine, which causes a proportional reduction in the HC and CO emissions (*i.e.* the exhaust concentrations of HC and CO are not necessarily reduced by adding EGR, but the amount of gas exiting the exhaust pipe is significantly reduced since a large fraction of the exhaust gas is recirculated back into the intake). An alternative view point is that the HC and CO in the EGR gases get a second chance to burn.

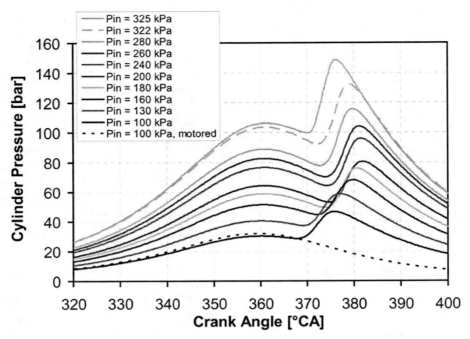

Figure 5. Cylinder pressure traces corresponding to the maximum-IMEP$_g$ points in <u>Fig. 3</u>*.*

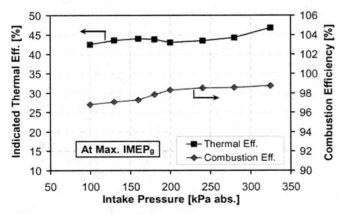

Figure 6. Indicated thermal efficiency and combustion efficiency for the maximum-IMEP$_g$ points in <u>Fig. 3</u>

= 325 kPa vs. C/F = 31 for P_{in} = 100 kPa). However, the CO in percent fuel carbon actually decreases since the net throughput of gases is reduced by more than a factor of two as P_{in} is increased up to 325 kPa. Similarly, the HC concentration drops only slightly, most likely because higher wall temperatures promote reaction closer to the wall or crevices where HC emissions tend to arise [19,20]. However, HC in percent fuel carbon shows a much greater drop due to the lower net flow through the engine. It is noteworthy that the combustion efficiency shows a particularly steep rise from $160 \leq P_{in} \leq 200$ kPa as the EGR levels are ramping up steeply (<u>Fig. 4</u>).

Evidence of this latter effect can be found in an analysis of the HC and CO emissions plotted in <u>Fig. 7</u>. This plot shows HC and CO emissions in both concentration (ppm) and normalized by the amount of fuel supplied (percentage of fuel carbon). As can be seen, for $P_{in} \leq 160$ kPa, the two curves for each emission-type track almost perfectly, but with EGR addition for $P_{in} \geq 180$ kPa, the curves diverge due to the reduction of gases moving though the engine (reason 2 above). Exhaust CO concentration increases with increasing P_{in} indicating less complete combustion due to lower combustion temperatures (see <u>Fig. 8</u>) resulting from the reduced intake temperature, lower γ of the EGR gases, and the increase in C/F at the higher boost levels (C/F = 38 at P_{in}

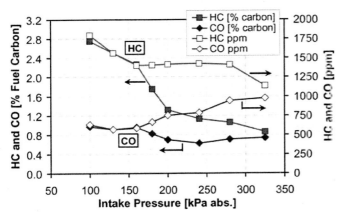

Figure 7. HC and CO emissions in both concentration (ppm) and normalized by the amount of fuel supplied (% of fuel carbon), for the maximum-IMEP$_g$ points.

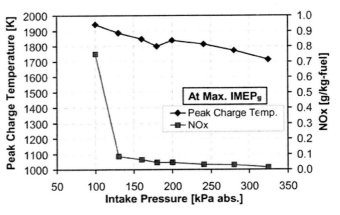

Figure 8. Peak charge temperature (mass-averaged) and NOx emissions as a function of P$_{in}$ for the maximum-IMEP$_g$ points. NOx = 1 g/kg-fuel is approximately the US 2010 emission standard.

NOx Emissions and Temperature

NOx emissions and the peak, mass-averaged combustion temperatures are presented in Fig. 8. NOx is given in g/kg-fuel, which has the convenience of 1 g/kg-fuel being representative of the US-2010 NOx standard. The figure shows that for all boosted conditions, NOx emissions are more than a factor of ten below the US-2010 standard. The only significant NOx emission occurs for the baseline, naturally aspirated point (P$_{in}$ = 100 kPa), but even this point is comfortably below the standard. In practice, NOx emissions at this load IMEP$_g$ = 5.06 bar could be reduced by increasing the C/F slightly and compensating with a small amount of boost.

The temperature curve in Fig. 8 shows the reason for the low NOx emissions. Peak mass-averaged temperatures are below 1900 K for all boosted conditions, falling as low as 1712 K at P$_{in}$ = 325 kPa, where NOx emissions are almost a factor of 70 below the US-2010 standard. The higher NOx levels for Pin = 100 kPa correspond to the peak temperature rising above 1900 K to a value of 1943 K.

Ringing Intensity and COV of IMEP$_g$

Figure 9 shows the ringing intensity and coefficient of variation (COV) of the IMEP$_g$. The COV provides a quantitative measure of the cycle-to-cycle variation in the IMEP$_g$, *i.e.* the combustion stability. The data show that the stability is very good for all the maximum IMEP$_g$ points, with the COV never exceeding 1.5%. Above P$_{in}$ = 240 kPa, the COV becomes smaller again due to the more advanced combustion phasing. As can be seen by a comparison of the COV curve with the CA50 curve in Fig. 10, the changes in COV with P$_{in}$ correspond closely with changes in the amount of CA50 retard.

As discussed previously, the ringing intensity was kept ≤ 5 MW/m^2 to prevent engine knock. However, as Fig. 9 shows, the knock/stability limit was reached at a ringing intensity of 2 MW/m^2 for P$_{in}$ = 100 kPa. The reason is that although this point was stable, with this intake pressure, the combustion phasing was very sensitive to small variations in the compressed-gas temperature. Any small increase in fueling or small increase in retard (initiated by a reduction of T$_{in}$), caused the engine to work its way to run-away knock or decay to misfire. This behavior is discussed in greater depth in Refs. [1,2]. As boost was increased, the propensity for run-away knock progressively decreased, and by 180 kPa, the load could be increased to the ringing limit of 5 MW/m^2. This corresponds to a large increase in CA50 retard as shown in Fig. 10. This behavior indicates that despite the small increase in cycle-to-cycle variation (COV) with increased retard, the likelihood of run-away knock is greatly diminished. For P$_{in}$ > 180 kPa, the ringing was held constant at 5 MW/m^2 to prevent audible knock; however, higher ringing levels did not lead immediately to run-away knock as they did at lower P$_{in}$. Keeping the ringing below this limit was accomplished by maintaining sufficient combustion phasing retard.

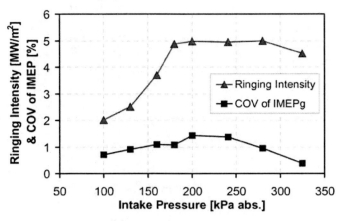

Figure 9. Ringing intensity and COV of the IMEP_g for the maximum-IMEP_g points.

Combustion Phasing

As indicated in the above discussion, combustion phasing retard is critical to achieving stable high-load combustion without knock under boosted conditions. Figure 10 shows that to achieve the maximum-IMEP_g points discussed above, CA50 was retarded substantially, as far as 379.2° CA (19.2° aTDC). This is not normally possible under naturally aspirated conditions, where the maximum CA50 retard with acceptable stability is typically found to be about 373° CA, similar to the P_{in} = 100 kPa high-load point presented here. Nevertheless, as boost was increased, more retarded combustion with good stability could be achieved in a straightforward manner, as discussed above.

As Fig. 10 shows, the increase in CA50 retard is nearly linear with increasing boost up to P_{in} = 200 kPa. However, the amount of CA50 retard with increased boost seems to be limited to about 379° CA. For the P_{in} = 240 kPa data point, CA50 is essentially the same, 378.8° CA. The small difference is likely not significant considering the uncertainty in achieving the "true" high-load limit and in the repeatability of CA50 from one 100-cycle average to the next at these high-load conditions. This apparent limit to the CA50 retard for these operating conditions is thought to be one of the main reasons that the slope of the maximum IMEP_g vs. P_{in} line in Fig. 3 becomes a little less steep for $P_{in} \geq$ 200 kPa. (See also the discussion of this subject with respect to Fig. 3.) For $P_{in} \geq$ 260 kPa, the mixture is nearly stoichiometric due to the high levels of EGR, and less retard is required due to the decreased fraction of reactants (fuel and air) in the charge. It should be noted, however, that stability remains very good for these high intake pressures. Therefore, more retard can be applied if desired, but reactants must be displaced with

additional EGR to achieve this, and thermal efficiency will be reduced slightly due to the greater retard. (Additional discussion of this is given above with respect to the second-highest load point in Fig. 5.) The second-highest load point is also plotted in Fig. 10 to demonstrate this ability to further retard CA50 at these high-boost conditions.

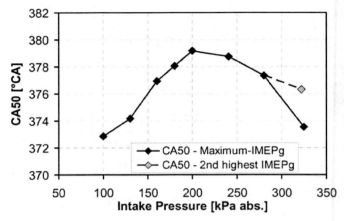

Figure 10. Combustion phasing as measured by the 50% burn point (CA50) for maximum-IMEP_g points.

RESULTS AND DISCUSSION - HRR ANALYSIS

The performance data presented in the previous section demonstrate that boosted HCCI can provide high loads up to IMEP_g = 16.34 bar with ringing $\leq$ 5 MW/m^2, and a low COV of IMEP_g. Achieving this performance requires substantial combustion-phasing retard to control knock, and this must be done while maintaining good stability. As discussed above, under boosted conditions, CA50 could be retarded as late as 379° CA with good stability, whereas for naturally aspirated conditions, it is difficult to maintain stability for CA50 retard > 373° CA. To understand why this was possible a further investigation was undertaken.

IN-CYLINDER TEMPERATURES AND ITHR

As originally described by Sjöberg and Dec [4], and mentioned briefly in the discussion of the knock/stability point above, the stability of HCCI combustion depends on the rate of temperature rise prior to the hot ignition point. For retarded combustion, this temperature rise rate is affected by both the magnitude of early combustion reactions and the rate of expansion cooling due to piston motion. Since these early combustion reactions occur at temperatures above those of low-temperature heat release and below those required for hot ignition, the term intermediate-temperature heat release (ITHR) will be used here, as introduced in Ref. [21]. In order to reach the temperatures required for hot autoignition, the

ITHR must be sufficiently high for temperatures in the hottest regions of the charge to continue rising despite the expansion. With greater combustion retard, the rate of expansion increases, which slows the rate of temperature rise for a given operating condition and makes it increasingly difficult to obtain stable hot ignition. Thus, the amount of allowable combustion phasing retard is limited unless the magnitude of the ITHR increases.

To understand how this relates to the current study, Fig. 11 presents in-cylinder temperatures for the maximum-$IMEP_g$ points and for motored operation. To facilitate comparison of the traces, the temperatures have been offset to match the value at 345° CA (prior to any ITHR) for the $P_{in} = 160$ kPa case, which is in the middle of the temperature range (maximum adjustment was 40 K). This is necessary since the compressed-gas temperatures vary significantly between these data points due to changes in T_{in} and the amount of EGR.[6]

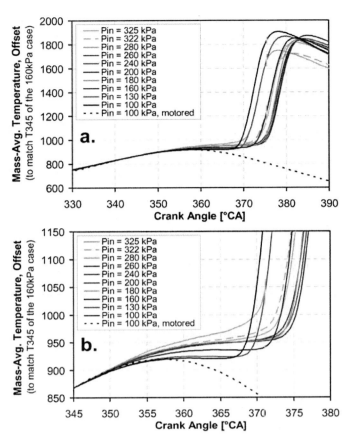

Figure 11. Normalized mass-averaged in-cylinder temperature corresponding to the maximum-IMEPg points in Fig. 3; (a) shows the full temperature rise with combustion, and (b) has an amplified scale to better view the early temperature rise.

As Fig. 11a shows, the motored-engine temperature drops after TDC, but for the fired cases, temperatures rise above the motored temperature even before TDC. Then they continue to rise slowly or to be essentially flat until the point of hot ignition where the temperature curves turn sharply upward. Note that with careful examination using the amplified scale in Fig. 11b, some of the curves show a slight temperature drop just prior to hot ignition; however, this is thought to be only an artifact of using mass-averaged temperatures, which will be colder than the hottest parts of the charge that must continue to rise to reach hot ignition. In any case, it can be seen that as P_{in} is increased from 100 - 130 - 160 kPa, temperatures progress to hot ignition at increasingly retarded crank angles. Thus, the ITHR reactions must be increasing with boost to counteract the increasing rapid expansion cooling at later crank angles. For further increases in P_{in} above 160 kPa, even the mass-averaged temperatures show a continuous upward trend despite most of them being even

---

[6] Aligning the temperature plots by scaling them to match the temperature at 345° CA for the $P_{in} = 160$ kPa case was also examined, and the results are nearly identical to those presented. However, offsetting is preferred since it preserves the slope of the temperature rise which is central to this comparison.

more retarded. For the maximum load case, combustion phasing was more advanced again, and the mass-averaged temperature shows a very strong upward trend throughout the ITHR period.

The increasing magnitude of the ITHR with boost is more clearly evident in Fig. 12, which presents the HRR traces for the maximum-$IMEP_g$ points on two different scales. Since the magnitude of the heat release and combustion phasing vary greatly between conditions, these HRR traces have been normalized by the total heat release and offset to align the peak HRR. With this normalization and offset, the overall trends in the HRR appear quite similar, as can be seen in Fig. 12a. However, examining the ITHR on an amplified scale in Fig. 12b shows that there are distinct differences with boost. In agreement with the discussion of the temperature traces in Fig. 11, the naturally aspirated case has the lowest ITHR. Then, there is a progressive increase in the magnitude of the ITHR as P_{in} is increased from 100 - 130 - 160 - 180 kPa. However, for $P_{in} \geq 180$ kPa, the enhancement appears to saturate and there is little change from 180 to 325 kPa. This is likely the reason that boost levels above $P_{in} = 200$ kPa did not allow CA50 to be retarded beyond limit of 379° CA. The reason that $P_{in} = 200$ kPa could be retarded 1° CA more than $P_{in} = 180$ kPa is unclear, but the difference is small in any case.

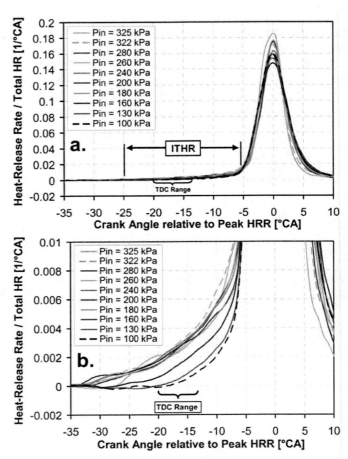

Figure 12. Normalized heat-release rates (HRR) for the maximum-IMEP$_g$ points in Fig. 3. The curves are offset to align the points of peak HRR; (a) shows the full HRR plots, and (b) has an amplified scale to better view the ITHR. "TDC range" indicates the variation in combustion phasing for these operating points.

PRESSURE AND TEMPERATURE EFFECTS ON ITHR

The HRR data in Fig. 12 suggest that increasing boost enhances the ITHR, which is critical for the required increase in CA50 retard. However, it should be noted that the intake temperature was reduced from 131°C → 60°C as P_{in} was increased from 100 → 180 kPa. Then, for $P_{in} \geq 180$ kPa, T_{in} was held constant at 60°C, and EGR was added to control combustion phasing. Since the reduction in T_{in} corresponds to the intake pressures for which a large increase in ITHR occurred, and the constant $T_{in} = 60$°C corresponds to the intake pressures for which the ITHR did not change significantly, the changes in ITHR may result from the temperature change, the pressure change, or a combination of the two.

To clarify this, additional tests were conducted, as shown in Fig. 13. For these data, the 10% burn point (CA10), which is commonly taken as being representative of the ignition point, was held nearly constant by adjusting the amount of EGR to avoid the significant offset required for the data in Fig. 12. This is reflected in the smaller "TDC range" noted near the bottom of the figure. To separate the effects of temperature and pressure, data were acquired for a baseline P_{in} = 100 kPa, T_{in} = 131°C condition, for a boosted condition with the same hot intake temperature, P_{in} = 200 kPa, T_{in} = 130°C, and for a boosted condition with lower intake temperature, P_{in} = 200 kPa, T_{in} = 60°C. As can be seen, boosting P_{in} to 200 kPa with a hot T_{in} = 130°C changes the nature of the ITHR compared to T_{in} = 60°C. However, the ITHR is still significantly enhanced relative to P_{in} = 100 kPa, particularly for the period after TDC which is the most critical for stable retarded combustion. In fact, for the last few crank angle degrees shown, the ITHR for P_{in} = 200 kPa is nearly the same for both the 130°C and 60°C intake temperatures.

Reducing T_{in} to 60°C at the boosted condition mainly increases the early portion of the ITHR. This increase in the early ITHR can only help maintain stability for the highly retarded operation. However, it is less significant than the late ITHR enhancement, which occurs when the expansion rate is higher. Thus, it appears that the pressure-induced increase in the ITHR is the more important effect for stability at late CA50, but the combined effect of temperature reduction likely contributes as well.

For the lower intake-temperature condition (P_{in} = 200 kPa, T_{in} = 60°C), the earliest part of this heat release, beginning 25° CA before CA10 (15° CA bTDC) is likely indicative of a small amount of LTHR, which then leads into enhancement of the early ITHR reactions [4]. To clarify this, additional P_{in} = 200 kPa data points were taken for a range of intake temperatures, also with CA10 held nearly constant by varying the amount of EGR. The early HRRs for this temperature sweep are presented in Fig. 14. Note that the T_{in} = 130°C and 60°C curves are the same as those presented in Fig. 13. As shown, reducing T_{in} from 80 to 60°C causes the onset of heat release to shift several degrees earlier. Further reduction in T_{in} to 45°C results in a distinct LTHR peak separated from the later ITHR by a local minimum in the HRR curve. Since the onset of the heat release for T_{in} = 45°C is nearly the same as for T_{in} = 60°C, it is likely that the early heat release for T_{in} = 60°C involves LTHR reactions.

Finally, it should be noted that maintaining a constant CA10 over the range of temperatures presented for P_{in} = 200 kPa requires substantial variation in the amount of EGR. The effect of this on the ITHR is uncertain. However, it seems most likely that it would reduce the ITHR, so the P_{in} = 200

kPa, T_{in} = 130°C curve in Figs. 13 and 14 should represent a "worst case" in terms of the magnitude of the pressure enhancement of the ITHR.

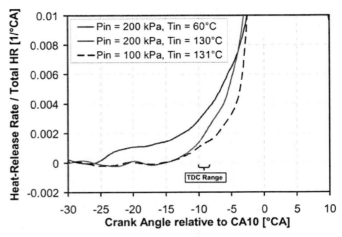

Figure 13. Normalized heat-release rates (HRR) showing the effect of boost on the ITHR both with and without reducing the temperature relative to the baseline P_{in} = 100 kPa case.

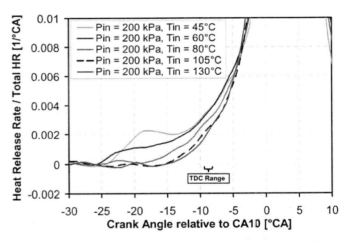

Figure 14. Normalized heat-release rates (HRR) showing the effect on the ITHR of varying T_{in} for P_{in} = 200 kPa.

A detailed HRR analysis on data from tests conducted under more well-controlled conditions was also performed to examine the apparent saturation effect of the ITHR enhancement for P_{in} ≥ 180 kPa. The results are presented in Fig. 15. These data were acquired with a nearly constant CA10, as indicated in the figure, and the C/F was nearly constant as well (C/F = 41 - 42). For these well-controlled conditions the ITHR curves align closely, showing that varying P_{in} from 180 - 280 kPa has virtually no effect on the ITHR. This verifies the conclusion drawn from the maximum-IMEP$_g$ data in Fig. 12 that indeed the ITHR is no longer enhanced with boost levels above 180 kPa for this engine and operating condition.

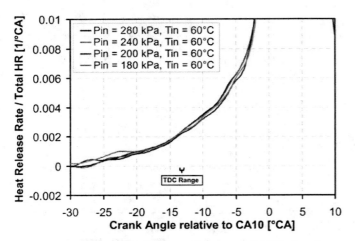

Figure 15. Normalized heat-release rates (HRR) showing that increasing P_{in} from 180 → 280 kPa at a constant $T_{in} = 60°C$ has little effect on the ITHR.

FUEL EFFECTS ON ITHR

As has been shown, the reference gasoline used in this study allowed a significant increase in CA50 retard as intake boost was increased, due to an increased magnitude of the ITHR. As a first step toward investigating the role of fuel composition on this enhancement, Fig. 16a and 16b compare the ITHRs of gasoline with those of ethanol, the latter of which were acquired as part of a concurrent study of the potential of ethanol as an HCCI fuel [22]. For both fuels, the C/F was held constant over the pressure sweep, and fueling rates were similar between the two fuels. For ethanol $\phi = 0.4$, and for gasoline $\phi = 0.38$. For the $P_{in} = 180$ kPa gasoline point, EGR addition was required, but the C/F was kept the same as that of the lower-P_{in}, $\phi = 0.38$ points.

Figure 16a shows that the ITHR for gasoline increases with boost up to $P_{in} = 180$ kPa, similar to the changes noted for maximum-IMEP$_g$ points presented in Fig. 12. In contrast, ethanol shows no change in ITHR with boost from $P_{in} = 100$ → 247 kPa. The lack of ITHR enhancement with ethanol is thought to be related to its relatively simple molecular structure, which apparently does not have pressure-dependent reactions during the early stages of autoignition. Nevertheless, these data show that the ITHR enhancement with boost can vary widely between fuels.

The lack of ITHR enhancement with boost for ethanol indicates that the combustion-phasing retard with boost will probably be limited to values much less than those possible with gasoline. This will likely limit the high-load capability under boosted operation. In fact, the ITHR with ethanol is

even a little less than that of gasoline for naturally aspirated conditions, as can be seen by comparing the $P_{in} = 100$ kPa ITHR of gasoline with that of ethanol, which is included in Fig. 16a for reference. It is unclear whether this small difference is significant, but it might slightly reduce the allowable retard, and therefore load, at naturally aspirated conditions as well.

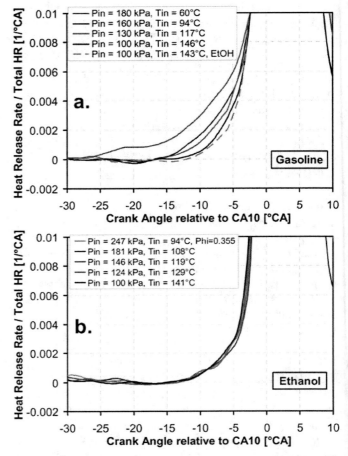

Figure 16. Normalized heat-release rates (HRR) showing a comparisons of the ITHR for (a) gasoline and (b) ethanol over a range of intake pressures. The ethanol data is reproduced from Ref. [22].

SUMMARY AND CONCLUSIONS

The use of intake-pressure boosting to obtain high loads with well-mixed HCCI has been systematically investigated. Intake pressures (P_{in}) were progressively increased from 100 kPa (naturally aspirated) to 325 kPa absolute, and for each P_{in} the fueling was gradually increased up to the knock/stability limit. To compensate for the pressure-induced enhancement of autoignition and to retard the combustion phasing sufficiently to prevent knock, a combination of intake-temperature reduction and cooled-EGR addition was used. Intake temperatures ranged from the 130°C required for $P_{in} =$

100 kPa to 60°C for $P_{in} \geq 180$ kPa. Above 180 kPa, intake temperatures were held constant at 60°C, and EGR was added to further retard the combustion phasing. The amount of EGR addition increased with increasing boost, and for $P_{in} \geq 260$ kPa, the charge was stoichiometric (*i.e.* all the excess air was replaced with EGR). For the highest $P_{in} = 325$ kPa, the EGR mole fraction reached 60%. The fuel was a research-grade 87-octane gasoline (RON = 91, MON = 83). All data were taken at 1200 rpm, and the HCCI research engine was equipped with a compression-ratio 14 piston. The study produced the following results:

1. A high load of 16.34 bar $IMEP_g$ was obtained with no ringing (*i.e.* no engine knock), very good stability, high efficiency, and ultra-low NOx, at $P_{in} = 3.25$ bar.

2. The maximum attainable $IMEP_g$ increased almost linearly with boost up to $P_{in} = 200$ kPa, and above 200 kPa at a slightly lesser rate. However, the rate of increase was still substantial at the highest load, indicating the even higher loads could be achieved at greater boost levels.

3. For the maximum-$IMEP_g$ points at the various P_m, indicated thermal efficiencies ranged mostly from 43 - 44%, but reached 47% for the highest-load point. Combustion efficiencies ranged from 97 to 99%.

4. NOx emissions were extremely low (more than a factor of 10 below US-2010 standards) for all boosted conditions, due to low peak combustion temperatures. For the naturally aspirated condition, NOx was higher, but still comfortably below US-2010 standards.

5. The COV of the $IMEP_g$ was less than 1.5% for the maximum $IMEP_g$ points at all boost levels.

6. The ability to substantially retard the combustion phasing under boosted conditions to control knock was key to achieving these results. CA50 was retarded as late as 379° CA (19° aTDC) with good stability.

7. A detailed examination of the heat-release rates showed that this highly retarded combustion was possible because the early autoignition reactions (*i.e.* the intermediate-temperature heat release, ITHR) are significantly enhanced for boosted operation. This keeps the bulk-gas temperatures rising despite the late CA50.

8. Experiments with intake pressures and temperatures controlled independently showed that increased P_{in} is the most important factor for enhancing the ITHR, but that reduced intake temperatures also contribute.

9. The ITHR is progressively enhanced as boost is increased from 100 → 180 kPa. However, the enhancement appears to reach a limit at $P_{in} = 180$ kPa, and it does not increase further as P_{in} is increased up to 325 kPa.

10. The ITHR enhancement is fuel dependent for fuels whose molecular structure varies substantially. A comparison of these gasoline data with ethanol data from a concurrent study [22] showed that ethanol has no enhancement of the ITHR with intake boost.

Overall, the results of this study show that well-controlled, boosted HCCI has a strong potential for achieving high power levels, close to those of turbo-charged diesel engines.

REFERENCES

1. Olsson, J-O., Tunestål, P., Johansson, B., Fiveland, S., Agama, J. R., and Assanis, D. N., "Compression Ratio Influence on Maximum Load of a Natural Gas-Fueled HCCI Engine," SAE Technical Paper 2002-01-0111, 2002.

2. Sjöberg, M., Dec, J. E., Babajimopoulos, A., and Assanis, D., "Comparing Enhanced Natural Thermal Stratification against Retarded Combustion Phasing for Smoothing of HCCI Heat-Release Rates," *SAE Transactions*, 113(3), pp. 1557-1575, Paper 2004-01-2994, 2004.

3. Sjöberg, M. and Dec, J. E., "Influence of Fuel Autoignition Reactivity on the High-Load Limits of HCCI Engines," *SAE Int. J. of Engines* 1(1):39-58, 2008.

4. Sjöberg, M. and Dec, J. E., "Comparing Late-cycle Autoignition Stability for Single- and Two-Stage Ignition Fuels in HCCI Engines," *Proceedings of the Combustion Institute*, Vol. 31, pp. 2895-2902, 2007.

5. Christensen, M., Johansson, B., Amnéus, P. and Mauss, F., "Supercharged Homogeneous Charge Compression Ignition," SAE Technical Paper 980787, 1998.

6. Christensen, M. and Johansson, B., "Supercharged Homogeneous Charge Compression Ignition (HCCI) with Exhaust Gas Recirculation and Pilot Fuel," SAE Technical Paper 2000-01-1835, 2000.

7. Olsson, J.-O., Tunestål, P., Haroldsson, G. and Johansson, B., "A Turbo-Charged Dual Fuel HCCI Engine," SAE Technical Paper 2001-01-1896, 2001.

8. Bessonette, P. W., Schleyer, C. H., Duffy, K. P., Hardy W. L., and Liechty, M. P. "Effects of Fuel Property Changes on Heavy-Duty HCCI Combustion," SAE Technical Paper 2007-01-0191, 2007.

9. Kalghatgi, G. T., Risberg, P., and Ångström, H.-E., "Partially Pre-Mixed Auto-Ignition of Gasoline to Attain Low Smoke and Low NOx at High Load in a Compression Ignition Engine and Comparison with a Diesel Fuel," SAE Technical Paper, 2007-01-0006, 2007.

10. Sjöberg, M. and Dec, J. E., "Influence of EGR Quality and Unmixedness on the High-Load Limits of HCCI Engines," *SAE Int. J. of Engines*, 2(1):492-510, 2009.

11. Dec, J. E., Sjöberg, M., and Hwang, W., "Isolating the Effects of EGR on HCCI Heat-Release Rates and NOx Emissions," *SAE Int. J. Engines* 2(2):58-70, 2009.

12. Sjöberg, M., Dec, J. E., and Hwang, W., "Thermodynamic and Chemical Effects of EGR and Its Constituents on HCCI Autoignition," *SAE Transactions* 116(3), SAE Technical Paper 2007-01-0207, 2007.

13. Dec, J. E. and Sjöberg, M., "Isolating the Effects of Fuel Chemistry on Combustion Phasing in an HCCI Engine and the Potential of Fuel Stratification for Ignition Control," *SAE Transactions*, 113(4), pp. 239-257, SAE Technical Paper 2004-01-0557, 2004.

14. Dec, J. E. and Sjöberg, M., "A Parametric Study of HCCI Combustion - the Sources of Emissions at Low Loads and the Effects of GDI Fuel Injection," *SAE Transactions*, 112(3), pp. 1119-1141, SAE Technical Paper 2003-01-0752, 2003.

15. Heywood, J. B., Internal Combustion Engine Fundamentals, McGraw-Hill, New York, 1988.

16. Sjöberg, M., Dec, J. E., and Cernansky, N. P., "The Potential of Thermal Stratification and Combustion Retard for Reducing Pressure-Rise Rates in HCCI Engines, based on Multi-Zone Modeling and Experiments," *SAE Transactions*, 114(3), pp. 236-251, SAE Technical Paper 2005-01-0113, 2005.

17. Eng, J. A., "Characterization of Pressure Waves in HCCI Combustion," SAE Technical Paper 2002-01-2859, 2002.

18. Sjöberg, M. and Dec, J. E., "EGR and Intake Boost for Managing HCCI Low-Temperature Heat Release over Wide Ranges of Engine Speed," *SAE Transactions*, 116(3), pp. 65-77, SAE Technical Paper 2007-01-0051, 2007.

19. Christensen, M., Johansson, B., and Hultqvist, A. "The Effect of Piston Topland Geometry on Emissions of Unburned Hydrocarbons from a Homogeneous Charge Compression Ignition (HCCI) Engine," SAE Technical Paper 2001-01-1893, 2001.

20. Dec, J. E., Davisson, M. L., Leif, R. N., Sjöberg, M., Hwang, W., "Detailed HCCI Exhaust Speciation and the Sources of Hydrocarbon and Oxygenated Hydrocarbon Emissions," *SAE Int. J. of Fuels and Lubricants*, 1(1):50-67, 2008.

21. Hwang, W., Dec, J. E., and Sjöberg, M., "Spectroscopic and Chemical-Kinetic Analysis of the Phases of HCCI Autoignition and Combustion for Single- and Two-Stage Ignition Fuels," *Combustion and Flame*, 154(3), pp. 387-409, 2008.

22. Sjöberg, M. and Dec, J. E., "Ethanol Autoignition Characteristics and HCCI Performance for Wide Ranges of Engine Speed, Load and Boost," *SAE Int. J. Engines* 3(1): 84-106, 2010, doi:10.4271/2010-01-0338.

CONTACT INFORMATION

Corresponding author

John E. Dec
Sandia National Laboratories
MS 9053, PO Box 969
Livermore, CA 94551-0969, USA.

ACKNOWLEDGMENTS

The authors would like to thank the following people from Sandia National Laboratories: Magnus Sjöberg for many helpful discussions and for use of the ethanol data, Nicolas Dronniou for help with filtering the detailed HRR curves, and Gary Hux, Kenneth St. Hilaire, David Cicone, Christopher Carlen and Gary Hubbard for their dedicated support of the HCCI engine laboratory.

This work was performed at the Combustion Research Facility, Sandia National Laboratories, Livermore, CA. Support was provided by the U.S. Department of Energy, Office of Vehicle Technologies. Sandia is a multiprogram laboratory operated by the Sandia Corporation, a Lockheed Martin Company, for the United States Department of Energy's National Nuclear Security Administration under contract DE-AC04-94AL85000.

Measuring Near Zero Automotive Exhaust Emissions - Zero Is a Very Small Precise Number	2010-01-1301 Published 04/12/2010

Wolfgang Thiel, Roman Woegerbauer and David Eason
BMW Group

ABSTRACT

In the environmentally conscious world we live in, auto manufacturers are under extreme pressure to reduce tailpipe emissions from cars and trucks. The manufacturers have responded by creating clean-burning engines and exhaust treatments that mainly produce CO2 and water vapor along with trace emissions of pollutants such as CO, THC, NCx, and CH4. The trace emissions are regulated by law, and testing must be performed to show that they are below a certain level for the vehicle to be classified as road legal. Modern engine and pollution control technology has moved so quickly toward zero pollutant emissions that the testing technology is no longer able to accurately measure the trace levels of pollutants. Negative emission values are often measured for some pollutants, as shown by results from eight laboratories independently testing the same SULEV automobile. The negative emission values are shown to be caused by actual values that are very near zero, physically imperfect testing equipment and practical testing issues.

INTRODUCTION

BMW has been at the forefront of automotive vehicle emission testing since it began in the early 70's, helping to improve much of the testing equipment, conducting research into emissions theory, and assisting in developing the mathematical equations used in vehicle emissions testing. As engines become cleaner and cleaner, emissions testing must become equally more precise and accurate. This requires engineers to investigate the measurement limits of current equipment. Engines today produce mostly CO_2 and H_2O emissions as a result of burned hydrocarbons, but they also produce less desirable pollutant emissions that are a result of imperfect combustion. As less of these pollutant emissions are produced, the measurement limit of detection for each

pollutant must also approach zero or it will not be possible to measure what is produced.

The United States government has had emission regulations since 1972, enforced by the EPA (Environmental Protection Agency). The standard test for emissions is known as the FTP 75 (Federal Test Procedure 1975). This is a test cycle where the car is driven on a chassis dynamometer following a simulated driving route while the tailpipe emissions are measured. It consists of three phases. The first simulates city driving when the vehicle is "cold," meaning the engine has not yet reached operating temperature since it has not been running previously. The second phase is known as stable phase, as the engine has now reached normal running temperature. This phase is where most engines will produce the least emissions. The engine is running relatively low load and there are not so many dynamic throttle changes as compared with the first phase. The catalysts in the exhaust system have also reached operating temperature. The third phase also produces low emissions since the engine is warm and the driveline and exhaust are still at running temperature. There is a 10-minute pause between Phase 2 and 3, which is meant to simulate a "hot" start.

Engines now are cleaner than ever and produce near zero pollution component emissions in most cases This presents a problem for the current emission testing method. Measuring zero exactly is impossible with physical testing, as is shown in this paper by results of eight independent tests performed by six laboratories independent of BMW and by two BMW testing facilities. These tests were run on the same car to see what happens near the limit of zero emissions.

There are many ways to get negative results for emissions measurement. Three common ways are "faulty" analyzer readings, subtraction between two exhaust bags, and subtraction of THC and CH_4 to obtain NMHC. "Faulty" analyzer readings can be the result of subtraction, quenching, noise, or other impacting factors, and can yield negative measurement values. Subtracting two different exhaust gas bag readings that are both near zero (according to the bracketed term of Eq. (1)) is another common source for negative values. Finally, subtracting CH_4 from THC to get NMHC can yield negative measurement values for sufficiently small concentrations of CH_4 and THC [1].

One cause of the possible negative values is the bracketed term in Eq. (1).

$$m_E = \left[c_{Emix} - c_{Air} \bullet (1 - \frac{1}{DF}) \right] \bullet \rho \bullet V_{mix}$$

(1)

How close these values are to zero is something that should be clarified. The mass values that are important for the EPA and other agencies are given in g/mile, grams of component per mile. This is done through calculations that involve factors and conversions. The physical measurement is done, however, by measuring concentration in ppm (parts per million) or in volume percentage. For this reason this report only speaks of concentrations in ppm based on Eq. (1), which are the diluted net concentration values.

Of the three components, THC generally exhibits the highest concentration levels. During the 24 tests described below, the highest value of THC, and therefore the highest measured component concentration, was 4 ppmC, or 0.0004 % of everything that came from the exhaust pipe mixed with background air. Despite the miniscule volume percentages, such levels of pollutant have been accurately measured for decades with little error.

In contrast to THC, NO_x concentrations in the second phase could be as close to zero as 0.001 ppm, or 0.0000001% of total exhaust volume. This is such a low amount of substance that any measuring device will have problems distinguishing 0.0000001% of volume from zero, no matter how precise the equipment. This is the basis of this report, to discuss the ability to measure such low concentrations and to determine if and when real negative values are measured or calculated from diluted vehicle exhaust.

Why this is important can be illustrated in Table 1. This data table is calculated with Eq. (1) using hypothetical but realistic data. The arrows show how the NMHC concentration is calculated in the EPA and CARB methods, as described in [2]. Despite the different calculation methods, mathematically the end result should be identical. This is not the case, however, when a vehicle produces extremely low concentrations of THC or CH_4, such as a SULEV or Hydrogen IC-powered car. Two cases are presented that explore this issue.

<table 1 here>

Case 1 shows the problem with the different EPA and CARB measuring methods. In the EPA method, a difference is calculated between the diluted air and exhaust gas measurements for each species. The resulting THC and CH_4 differences are then subtracted to obtain the NMHC measurement. In the CARB method, a difference is calculated between the THC and CH_4 diluted air, and then again for the diluted exhaust gas. These two differences are then used to calculate the total NMHC.

This practice works well in most cases, but what happens when the diluted air or diluted exhaust gas THC and CH_4 concentrations are close to the same value? According to the CARB methodology, a negative result can occur, which is then corrected to zero [3]. The Control column shows what the difference value normally would be if it was not corrected to zero. This negative difference will cause the NMHC result for the CARB method to be markedly different from the EPA method.

This discrepancy is especially important for hydrogen vehicles, as their CH_4 and THC emissions can be incredibly close to zero ppm. The majority of this THC and CH_4 come from the vehicle intake and from the ambient air. In some cases, the emissions are so low such that the vehicle actually cleans the air as it drives.

Case 2 is simply a control case to show that this EPA - CARB discrepancy occurs only for near zero emissions results. For the most part, the EPA and CARB methodology agree quite well.

Because the analyzer may return negative values, or display negative values initially, that issue is discussed further below. However, as there are a wide range of possible causes for this analyzer error, it is only possible to briefly touch upon this effect within the scope of this paper. This will be presented as so-called "food for thought" in a separate subsection of this report. This analyzer issue has also been shown to affect emissions measurements from hydrogen vehicles when testing for CO and CO_2 concentrations [4].

TESTS ON A SINGLE SULEV VEHICLE AT EIGHT LABORATORIES

TEST DESCRIPTION AND OVERALL RESULTS

Eight independent testers ran three official FTP 75 tests each. This means a total of 24 tests were studied for negative or near negative values in THC, NO_x, and CH_4. The tests were conducted following the FTP 75 official guidelines. The locations and testers were changed and each set of tests were conducted on a different chassis dynamometer. The emissions of THC (Total Hydrocarbon), NO_x, and CH_4 were studied to explore the precision of the CVS (Constant Volume Sampling) system and to find out if negative values that occur from time to time are truly negative or just an error of testing.

The vehicle used was a 2005 BMW 325i, with a SULEV rated engine. In twenty four tests there were a total of twenty two negative values for THC, CH4, and NOx, from a total of 24*3*3 =216 values.

Figure 1 shows the break down of negative values of the bracketed term in the different lab tests. Lab 5 had the most negative values with six while Lab 6 found no negative values in testing.

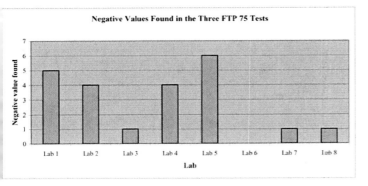

Figure 1. Negative values found in FTP 75 testing.

Figure 2 shows which pollution component produced the negative values. NO_x has the greatest number of negative values. Five of the eight labs had problems only with NO_x, while two other labs had negative values in CH_4 and THC. Individual results are shown in Appendix A. It is important to note that each lab had negative values of the bracketed term for only one of the three components.

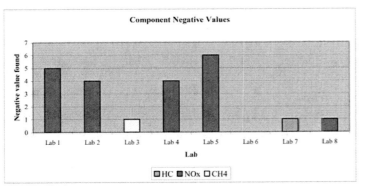

Figure 2. Component negative values found.

As discussed before, Phase 2 is the most stable phase since the engine is at running temperature and the driving cycle is less dynamic. As expected, Phase 2 has the most negative values. It has more than double Phase 1 or 3 values. This is shown in Fig. 3 and 4. Figure 4 shows in which phase the different labs experienced their negative values. For instance, Lab 1 had three of its negative values occur in Phase 1, two in Phase 2, and none in Phase 3. Lab 5 had an equal amount in each of the three phases. While it seems negative values are more likely in Phase 2, they can still occur in the other phases as well. This may indicate that the problem is not entirely caused by the low emissions from the vehicle, but instead may be partly caused by issues with the testing equipment.

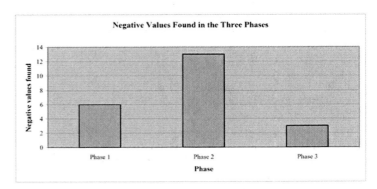

Figure 3. Negative values in the three phases.

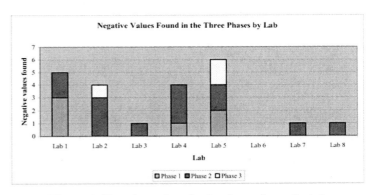

Figure 4. Negative values in the three phases by lab.

It is also important to know the results expressed as pollutant g/mi emissions. The pollutant mass per distance driven was measured and recorded for each lab. The three test results per lab were averaged together and are displayed in Fig. 5. The results are displayed per lab per species, with the Overall Average of all results displayed at the end.

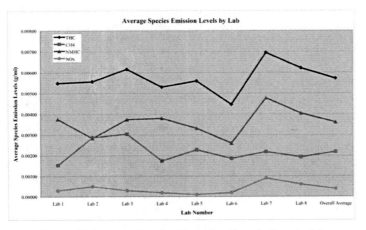

Figure 5. Average Species Emission Levels (in g/mi) by Lab.

It can be seen from Fig. 5 that although there are some deviations from the overall average for THC, CH₄, NMHC, and NO$_x$, few of the deviations occur simultaneously or to the same degree as that of another species for a specific testing lab. This suggests that these deviations may simply be outlying values from a faulty test or from an outside influencing factor, or that they are just the result of deviations stemming from the lowest possible variation.

The averages for each species for all twenty four tests are collected in Table 2, shown below.

Table 2. Test Pollutant Emission Averages by Species

	THC (mg/mi)	CH₄ (mg/mi)	NMHC (mg/mi)	NO$_x$ (mg/mi)
Test Average	5.72	2.18	3.61	0.39

These values are very low, as befits a SULEV vehicle. The uncertainty for the above pollutant values is ±1 mg/mi, or 0.001 g/mi. The NO$_x$ average value is so small as to be statistically zero. These values and graph are not the focus of this paper, but are simply given to show how close we are to "zero" emissions.

PHASE 1 TOTAL HYDROCARBON (THC) RESULTS

In order to understand where the negative values are coming from, it is necessary to discuss null tests. Null tests are tests where the vehicle is not hooked up to the CVS testing system, but the test is conducted as if it were. The CVS system takes in and analyzes room air and compares it to the same room air. This is a way to test for calibration and drift errors in the testing system. It is common to run this type of test before and after an official test to make sure the values obtained are accurate.

Each of the labs ran a null test before the official test for better precision. These tests are important because they allow us to see how accurate a machine can measure these concentrations on a given day under given circumstances. The null test values should be zero since the room air that is analyzed like vehicle exhaust is compared to the same room air. The null test results for THC in Phase 1 are shown in Fig. 6, with many hovering around the +0.02 ppm mark. This shows that the total measurement process is not perfect. If the system were perfect, the values should all be close to zero with equal amount slightly above or below zero.

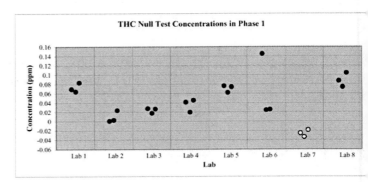

Figure 6. Null THC test concentration in Phase 1 (negative values marked in white).

At first glance it would seem the above values would throw off the test since they are clearly not zero. However, this is not the case when the actual test values are shown with the null test values and plotted to the same scale as in Fig. 7.

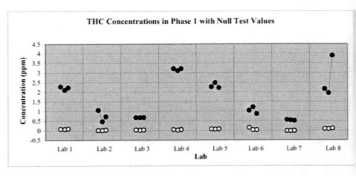

Figure 7. Null and test THC concentration in Phase 1 (null test values in white, test values in black)

It is obvious in Fig. 7 that the slight error in the null test is dwarfed in comparison to the actual value of THC concentration. In most cases the test value is over an order of magnitude higher than the null value. Figure 7 also shows that there appears to be no apparent connection between null value and test value, as the pattern of high or low null values does not follow the pattern of test values.

All of the results for experimental results and null tests can be found in Appendices A and B.

PHASE 2 NITROGEN OXIDES (NOX) RESULTS

NO_x concentrations can be evaluated in much the same way as THC. Results for CH_4 are similar to NOx and can be found in the Appendices. The method of analysis is the same for all tests. The NO_x concentrations had the greatest amount of negative values and most of those came in Phase 2. Looking at the null test in Fig. 8, the results appear better than the THC null results, as many of them are within 0.01 ppm of zero. The negative values indicate actual zero values. Lab 1 for instance has three values under 0.004 ppm difference from zero, and one of them is negative.

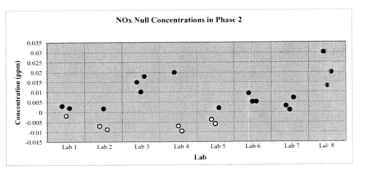

Figure 8. Null NO_x test concentration in Phase 2 (negative values in white).

The test values can now be plotted alongside the null values for reference, just like in the THC test earlier. There is, however, a very obvious difference: the test values for NO_x do not dwarf the null values. This is shown in Fig. 9.

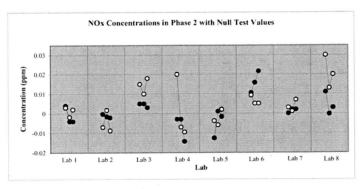

Figure 9. Null and test NO_x concentration in Phase 2 (Null values in white, test values in black).

The test concentrations of NO_x coming from the vehicle are comparable to the null test results. Now the question is: what is a true negative value and what is simply zero or just above zero? This is not a simple question for many reasons, as discussed below. If the two highest and the two lowest values in Fig. 9 are removed, the remaining data points appear to be evenly distributed around zero or slightly above zero. This is shown in Fig. 10.

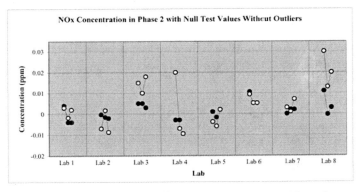

Figure 10. Null and test NO_x concentration in Phase 2 excluding highest and lowest two test values (Null values in white, test values in black).

DISCUSSION

INTERPRETING NEGATIVE EMISSIONS CONCENTRATIONS

Perhaps the most important question to ask before going any further in the analysis would be: is it possible to have true negative values for THC, NO_x and CH_4? In the past the answer was a simple no; however, now things are a little more complicated. With super effective catalytic converters and precision fuel injection, the possibility of negative calculated emissions, according to Eq. (1), is just that, possible. In some cases, such as for NO_x, some bag readings already were below zero.

Of course, real negative values for emission concentrations are impossible. This situation arises when diluted exhaust bag concentrations are subtracted from diluted ambient air bag concentration, and the exhaust bag has a lower concentration of pollutants than the ambient air. The analyzer would return a negative value for the emissions result but in reality the vehicle still is emitting minute quantities of pollutants. It is, however, essentially cleaning the air as it drives.

With cleaner hybrid engines and alternative fuel vehicles like hydrogen, negative values may start to be the norm instead of an anomaly. Despite this, BMW has never tested a vehicle with known negative emissions. For that reason, any statistical analysis cannot be based on a normal distribution as a normal distribution must range from negative to positive infinity. If statistics could be applied to the data it would only provide a probability that an apparent negative value is actually negative. In the end, the situation is too complex to rely on statistics alone to draw conclusions.

An interesting thing happens in the previous example of NO_x: there are more negative values in the emission test than the null test. This, however, does not prove that there are actual negative values. Many of the test results are within 0.005 ppm from zero, and even the outlying values are at most 0.03 ppm above and 0.015 ppm below zero. In the end it comes down to an accuracy issue: 0.005 ppm is such a small amount of substance that when zero is measured, noise alone could create such a signal.

There are two values, one from Lab 4 and one from Lab 5, which are more than 0.10 ppm below zero. The rest of the negative values are within 0.005 ppm below zero, or 0.0000005 % of total exhaust volume. At some point one must call such a small value zero. For any given test and for any given pollution component there are always outliers. These outliers can be positive or negative. However, having a negative calculated value does not prove that the actual value is negative, only that the calculated value is not valid. This is the case with many of the negative values. This means that even if the actual value of NO_x is positive, such as 0.01 ppm for example, statistically there will be some outliers that will end up negative, even very negative, just as there will be similarly large inaccurate positive values.

The data variability near zero was almost completely random. There were no obvious systematic errors, only apparent causes of negative values. The negative values were observed in Phase 2 more than the other phases simply because there was less load on the engine and therefore more complete combustion, so the values were closer to zero. NO_x had more negative values because it had the lowest concentrations, again closer to zero. In the end it is impossible to say whether negative values for this particular automobile are correct. One can only say that the current testing equipment has some

error, and engines have come to a point where the emission component is as small as the measurement error.

There is a trend in the data involving NO_x, such that the greater the standard deviation, the more negative values appear. This is shown in Fig. 11. This would seem to indicate that precision alone is the biggest influencing factor in measuring negative values. This means that the negative values are not a result of the car, but are rather due to measurement imprecision. The negative THC and CH4 values did not share this trend, nor did the NO_x values in the other two phases. This would seem to indicate the other negative values are random error or outlying data points.

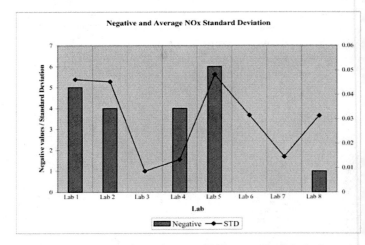

Figure 11. Negative values and NO_x standard deviation.

This brings us to an important question: what else can cause negative results to be measured instead of zero, and what limits our measurement precision? The answer for this question requires a detailed explanation of how the measurement equipment functions and a discussion on the limiting factors that affect the ability of each machine to measure low concentrations consistently and precisely.

PHYSICAL DIFFERENCES AMONG EXHAUST ANALYZERS

FLAME IONIZATION DETECTOR

The FID, or Flame Ionization Detector, is a simple device used to measure the concentration of hydrocarbon species in the exhaust gas.

In its simplest form, the FID is essentially a hydrocarbon detector consisting of a burner, electric field, detector, fuel, and sample. The sample gas enters the hydrogen flame inside the FID. Hydrocarbons in the sample gas become oxidized by reacting with the synthetic air in the detector chamber. Some of the oxidized molecules release an electron and become CHO+ ions. An electric field is applied across the hydrogen flame which captures the ions and electrons and makes them flow to the corresponding electrodes of the field. This movement generates a small current which is then amplified to be measured by the FID itself and displayed as a ppmC concentration.

A detailed description and explanation of the functional parts of an FID can be found in [5]. A gas cutter mechanism for the separate measurement of CH_4 is also described therein.

The FID calculates the concentration of THC and CH_4 by comparing the number of ionized molecules of either exhaust or ambient air with the number of background ions present in zero gas. When no sample gas is flowing into the FID, there is still a current generated. This must be offset in the device calibration, or else the measured concentration is nonsensical. The number of background ions in zero gas is subtracted from the number of ions of "unknown" gas, such as exhaust or ambient air. This quantity is then multiplied by a calibration factor (Cf) according to Eq. (2).

$$[THC] = [Ion\#_{unknown} - Ion\#_{background}] * C_f$$

(2)

When the vehicle undergoing testing is producing near-zero levels of THC and CH_4, Eq. (2) can return negative concentrations. While that is not physically possible, it is mathematically possible. Because there are very few ions produced when zero gas is measured, a vehicle emitting near-zero levels of THC can cause this equation to be negative.

The FID measures the concentration by summing the current generated by each ion captured in the electric field. Even in the absence of THC, this current is not zero. This is why it is necessary to span and zero the analyzer before each test, as the calibration factor may change from day to day, analyzer to analyzer.

CHEMILUMINESCENT DETECTOR

The Chemiluminescent Detector, or CLD, is used to measure total NO_x concentration according to the principle of chemiluminescence.

Chemiluminescence is the emission of light with a limited emission of heat as the result of a chemical reaction. In our case, nitric oxide exhibits chemiluminescent behavior when combined with ozone, according to the following reaction:

$$NO + O_3 \Rightarrow NO_2^* + O_2$$

(3)

The asterisk indicates that the NO_2 exists in an excited energy state, and releases photons of infrared light energy, which can then be counted. This infrared light is detected by a pair of photodiodes which generate a current that is measured after amplification to determine the concentration of NO_x.

A CLD is a rather simple device in comparison to the FID. It consists of a catalyst, ozone feed, reaction chamber, a pair of photodiodes, and measurement electronics. The workings of a CLD system are well known, and therefore describing it is not necessary for the scope of this paper.

A CLD functions in much the same way as the FID does when it comes to measuring the concentration of NO_x in the sample gas. Like the FID, when zero gas is used there are still a few molecules that give off photons, which generate a current that must be calibrated out. However, unlike the FID, the CLD is subject to a phenomenon known as "dark current" where the photodiodes generate a slight current even without any excited molecules. This dark current and the background zero gas current must be calibrated out for the concentration results to be meaningful at all. The calculation for this is shown in Eq. (4).

$$[NO_x] = [Light_{unknown} - Light_{back} - Light_{dark}] * C_f$$

(4)

The CLD compares the measured amount of light from the unknown sample gas and subtracts away the known background concentration from the zero gas and the known dark current level. This quantity is then multiplied by the calibration factor.

Where problems arise in the CLD method is when the ambient air and exhaust gas are similar enough that the difference between the two measured concentrations is zero or negative, such as in SULEV and Hydrogen IC engines. It is possible for the measured NO_x to be close enough to the ambient air concentration that the end result is a negative value or zero.

NOISE, QUENCHING, AND CROSS SENSITIVITY ISSUES

NOISE ISSUES

The FID is very susceptible to electronic noise, which reduces the accuracy to which measurements can be made.

As a result of using a very high impedance operational amplifier to boost the ionization current a correspondingly large noise signal is introduced.

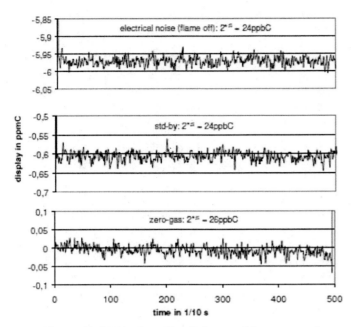

Figure 13. Noise signal levels in stand-by, zero, and flame-off operational modes.

Figure 13 shows a much less sensitive detector than is normally used. This detector has a noise signal of approximately 24ppbC, taken as twice the standard deviation. For a Pierburg FID with a sensitivity of 1 pA per ppmC this leads to a noise signal and corresponding limit of detection, or LOD, in the range of 20ppbC.

Depending on the sensitivity of the detector a noise signal in the range of 1 to 5 ppbC (twice the standard deviation) can be achieved. The sensitivity of the detector to superfluous noise is therefore a key influencing factor on the ability of the detector to measure zero and near-zero values.

The CLD also has an electronic noise issue, but due to the much simpler nature of the device, this noise effect occurs at significantly lower concentrations. The source of the noise signal is the same. The amplification required to measure the current generated by the photodiodes is quite large and introduces a non-negligible noise signal. This noise signal is orders of magnitude smaller than that of the FID, and as a result, it is possible to average the CLD results over time for much lower concentration values.

QUENCHING ISSUES

Oxygen causes a quenching effect in the hydrocarbon FID. Oxygen concentrations in the exhaust sample bags are typically lower than in the ambient air samples. The comparatively higher oxygen concentration in the ambient air bag results in a larger viscosity and a proportionally lower volume flow rate, resulting in a lower measured concentration.

Changing oxygen concentration also affects the sensitivity of the FID detector. It is therefore important to adjust the flow ratio of fuel to combustion air to keep the increasing detector sensitivity and decreasing signal balanced [6]. For measurements near zero, a mathematical correction for the FID oxygen quench effect becomes nonsensical especially for data from Phase 2 and 3 collection bags. A better correction for such low concentrations is one that takes the physical geometry of the flame into account.

Oxygen also introduces a quenching effect for the CLD. Since the CLD relies on the excitement of NO_2 to measure the NO_x concentration in the sample gas, excess oxygen can oxidize the NO concentration before it has a chance to react with the ozone in the chamber and give off light. This oxygen oxidation reaction significantly reduces the amount of light measured by the detector, as the NO_2 produced via this method is not in an excited state. To avoid the quench effect impacting the results in a detrimental way, it is imperative that the oxygen content in the detection chamber be controlled.

CROSS SENSITIVITY ISSUES

The FID is susceptible to cross sensitivities with other gas species. The most common are water and carbon monoxide. The influence of water on FID hydrocarbon detection is shown below in Fig. 14.

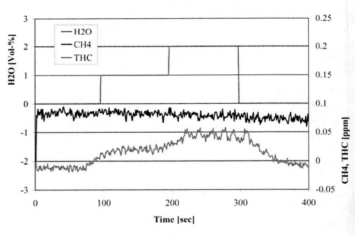

Figure 14. Water cross sensitivity for the FID.

The data for Fig. 14 were generated by using a Hovacal humidifier in conjunction with a double FID system. One side of the FID was unaltered, measuring THC. The other side had a methane cutter mechanism attached which eliminates most of the NMHC molecules present in the sample gas, allowing the FID to measure CH_4 only. Null gas was passed through both, then null gas with one percent H_2O, then null gas with two percent H_2O.

The results show that as the H₂O percent increases, so does the THC concentration. The CH$_4$ data is not affected from the H$_2$O concentration due to the methane cutter operating with a constant humidity of 1.6%.

When there is a surplus of water, or the FID is uncorrected for the water concentration, the FID will return greater concentrations of hydrocarbons than are actually present. This cross sensitivity may stem from the ionization of the water but the full extent and cause of this effect is unknown. Water is ionized into H_3O^+ ions which are in equilibrium with the CHO^+ ions. This equilibrium is biased towards the H_3O^+ ions by a very large degree [7].

Excess water will create excess hydronium (H_3O^+) ions, which will revert to CHO^+ ions to maintain their equilibrium. This increased level of CHO^+ ions generates a larger measured current and therefore a greater concentration is measured.

It is hypothesized, however, that the increased hydronium content has less of an effect on the ions themselves than on the overall mass flow of the system, and that it is this mass flow that has the most influence on the increased hydrocarbon concentration [8]. The oxidation of CO is also affected by the humidity of the sample gas.

A stable mass flow which accounts for greater water gas content will produce more accurate results. This is especially important for hydrogen combustion vehicles in which water is a significant portion of the exhaust.

This changing mass flow is caused by the changing viscosity of the humidified air. As the viscosity of moist air is less than that of dry air, the mass flows of dry and wet sample gas into the detector will be different. With lower viscosity the mass flow rate is higher and therefore more sample gas enters the detector chamber. This increases the probability of impurities contained in the sample gas interfering with the measurement results.

Furthermore, the lower viscosity of humid sample gas causes greater ion mobility and therefore better ion conductivity. As shown in Figure 14, THC increased significantly for a water concentration of one percent by volume for this specific detector. It is not possible to say definitively that all detectors will exhibit this behavior, but rather that all detectors show some degree of water cross-sensitivity with regards to measurement results.

These noise signals, quenching factors, and cross sensitivities all limit the ability of the measurement equipment to accurately measure zero. It is important to note, however, that only when measuring SULEV, SULEV+, or Hydrogen combustion vehicles are the measured emissions results as small as these factors.

PROPOSED MEASUREMENT ZERO LEVELS

There comes a point where one must call a value that is extremely close to zero simply zero. Each of the components measured in exhaust testing is measured with a different system. Therefore the accuracy to which a particular component is measured may not be the same as another. For this reason there cannot be one range of values that works for each of the components. Looking at the individual components, a reasonable range to call THC zero would be somewhere between ±0.05 ppm. This is not based on the standard deviation since the normality of the data is questionable. It is more of a reasonable estimation rather than an actual sigma range. NO$_x$, which has smaller values, could reasonably be called zero between ±0.01 ppm. CH$_4$ lies somewhere in-between, and ±0.03 ppm can be effectively called zero. When one puts this theory into practice, the number of negative values falls considerably. In fact, there are now only five possible negative values left from the entire NO$_x$ results, as shown in Fig. 15.

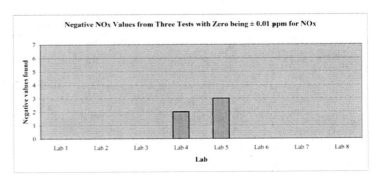

Figure 15. Negative values when values between ±0.01 ppm NO$_x$ are considered to be zero.

Whether these remaining "negative" values are actually negative is impossible to say, but five total negative outliers out of 72 data points is a reasonable number in this situation. To be even more stringent with NO$_x$, one could call ±0.005 ppm zero. This results in seven total outliers, as shown in Figure 16.

The difference in the number of negative values between the ±0.01 ppm and ±0.005 ppm limits is only two data points. This means that the rest of the negative NO$_x$ values (13 data points) are below 0.005 ppm from zero on the negative side.

From this data, one could reasonably hypothesize that this particular vehicle is incapable of negative values. It is, however, capable of zero emissions in NO$_x$ according to our new standards. Other than a few outlying values, the majority are so close to zero that they are effectively zero.

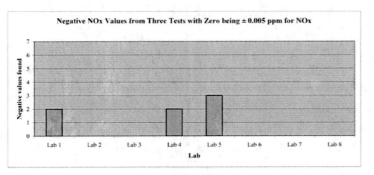

Figure 16. Negative values when values between ±0.005 ppm NO_x are considered to be zero.

CONCLUSIONS

NO_x emissions have reached a point where the values coming from the exhaust are at the limit of measurement capability. Negative calculated values are observed, but this does not prove that the emissions are negative; rather, only that measurement equipment fails to recognize true zero values. This is true of all testing systems, since noise and error are inherent in every form of testing equipment whether using a CVS or a BMD (Bag Mini Dilution) or another measurement system [9, 10].

No measurement can ever be perfect and so there will always be an uncertainty. With the advent of SULEV, SULEV+, and hydrogen vehicles, the measured exhaust emission can be well within the machine uncertainty. A perfect emissions result of zero is not physically possible and so it is necessary to define a range within which the result can effectively be called zero.

It is important to set zero standards now before new vehicles that actually clean the air are found in the mainstream market. A reasonable range, based on extended error testing of the CVS system and other testing methods, should be decided for all components under all regulation in preparation for zero emission vehicles.

It is difficult or impossible to distinguish NMHC below 3 mg/mi or NO_x below 1.5 mg/mi with current technology. Everything below that border is effectively zero. Furthermore, SUELV+ is not really measureable. AIGER criteria for BMD vs CVS correlations state that THC and NO_x results must match within 4 and 2 mg/mi, respectively between the two systems [11]. The 1.5 mg/mi and 3 mg/mi limits are more of an experimentally discovered limit rather than any hard mathematical lower bound. These limits are fluid and are not absolute by any means: there is some so-called 'wiggle room' inherent in those values. These values are a direct result of the uncertainties of the FID and CLD, respectively.

Because we currently can only measure down to 3 mg/mi for NMHC and 1.5 mg/mi for NO_x, in order to compare BMD and CVS device systems we now have a range of 3 ± 4 mg/mi and 1.5 ± 2 mg/mi for THC/ NMHC and NO_x, respectively. For SULEV+, the NMHC limit is 5 mg/mi. While this is above the detection level minimum of 3 mg/mi, it is inside the 3 ± 4 mg/mi range. Statistically speaking, this is zero, and not definitively measureable. Please refer back to Table 2 for our SULEV-rated vehicle results to see how these ranges effectively define statistical zero.

The noise signals, quenching factors, and cross sensitivities provide physical reasons limiting the accuracy of the equipment to measure emissions components at near zero and zero concentrations. This becomes most important when testing SULEV, SULEV+, and Hydrogen combustion vehicles, as their emissions are of the same magnitude as the measurement uncertainty.

FUTURE WORK

Areas for more investigation include the additional influences that make it difficult if not impossible to measure near-zero emissions. Along those lines, sensible limits must be found for CO, CO_2, THC, and other pollutant species to define a range for "zero" values. Exhaust gas bag stabilization was found to play a role in our results. Research into the optimum method for bag stabilization, and subsequent standardization thereof, is needed.

The effects of water in exhaust gas in these near-zero measurements should be investigated. All zeroing and spanning is performed with dry gases, and as a result the permeation of sample bags is not really understood [12, 13]. Investigating how moisture affects FID flame temperature would add to our understanding of the influence of water on hydrocarbon measurement. The lower bounds for NMHC and NO_x, 3 mg/mi and 1.5 mg/mi respectively, are believed to be affected by span gas humidity. Calibrating the FID and CLD devices with dehumidified span gas may lower these bounds and allow lower emissions to be measured more precisely as a result of decreasing a source of uncertainty in the machine. Further investigation into this effect is needed.

Running a test series with both an FID and a Gas Chromatograph (GC) would be very beneficial to understanding the effect of the response factors of CH4 and other hydrocarbons on the measurement uncertainties and accuracy.

Finally, additional work could investigate how near-zero emissions levels affects particulate measurements and CO emission concentrations for both gasoline and hydrogen engines.

REFERENCES

1. Krough, B., "Negative Numbers Report," BMW internal documentation.

2. California Air Resources, California Non-Methane Organic Gas Test Procedures, Amended July 30, 2002.

3. Letter to Manufacturers, CCD-01-01, US EPA, February 8, 2001.

4. Thiel, W., Hartmann, K., "Possible Influences on Fuel Consumption Calculations While Using the Hydrogen-Balance Method," SAE Technical Paper 2008-01-1037, 2008.

5. Garthe, C., Ballik, R., Hornreich, C., Thiel, W., "HC Measurements by Means of Flame Ionization: Background and Limits of Low Emission Measurement," SAE Technical Paper 2003-01-0387, 2003.

6. Sun, E., McMahon, W., Peterson, D., Wong, J., Stenerson, E., "Evaluation of an Enhanced Constant Volume Sampling System and a Bag Mini Diluter for Near Zero Exhaust Emission Testing," SAE Technical Paper 2005-01-0684, 2005.

7. Baronick, J.D. et al., "Final Report on the Research Project: Improving the Method of Hydrocarbon Analysis," Volkswagen AG Research and Development, 1980.

8. Wallner, T, et al., "Fuel economy and emissions evaluation of BMW Hydrogen 7 Mono-Fuel demonstration Vehicles," International Journal of Hydrogen Energy (2008), doi:10.1016/j.ijhydene.2008.08.067.

9. Manufacturers Advisory Correspondence, MAC #2002-02, California Air Resources Board, May 14, 2002.

10. Letter to Manufacturers, CCD-01-23, US EPA, December 6, 2001.

11. Nevius, T., Porter, S., Rooney, R., Rauker, D., "Techniques for Improved Correlation Between Constant Volume and Partial Flow Sample Systems," SAE Technical Paper 2009-01-1351, 2009.

12. U.S. Code of Federal Regulations, "Analytical Gases," Title 40, Part 86.114-94, Rev. July 2008.

13. U.S. Code of Federal Regulations, "Hydrocarbon Analyzer Calibration," Title 40, Part 86.331-79, Rev. July 2008.

CONTACT INFORMATION

Wolfgang Thiel
Wolfgang.Thiel@bmw.de

Roman Woegerbauer
Roman.Woegerbauer@bmw.de

David Eason
David.Eason@Colorado.edu

ACKNOWLEDGEMENTS

The authors would like to acknowledge Mr. C. Garthe from AVL for his detailed discussions and Mr. B. Krough, a former internship student at BMW, for his preliminary work. The authors appreciate the efforts of friends and colleagues at ANL and especially Mr. A. Fredrich of AVL for their help and actions on this project.

APPENDIX A

TEST GRAPHS

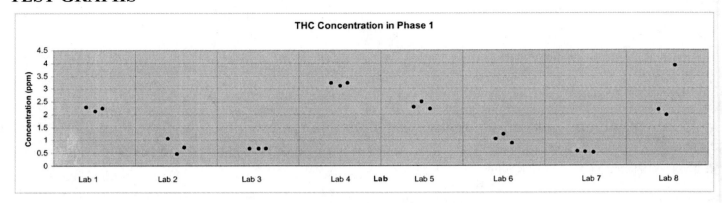

Appendix A - Phase 1.

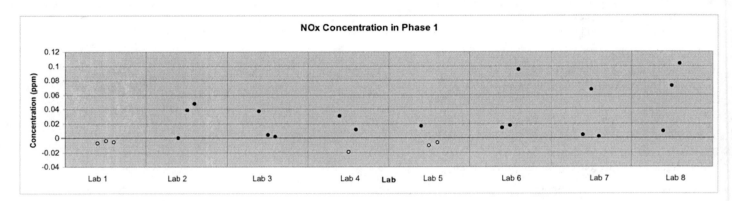

Appendix A - Phase 1.

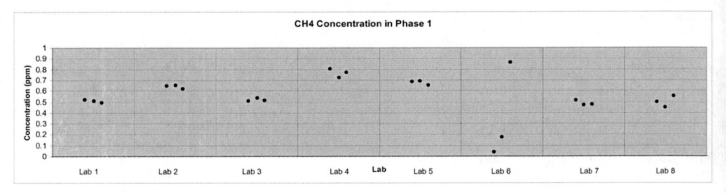

Appendix A - Phase 1.

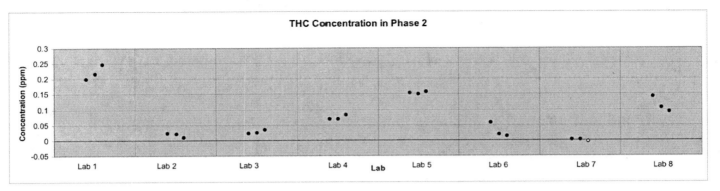

Appendix A - Phase 2.

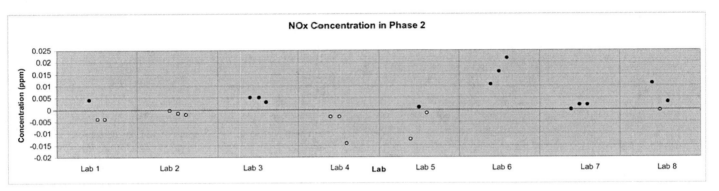

Appendix A - Phase 2.

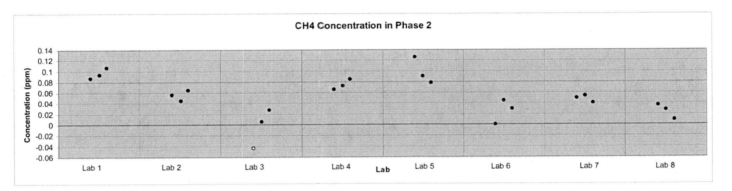

Appendix A - Phase 2.

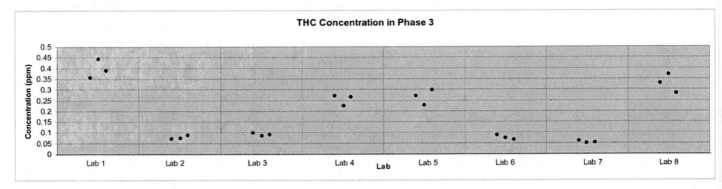

Appendix A - Phase 3.

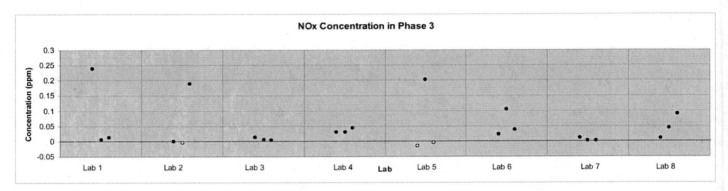

Appendix A - Phase 3.

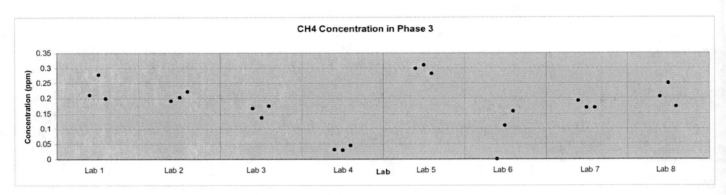

Appendix A - Phase 3.

APPENDIX B

NULL TEST GRAPHS

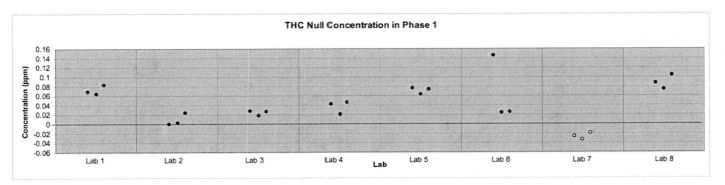

Appendix B - Phase 1.

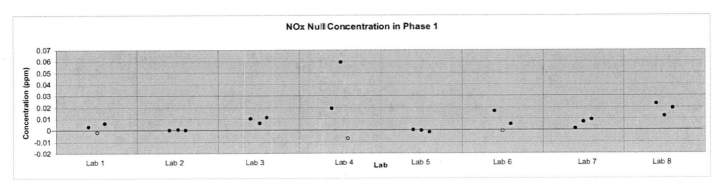

Appendix B - Phase 1.

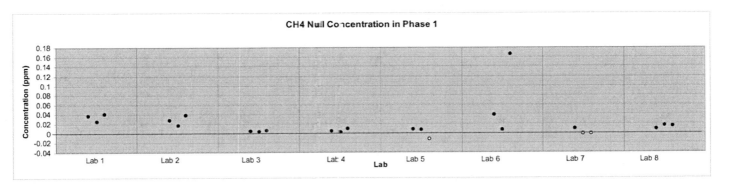

Appendix B - Phase 1.

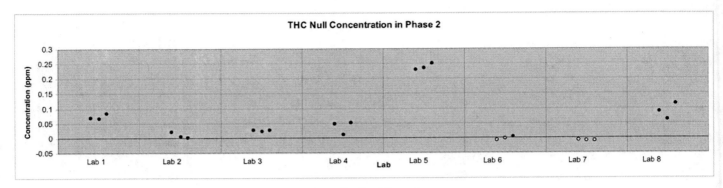

Appendix B - Phase 2.

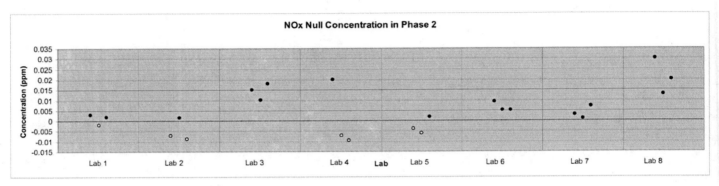

Appendix B - Phase 2.

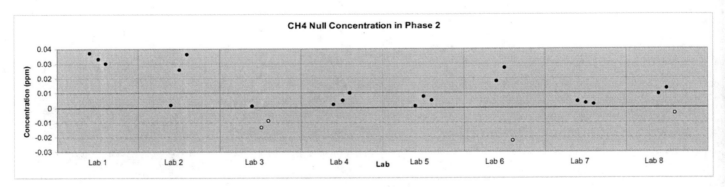

Appendix B - Phase 2.

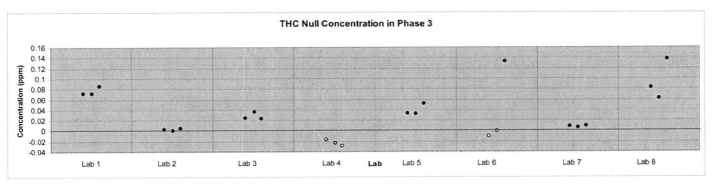

Appendix B - Phase 3.

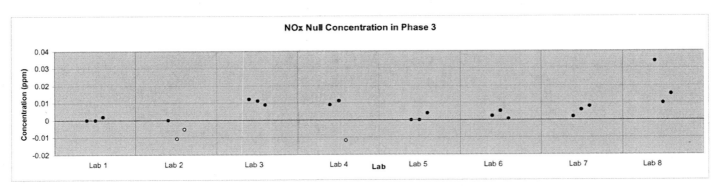

Appendix B - Phase 3.

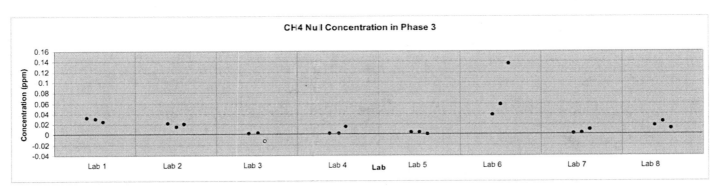

Appendix B - Phase 3.

Table 1. Two cases exploring the use of Eq. (1).

Concentrations			THC	NOX	CH4	NMHC (EPA)	NMHC (Carb)	Control
Item			PPM C	PPM	PPM C	PPM C	PPM C	PPM C
Case 1	Diluted Air	Yd	3.00	0.0420	2.80		0.00	-0.19
	Diluted Exhaust Gas	Ye	4.00	0.0370	2.57		1.08	1.08
	Difference	Kon	1.287	-0.0010	0.033	1.250	1.076	1.250
Case 2	Diluted Air	Yd	3.00	0.0420	2.60		0.04	0.04
	Diluted Exhaust Gas	Ye	4.00	0.0390	2.57		1.07	1.07
	Difference	Kon	1.287	0.0010	0.219	1.038	1.038	1.038

CH4-Response	EPA
1.14	CARB

Nanotechnology Applications in Future Automobiles	2010-01-1149 Published 04/12/2010

Edward Wallner, D.H.R. Sarma, Bruce Myers, Suresh Shah, David Ihms, Suresh Chengalva, Richard Parker, Gary Eesley and Coleen Dykstra
Delphi Corp.

ABSTRACT

It is rare for a single technology to have the power to dramatically influence almost every major industry in the world. Nanotechnology falls into this category and offers fundamentally new capabilities to architect a broad array of novel materials, composites and structures on a molecular scale. This technology has the potential to drastically re-define the methods used for developing lighter, stronger, and high-performance structures and processes with unique and non-traditional properties.

This paper focuses on some of the automotive applications for nanotechnology and showcases a few of them that are believed to have the highest probability of success in this highly competitive industry.

No discussion of nanotechnology is complete without touching upon its health and environmental implications. This paper addresses some of the safety issues and the precautions that we as an automotive industry need to take in the production, processing, storage and handling of such minute particles.

The goal of this paper is to raise the awareness on the promise of nanotechnology and the potential impact it will have on the future of the automotive industry.

INTRODUCTION

A huge amount of research and development activity has been devoted to nano-scale related technologies in recent years. The National Science Foundation projects nanotechnology related products will become a $1 trillion industry by 2015 [1]. Nano-scale technology is defined as any technology that deals with structures or features in the nanometer range or that are less than 100 nanometers, about one-thousandth the diameter of a human hair, and larger than about 1 nm, the scale of the atom or of small molecules. Below about 1 nm, the properties of materials become familiar and predictable, as this is the established domain of chemistry and atomic physics. It should be noted that nanotechnology is not just one, but many wide ranging technologies in many technical disciplines including but not limited to chemistry, biology, physics, material science, electronics, MEMS and self-assembly. Nano-structures have the ability to generate new features and perform new functions that are more efficient than or cannot be performed by larger structures and machines. Due to the small dimensions of nano-materials, their physical/chemical properties (e.g. stability, hardness, conductivity, reactivity, optical sensitivity, melting point, etc.) can be manipulated to improve the overall properties of conventional materials. At nanometer scales, the surface properties start becoming more dominant than the bulk material properties, generating unique material attributes and chemical reactions. More fundamentally, the electronic structure of materials becomes size-dependent as the dimensions enter the nanoscale. Delocalized electronic states as in a metal or a semiconductor are altered by the finite dimensions. Hence, the optical properties, including light absorption and emission behavior, will be altered, The fact that nanoscale features are smaller than the wavelength of visible photons also impacts light scattering, enabling the design of nanocrystalline ceramics that are as transparent as glass. Changes in the bonding at the surface of a nanoparticle will affect the electronic structure as well, and the implications for the reactivity of the surface can be significant. Beyond the electronic structure, nanostructuring can also affect transport properties markedly. Nanoscale features that are smaller than or comparable to the wavelengths or mean-free paths of phonons (quanta of lattice vibrations) or electrons permit the design of materials with thermal and electrical conductivity that may be outside the range accessible with ordinary materials. The most significant nano-structures investigated to date are made from single atomistic layers of carbon. These structures include hollow ball shaped "Buckyballs" (Fullerene - C60), carbon nano-tubes (CNTs) and graphene sheets which have a very interesting range of mechanical, thermal and electrical properties. It should also be noted that even though the environmental and health effects of nano-scale structures are poorly understood at this time, nano-scale-based technologies are already being used in some industrial applications. A series of nano-materials, including metal nanoparticles and nano-powders, magnetic fluids, nano-adhesives, nano-composite polymers, and nanocoatings (anti-fog, anti-reflective, wear and scratch resistant, dirt repellant, biocide, etc.) are being introduced for potential application in the automotive market.

Metal nanoparticles are being considered for potential use in catalytic converters since the catalytic reactivity is significantly enhanced due to the increased surface area and altered electronic structure of the metal nanoparticle. Coolants utilize nanoparticles and nano-powders to increase the efficiency of heat transfer and potentially reduce the size of the automotive cooling equipment. Some manufacturers are currently using nanomagnetic fluids in shock absorbers to increase vibration control efficiency. Wear-resistant, hard-surface nano-coatings are being investigated for applications in bearings, cylinders, valves, and other highly stressed areas.

Nano-layers of semiconducting materials provide high efficiency electronic components and systems with a longer lifetime. Sensors based on nanolayer structures are used in engine control, airbag, anti-lock brake and electronic stability program systems. Nanoparticles also support the optimization of conventional components like batteries, catalysts, solar cells or fuel cells.

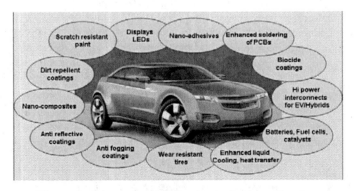

Figure 1. Automotive applications of nanotechnology

The most promising automotive applications of nanotechnology include the following:

- Improved materials with CNTs, graphene and other nanoparticles/structures

- Improved mechanical, thermal, and appearance properties for plastics

- Coatings & encapsulants for wear and corrosion resistance, permeation barriers, and appearance

- Cooling fluids with improved thermal performance

- Joining interfaces for improved thermal cycle and crack resistance

- Metal alloys with greater mechanical strength

- Metal matrix and ceramics with improved mechanical properties

- Solder materials with crack resistance or lower processing temperature

- Displays with lower cost and higher performance

- Batteries for electric vehicles and fuel cells with improved energy capacity

- Automotive sensors with nano-sensing elements, nano-structures and nano-machines

- Hybrid electric vehicles using electrical interconnects for high-frequency and high-power applications

- Electrical switching including CNT transistors, quantum transistors, nano-electro-mechanical switches, electron-emission amplification, and more efficient solar cells

- Self-assembly using fluid carriers

A few of these automotive applications and specific examples are reviewed in more detail in this paper.

NANOPARTICLE THERMAL MATERIALS

In spite of advances in efficiency of vehicle powertrain systems and electronics, the removal of waste heat continues to be an important challenge. With increasing focus on reduced component size and mass, the traditional approach of increasing the area available for heat exchange with a cooling fluid (air, water/ethylene glycol) to manage higher heat loads is not acceptable. Increasing thermal power densities requires innovations in new coolants and thermal coupling materials.

The concept of using nano-fluids as a means of improving coolant performance was proposed over a decade ago [2]. Reports of up to 100% increase in liquid thermal conductivity with the addition of nanometer scale particles motivated a large amount of scientific/technical inquiry in the ensuing years [3].

Nano-fluids are a solid-liquid composite containing nanoparticles with sizes in the 1-100 nm range dispersed and suspended in a liquid. A variety of nanoparticle solids have been used as additives, including metals such as copper and gold, alumina, SiC and CuO ceramics and carbon nano-tubes. The surprisingly large increases in liquid thermal conductivity have been reported for relatively small particle loadings (<10% by volume). In addition, there have also been reports of higher critical heat flux (dry-out) for nano-fluids used in liquid-vapor phase cooling applications. These observations have been made for a number of liquids, including water/ethylene glycol, alcohols and oils. The results defy conventional experience which requires much higher volume loading of larger particles to produce slurries with comparable increases in effective liquid thermal conductivity. These observations have stimulated numerous theories attempting to understand and describe the phenomena, but the nature of the thermal enhancement mechanism still remains controversial. This situation is further aggravated by inconsistent results from different laboratories, and some claims that if carefully measured, the enhancements are smaller and explained by established theories.

Nevertheless, the potential for significantly improved coolants may provide impetus for further improvements in engine efficiency and reduced size and weight of cooling system components. In addition, there are efforts to examine improvements in the thermal and rheological properties of lubricants with the addition of nano-scale particles [4].

In automotive electronics, the use of thermal interface materials (TIM) to thermally couple electronic devices to heat sinks for waste heat removal is common practice. Although the thermal resistance of TIM has been reduced over the years, these materials still represent a major bottleneck in the thermal stack-up between semiconductor die and the cooling medium. As a result, components capable of handling higher power densities often operate at de-rated performance levels to mitigate high temperatures and to compensate for the harsh automotive environment. This problem is especially critical in hybrid electric vehicle power control systems, where switching transistors can operate at power densities in excess of 300 W/cm^2.

Spurred by enhanced nano-fluid thermal properties, investigators translated the nano-composite ideas into the realm of TIM. It is common practice to boost the thermal conductivity of silicone oils, polymer gels, phase-change materials and thermoplastics by the addition of solid particles of micrometer scale size. Research has shown that optimal particle loading achieves improved thermal conductivity and low modulus (to accommodate thermal expansion mismatch of components) with a variety of materials and particle shapes/sizes [5]. Mixtures of nano- and micro- scale particles

add another dimension for controlling thermal, rheological and mechanical properties [6].

Of particular interest is the use of carbon nano-tubes for TIM applications. The CNT is essentially a single atomic layer of graphite (graphene) which is rolled up onto itself. There are single- and multi-walled versions of CNT which can exhibit thermal conductivity in excess of 1000 Watts/meter ° Kelvin (for comparison, Cu = 400W/mK) and high tensile strength along the axis of the tube. Applications to TIM have involved two basic approaches:

1. Simple addition of CNT to the TIM matrix (grease, gel, etc.)

2. Growth of vertically aligned CNT 'carpets' on the heatsink or device package.

In the former approach, CNT loading is increased until percolation of fibers provides a thermal path from mating surfaces. In the latter growth method, the individual CNT provide a direct high-conduction path between surfaces [6, 7]. In this case, tantalizing reports of low thermal impedance (~ 0.05 cm^{2}C/W) have motivated continuing development of growth methods more amenable to high volume, low cost electronics production. At this point, efficient growth of high quality CNT is still time consuming and requires temperatures in excess of 500°C on catalyzed surfaces.

In spite of the prospect that nano-composite materials offer improved thermal conduction, several issues need to be resolved. Dispersing nanoparticles to avoid aggregation can be crucial to improving performance. In many cases the dispersions are not stable and over time lead to degraded thermal performance. In the case of liquids, maintaining a time-stable suspension can be problematic, since many candidate particle materials are denser than the liquid and tend to settle out. As it turns out, it is the nanometer-sized particles that can mitigate this problem. The intrinsic Brownian motion of liquid molecules surrounding the particles can maintain a dispersion/suspension.

Although the thermal properties of CNT are impressive, the performance gains in CNT composites are not as large as anticipated. High-interface thermal resistance in both CNT fillers and vertically aligned CNT tips severely impedes coupling between the CNT and the matrix or mating surface. Work continues on materials and methods to functionalize the CNT surface to improve the thermal coupling.

As composite technology progresses, we would expect to see the eventual penetration of nanoparticles into the realm of thermal management materials. The final issues to be confronted will be the value of performance gains achievable in a high volume, low cost automotive market.

DISPLAYS USING NANOTECHNOLOGY

Displays with improved performance and unique features are made possible by nanotechnology. Additionally, lower cost light emission sources, such as lasers are possible in the near future. Display technology, under rapid development for consumer electronic devices and home entertainment systems, is also being pursued for automotive applications. Improved performance, longer life, higher energy efficiency, unique presentation features, reduced package size and innovation become the value proposition for implementing this new technology.

Automotive displays are expected to directly utilize nano-technology in a variety of ways. Light emitting devices, such as LEDs, OLEDs (Organic Light Emitting Diode), fluorescent or field-emissive displays, electro-luminescent and perhaps lasers, are utilizing nano-phosphors and nano-layers to improve their performance. For example, silver nanoparticles on the cathode surface allow surface plasmon localization. This provides a strong oscillator decay channel that generates a two-fold increase of intensity for flexible OLED displays. Optical thin films, non-linear holographic reflectors, micro-lenses, and light conversion films are examples of materials that modulate or redirect electromagnetic radiation. Light projection systems, flat-panel displays, including cameras and other optical detectors that provide the input signals are all expected to benefit from nanotechnology developments.

One particular area of interest is nano-phosphors, since these materials possess strikingly different absorption and emission characteristics while operating with better efficiencies and life times than their related bulk phosphors. Since the particle size determines the band-gap energy, coupling nano-phosphors with new semiconductor materials (with and without doping) means that a wide variety of designed phosphors and new devices will likely be developed. Although many materials under consideration are somewhat exotic and expensive, inexpensive materials, such as zinc oxide, and titanium dioxide are also used in the nano-world. Considerable work is being done but much of it is in the realm of industrial secrecy.

Most first generation nano-phosphors, Q-dots included, are based on toxic elements such as cadmium and lead. Alternative materials (manganese or copper-doped zinc sulfide, D-dots) are coming onto the market. Although these materials are still relatively expensive, the cost will reduce as applications are identified and escalate the demand for material. Today nano-phosphors have many applications in display devices and more are being discovered.

Photonic properties of these materials are indicative of their electrical properties. The arrangement of the electrons, dictated by energy states, sets the rules for how a material will interact with incident photons. In this regard, conductors, insulators, and semiconductors each have unique valance and conduction electron energy band arrangements. A dielectric or insulator material will absorb a photon when a valence band electron can be excited (interband) to a higher conduction band, the energy being greater than the band gap of the material. Most dielectrics are transparent to visible light since the energy of photons at these wavelengths are insufficient to promote the electrons. A conductive material is opaque since it will either absorb or reflect photons due to the many energy bands available for electrons to be promoted within the conduction band (intraband). It is the semiconductor materials (especially with doping) that allow controllable interaction with incident photons due to free electrons in the partially filled conductive band and the energy states available in the "adjustable" band gap energy.

Coupling these electrical properties with the dimensional size of the material, we now have the ability to break up the energy bands into discrete levels; that is, we can widen the band gap by controlling the physical size of the particle. Semiconductor particles at the size and scale where this is possible are known as quantum dots, and the smaller the quantum dot, the larger its corresponding band gap. Quantum dots can absorb photons over a broad wavelength interval. Conversely quantum dots emit photons over a very narrow, temperature insensitive wavelength band, since the quantum confinement of the energy states in three dimensions approximates that of an atom having discrete atomic levels. Quantum dots are also called artificial atoms.

In general, the area of nano-optics operates on different principles than bulk optics. Nano-optic elements consist of numerous nano-scale structures created in regular patterns on or in a material. Depending upon the optical function, they can be created with metals, dielectrics, non-metals and semiconductors, epitaxially grown crystals, glass and plastics. In whatever form, creating the nano-structured material is transformative. Nano-optic devices can perform their optical functions in very thin layers, often less than a micron in thickness. The optical effects can be achieved in a shorter focal length compared to bulk optics because the sub-wavelength-size structures of nano-patterns interact with light locally, involving quantum effects as well as classical optical performance. This feature of nano-optics allows for very compact form factors.

The ability to understand how a material will interact with photons for generating a display or display element is primarily dependent upon the energy states of the electrons. The nano-scale interaction of photons and materials, termed nano-photonics, is a field still in its infancy with plenty of room to grow. This term encompasses a very broad field of materials, processes, and potential applications. For example, a new emerging roadmap targeting development of concepts, technologies, and devices has been released within the framework of the Photonics21 strategic research agenda. This roadmap is promoted by the EU Network of Excellence on nano-photonics (PhOREMOST), composed of 34 partners and over 300 researchers. (www.phoremost.org)

The majority of the developing technologies referenced by PhOREMOST are not directly applicable to future automotive emissive optical displays, projection systems, or imagers. Many anticipated nano-photonic materials will be coupled with silicon-based wafer processing to generate digital information processing and communication light-based features (plasmonics) to increase processing speed while greatly reducing the power dissipation associated with today's electron-based metal and semiconductor materials. However, other processing developments such as material processing using sols and self-assembly techniques are expected to indirectly advance display technology as they provide the means to create these new properties economically.

Nanotechnology is engineering and it is all about practical applications of physics, chemistry, and materials science. Nano-photonics is that specialized region of study where the effects of light interacting with matter on a very small scale will be the engine to generate new products and features almost unimaginable today.

NANO-COMPOSITES

Nano-composites are materials that incorporate nano-sized particles into a matrix of standard material such as polymers. Adding nanoparticles can generate a drastic improvement in properties that include mechanical strength, toughness and electrical or thermal conductivity. The effectiveness of the nanoparticles is such that the amount of material added is normally only 0.5-5.0% by weight. They have properties that are superior to conventional microscale composites and can be synthesized using simple and inexpensive techniques. [8] A few nano-composites have already reached the marketplace, while a few others are on the verge, and many continue to remain in the laboratories of various research institutions and companies. The global nano-composites market is projected to reach 989 million pounds by the end of the 2010, as stated in a report published by Global Industry Analysts, Inc.

Nano-composites comprising nanoparticles such as Nano-clays (70% of volume) or nano-carbon fillers, carbon nano-tubes, carbon nano-fibers and graphite platelets are expected to be a major growth segment for the plastics industry.

HOW NANO-COMPOSITES WORK

Nanoparticles have an extremely high surface-to-volume ratio which dramatically changes their properties when compared with their bulk sized equivalents. It also changes the way in which the nanoparticles bond with the bulk material. The result is that the composite can be many times improved with respect to the component parts.

WHY NANO-COMPOSITES?

Polymers reinforced with as little as 2% to 5% of these nanoparticles via melt compounding or in-situ polymerization exhibit dramatic improvements in properties such as thermo-mechanical, light weight, dimensional stability, barrier properties, flame retardancy, heat resistance and electrical conductivity.

CURRENT APPLICATIONS OF NANO-COMPOSITES

Applications of nano-composite plastics are diversified such as thin-film capacitors for computer chips; solid polymer electrolytes for batteries, automotive engine parts and fuel tanks; impellers and blades, oxygen and gas barriers, food packaging etc. with automotive and packaging accounting for a majority of the consumption. [9] The automotive segment is projected to generate the fastest demand for nano-composites if the cost/performance ratio is acceptable. Some automotive production examples of nano-composites include the following: Step assist - First commercial application on the 2002 GMC Safari and Chevrolet Astro van; Body Side Molding of the 2004 Chevrolet Impala (7% weight savings per vehicle and improved surface quality compared with TPO and improved mar/scuff resistance); Cargo bed for GM's 2005 Hummer H2 (seven pounds of molded-in-color nano-composites); Fuel tanks (Increased resistance to permeation); under-hood (timing gage cover (Toyota) and engine cover (Mitsubishi).

KEY CHALLENGES FOR NANOCOMPOSITES FOR FASTER COMMERCIALIZATION

• Develop low cost and high production volume to meet fast to market needs.

• Develop fast, low cost analytical methods with small quantity of samples which can provide a degree of exfoliation and degree of orientation, (TEM, XRD, Rheology considered too expensive and time consuming) for example, IR can detect silicon-oxygen bond in clay, which can help to evaluate degree of clay dispersion.

• Develop in-line testing of nano-composites.

• Develop alternative nano-clay treatments for better adhesion of nano-filler to polymer.

- Improve understanding the effect on performance by blending nano-fillers with conventional reinforcements such as glass fiber.

- Prediction of orientation / flow modeling.

- Understand the rheology and chemo-rheology of the polymer composites.

- Cost/performance ratio to substitute HIPS (High impact polystyrene), PC/ABS (Polycarbonate/Acrylonitirile-Butadiene-Styrene) and PC (Polycarbonate) with TPO (Thermoplastics Polyolefins).

- Fine dispersion, full exfoliation and interfacial adhesion.

- High stiffness without affecting impact properties.

OPPORTUNITIES AND FUTURE TRENDS FOR NANO-COMPOSITES

Nano-fillers are expensive compared to conventional fillers, so one must use them wisely depending on the final part performance requirements. In many cases, it may be cost effective to use nano-filler where it is needed such as on the top layer of a part surface or middle layer of thickness or localized areas of the part (nano-composite pre-molded inserts).

New nanotechnology applications are being demonstrated by R&D engineers, but the commercial officers balk at increased costs. The nano-clays cost about $3/lb and are used in loadings of 3-4 percent. The conventional competitor material is talc, which costs 30 cents/lb and is used at loadings of 10-15 percent. Another issue: Widespread replacement with nano-composites may require extensive re-tooling because of differences in shrinkage rates.

Recent news of an innovative method of growing carbon nano-tubes may revolutionize the implementation of nanotechnology. Use of Nano-polypropylene (PP) for value added substitution such as high cost engineering plastics or development of molded-in-color nano-composites to replace glass-filled, painted PP for interior applications such as instrument panels will see major growth. Functional nano-composites development is underway such as functionalized clays which add properties to clay including anti-static and moisture repellent characteristics and selective chemical barriers. Ultraviolet-curable nano-composites (electronics) and foaming and nucleating effect of nano-fillers (improve properties, desirable cell size and density, use of microcellular processes such as MuCell) will be commercialized soon. There is potential for body panels and large moldings to substitute for steel, aluminum, magnesium and Sheet-Molding Compound (SMC), where thermoplastics are currently excluded due to inadequate physical or mechanical performance.

There is a need to develop a low cost, carbon nano-tube based composite for high-end engineered plastics. For designers, there is a need to develop flow simulation software with or without a hybrid fiber-filled system (including orientation effect and warpage) so output can be used directly for structural analysis.

There are also many opportunities for development of new fillers and improvements such as nano-composites of a new nano-ceramic fiber, titanium dioxide (TiO2), magnetic particles, carbon nano-tubes and other molecularly reinforced polymers. Mixtures of different nano-materials or combinations of nano-materials with traditional additives are increasingly being considered.

NANOTECHNOLOGY APPLIED TO SOLDERING SYSTEMS

Due to the European Union environmental legislation, the electronics industry is being forced to eliminate lead from the traditional solder alloy system of tin-lead (SnPb). The industry has developed new solder alloys to replace SnPb eutectic alloys, but the required processing temperature must increase by about 35°C to accommodate the alloy (tin silver copper family, SAC). This increase in processing temperatures to about 235°C to 245°C, results in additional unwanted thermal stress on the electronic components being assembled as compared to tin-lead assembly temperatures of about 210°C.

Research is being conducted in the realm of nano-particle-sized solder alloys. Metals undergo a melting point depression when the particle size is reduced to nano-scale. Preliminary work by an iNEMI Nano Solder Project Team [10] has worked towards demonstrating that a reduction in melting temperature of a solder alloy is feasible as a function of particle size. (see Figure 2)

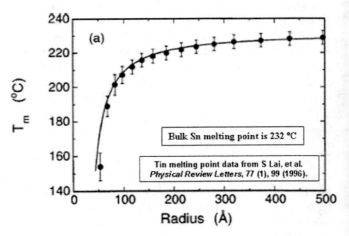

Figure 2. Melting point of solder as a function of particle size

It may be possible to develop a solder paste system using nano-sized solder particles (or alloying elements) to take advantage of this melting point suppression. This would be a technology enabler for upcoming high-density electronics that are heat sensitive, thereby improving reliability of the future electronic systems.

The first phase of the iNEMI project work was to produce nano-scale tin, silver and copper particles to test for the melting depression phenomena. The team used Differential Scanning Calorimetry (DSC) to demonstrate a reduction in the melting and subsequent solidification of the test materials. By repeating the temperature scan cycle many times, a record of the response is obtained. Figure 3 shows a typical DSC run on a sample of tin particles. The blue line includes the heat absorption (endothermic reaction) of the flux carrier that volatizes and reacts on the first thermal cycle combined with the melting of the nano-tin particles. By repeating the DSC thermal scan, one can demonstrate that the tin is no longer melting after the first cycle because it is no longer a nano-sized particle. This method has demonstrated the melting point suppression of the first cycle.

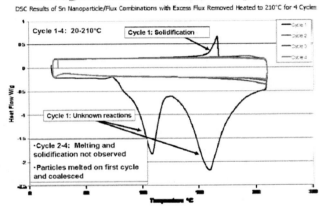

Figure 3. DSC scans depicting the melting depression for a Sn sample (courtesy of John Koppes, Purdue) [11]

Future work in Phase II of this iNEMI project will address increasing the metal density of the nano-solder paste and the development of a flux system that supports the coalescence of the particles. Solder paste printability and solder joint reliability tests would then follow.

NANO-REACTIVE FOILS

One promising application of nano-technology to the soldering process is in the use of nano-reactive foils. These foils are comprised of thousands of alternating nano-scale layers of aluminum and nickel that are placed between the two surfaces to be joined. For instance, a nano-reactive foil is placed between two solder performs to ultimately bond a semiconductor device to a Printed Circuit Board (PCB) as shown in figure 4.

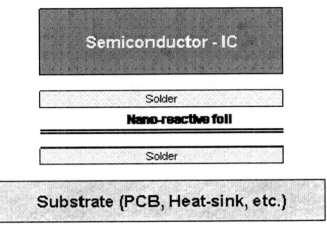

Figure 4. Cross-section showing nano-reactive foil

When activated with a small amount of energy, the nano-reactive foil rapidly reacts chemically in an exothermic reaction. The energy in the form of released heat, melts the adjacent solder and, in this example, bonds the IC to the substrate. The amount of energy released is directly proportional to the number of Al-Ni layers present in the nano-reactive foil.

There are several advantages to using this technology for the soldering process. It is not only quick, but it also eliminates the need for solder flux, thereby minimizing voiding and also eliminating the need for cleaning to remove any flux residue. Another significant benefit of this technology is that the reactive heat stays localized and does not subject the component to prolonged high temperature exposure found in the traditional PCB solder reflow process.

The superior quality of solder joints produced by using these nano-reactive foils is a big advantage when it comes to soldering high-power devices to heat sinks. This nano-reactive foil process is well suited for electric vehicle and hybrid power control applications. A small drop in thermal resistance goes a long way in increasing the performance of these components.

ENVIRONMENTAL HEALTH & SAFETY

The unconventional size, crystalline structure, large surface area and physical/chemical properties of nano-materials promise unprecedented technological advances; however these same properties also present significant challenges to understanding, predicting and managing potential health, safety, and environmental risks. The toxicological characteristics of familiar chemical compositions become uncertain when reconfigured at the molecular level in the form of nano-materials. Nanomaterials may differ from their larger particle counterparts with regard to viable routes of exposure, movement of the material once in the body and interaction of the materials with the body's biological systems; all data which are essential for predicting health risks.

Preliminary data indicate nanoparticles have the potential to be absorbed into the body via inhalation, ingestion, and through the skin. Occupational exposure is most likely to occur via inhalation or skin contact. While personal protective equipment (PPE) is often the preferred choice for minimizing employee exposures, the efficacy of traditional PPE toward specific nanomaterials is largely unknown.

A critical feature of nanoparticles is their high surface-area-to-mass ratio. This property provides additional sites for bonding or reaction with surrounding material and results in unique characteristics such as improved strength or heat resistance. Similarly, evidence suggests that when inhaled, the large surface area of insoluble nanoparticles creates the potential for greater biological activity [13, 14, 15]. Although dependent upon their effective size in the body and other physio-chemical characteristics, studies have also shown that when nanoparticles are inhaled they may penetrate cells allowing direct access to the bloodstream and possibly circumventing the blood-brain barrier or depositing in other organs of the body [16, 17].

Several studies have noted an increased risk of biological responses from exposure to carbon nano-tubes [18, 19, 20, 21]. Both single-walled (SWCNT) and multi-walled (MWCNT) carbon nano-tubes are non-biodegradable and resemble needle-like, carcinogenic asbestos fibers in size, shape and cellular persistence. Until recently several studies only suggested a potential link between inhalation exposure to long MWCNTs and cancer, but had not demonstrated that inhaled MWCNTs could actually pass from the lung and into the surrounding tissues [22, 23, 24]. The National Institute for Occupational Safety and Health (NIOSH) researchers have reported new data showing that MWCNTs can indeed migrate intact from the lungs of mice and into the tissue surrounding the lungs where asbestos induces a form of cancer known as mesothelioma [25].

Significant absorption of nano-material through the skin appears unlikely. Passive diffusion appears to be the primary mechanism for transport across the stratum corneum. The composition of this outer layer of the skin creates an effective barrier to dermal absorption of both chemicals. Although some studies have indicated that penetration of particles in low micron diameter size range is possible, penetration is likely to be slow and not present an acute hazard [26].

There is very limited information describing the behavior of nanoparticles once released to the environment. The fate of most nanoparticles in air, water, and soil is unknown particularly with respect to persistence and potential for bioaccumulation (wildlife toxicity). It is possible those microorganisms in soil or water could bioconcentrate nanoparticles within their cells, and that nanoparticles could consequently bioaccumulate up the food chain.

Because of these issues, it is important to adhere to safe practices for the handling, use and disposal of these materials. It is also essential that a comprehensive environmental and human health risk assessment accompany all new nano-materials and their proposed uses. Attention to each phase of an applications lifecycle is warranted due to the uncertainties and paucity of scientific data regarding toxicity and environmental impact.

NANO-MATERIAL REGULATIONS

The United States Environmental Protection Agency (EPA) Toxic Substances Control Act (TSCA) states that nano-formulations of existing chemicals do not require registration as a new chemical. However, in October 2008 the EPA announced that it considers carbon nanotubes (CNT) to be chemical substances distinct from graphite or other allotropes of carbon listed on the TSCA 8(b) Inventory [30]. As of March 1, 2009 the agency enforced filing of premanufacture notice (PMN) for carbon nano-tubes. This new chemical review action requires manufacturers or importers of CNTs not on the TSCA Inventory to submit a PMN or an applicable exemption at least 90 days before manufacture, unless the substance is excluded from premanufacture reporting. Any amount of CNTs triggers the notice requirement.

In November 2009, the EPA issued two proposed Significant New Use Rules (SNUR) under section 5(a) of TSCA for substances generically identified as single- and multi-walled carbon nanotubes [34]. The SNUR essentially binds all manufacturers, processors, and importers of CNTs to notify EPA at least 90 days before beginning any activity that EPA has designated as a significant new use. Significant new use of single-walled and multi-walled carbon nanotubes are deemed to occur in the absence of the corresponding TSCA section 5(e) consent orders which require protective measures to limit exposures and mitigate potential unreasonable risks. The consent orders require employees to use gloves and chemical protective clothing impervious to CNTs as well as a NIOSH-approved full-face respirator with an N-100 cartridge. The consent order also prohibits release of CNTs into water. This directive replaces the final SNUR issued in June 2009 because the prior directive did not apply to all variants of carbon nanotubes [33]. The former SNUR only applied to the specific carbon nanotubes that were the subject of the premanufacture notices (PMNs) submitted under Section 5 of TSCA and not to any other carbon nanotubes.

SNURs have also been issued for siloxane modified silica and siloxane modified alumina nanoparticles [31].

In early 2009 the EPA solicited comments on a petition to classify nano-scale silver as a pesticide. "In general, the petition requests that the agency require formal pesticide registration of all products containing nano-scale silver, analyze the potential human health and environmental risks of nano-scale silver, and take regulatory actions under the Federal Insecticide, Fungicide, and Rodenticide Act (FIFRA) against existing products that contain nano-scale silver."[32]

Under the European Union (EU) REACH Directive nano-materials are defined as "chemical substances". There are no special provisions for the nano-scale version of bulk chemicals. Unlike TSCA, the REACH one-ton threshold for registration might exclude many nano-scale materials. EU Parliament's Environmental Committee has requested removing the 1-ton threshold for nano-materials [29].

Under the provisions of the Canadian Environmental Protection Act of 1999, the New Substances Notification Regulations (Chemicals and Polymers) (NSNR) require that any nano-material not present on the Domestic Substance List (DSL) or defined as "new" to undergo a risk assessment of its potential effects on the environment and human health. Current policy considers the nano-scale form of a substance on the DSL to be a "new" substance if it has a unique structure or molecular arrangement. In February, 2009 Canada implemented a new rule requiring companies that manufacture or import more than 1 kg of a nano-material to report information on the quantity, usage and toxicity of nano-materials as well as any procedures, policies and technological solutions currently in place to protect environmental and human health [28].

The Japanese Ministry of the Environment released guidelines on March 10, 2009 with the intent of reducing the risk of environmental harm from nanomaterials [27]. Japan currently does not have any laws or regulations governing nano-materials and the new guidelines will be voluntary. The guidelines point out potential risks in nano-material manufacturing and urge companies to adopt policies that limit releases.

Although there is much debate as to whether or not nanotechnology should be regulated, it is becoming quite evident that it will be. Companies who are looking to reap the technological advantage that nano-materials can offer to their products will also have to factor in the effort and cost required. This cost and effort is needed to ensure employee and consumer safety through risk and life cycle assessments as well as meeting global regulations that will impact the sale of their product.

CONCLUSIONS

The automotive industry will be influenced by the development and implementation of nanotechnology. It is our hope to raise the awareness that nanotechnology will positively influence the business of the automotive industry over the next several years.

Due to the small size of nano-materials, their physical/chemical properties (e.g. stability, hardness, conductivity, reactivity, optical sensitivity, melting point, etc.) can be manipulated to improve the overall properties of conventional material.

Metal nanoparticles are being considered for potential use in catalytic converters since the catalytic reactivity is significantly enhanced due to the increased surface area of the metal. Coolants utilize nanoparticles and nano-powders to increase the efficiency of heat transfer and potentially reduce the size of the automotive cooling equipment. Some manufacturers are currently using nano-magnetic fluids in shock absorbers to increase vibration control efficiency. Wear-resistant, hard-surface nano-coatings are being investigated for applications in bearings, cylinders, valves, and other highly stressed components.

High efficiency nano-layers of semiconducting materials provide electronic components and systems with a longer lifetime. Sensors based on nano-layer structures find applications in engine control, airbag, anti-lock brake and electronic stability program systems. Nanoparticles also support the optimization of conventional components like batteries, catalysts, solar cells or fuel cells.

Nanotechnology is science and engineering, and it is all about practical applications of physics, chemistry and material properties. Nanotechnology will influence the auto industry initially on a very small scale, but will certainly be developed to deliver features, products and processes that are almost unimaginable today.

REFERENCES

1. J. of Nanoparticle Research, Kluwer Academic Publ., Vol. 3, No. 5-6, pp. 353-360, 2001 (based on the presentation at the symposium Global Nanotechnology Networking, International Union of Materials Meeting, August 28, 2001)

2. Choi S.U.S., in "Developments and Applications of Non-Newtonian Flows", edited by Singer D.A., Wang H.P., American Society of Mechanical Engineers, Vol.231/MD-Vol.66, p.99 (New York, 1995).

3. Keblinski P., Eastman J.A., Cahill D.G., Materials Today, p.36 (June 2005).

4. Marquis F.D.S., Chibante L.P.F., Journal of Materials, p. 32 (December 2005).

5. Elliot J.A., Kelly A., Windle A.H., J. Mat.Sci.Ltrs. Vol. 21, p.1249 (2002).

6. Prasher R., Proceedings of the IEEE Vol.94, No.8, p..1571 (August 2006).

7. HuX.J., Padilla A.A., Xu J., Fisher T.S., Goodson K.E., J. of Heat Transfer, Vol.128, p.1109 (November 2006).

8. http://www.nsti.org/press/PRshow.html?id=3203

9. www.AzoNano.com

10. iNEMI Nano Solder Project Work, 2008 www.inemi.org

11. Utilizing the Thermodynamic Nanoparticle Size Effects for Low Temperature Pb-Free Solder Applications Koppes John P., Grossklaus Kevin A., Muza Anthony R., Revur R. Rao, Sengupta Suvankar, Stach Eric A., and Handwerker Carol submitted to Acta Materialia.

12. Reactive NanoTechnologies (RNT) http://www.rntfoil.com

13. Tran, C. L., et al (2000). Inhalation of poorly soluble particles. II. Influence of particle surface area on inflammation and clearance. Inhal. Toxicol. 12, 1113-1126.

14. Donaldson, K., and Tran, C. L. (2002). Inflammation caused by particles and fibers. Inhal. Toxicol. 14, 5-27.

15. Duffin, R., et al (2002). The importance of surface area and specific reactivity in the acute pulmonary inflammatory response to particles. Ann Occup. Hyg 46 (Suppl.1), 242-245.

16. Kreuter, J., et al (2002). Apolipoprotein-mediated transport of nanoparticle-bound drugs across the blood-brain barrier. J. Drug Target. 10,317-325.

17. Oberdorster, G., et al (2004). Translocation of inhaled ultrafine particles to the brain. Inhal.Toxicol. 16, 437-445.

18. Lam, C.-W., et al (2004). Pulmonary toxicity of single-wall carbon nano-tubes in mice 7 and 90 days after intratracheal instillation. Toxicol. Sci. 77:126.

19. Shvedova, A.A., et al (2005). Unusual inflammatory and fibrogenic pulmonary responses to single walled carbon nano-tubes in mice. Am. J. Physiol.Lung Cell. Mol. Physiol., 289 (5), L698

20. Warheit, D.B., et al (2004). Comparative pulmonary toxicity assessment of single-wall carbon nano-tubes in rats. Toxicol. Sci.77:117.

21. Muller, J., et al (2005). Respiratory toxicity of multi-wall carbon nano-tubes. Toxicol. Appl. Pharmacol.207(3):221.

22. Poland, C.A., et al. (2008) Carbon nano-tubes introduced into the abdominal cavity of mice show asbestos-like pathogenicity in a pilot study. Nat. Nanotechnol. 3:423-428

23. Sakamoto, Y., et al. (2009) Induction of mesothelioma by a single intrascrotal administration of multi-wall carbon nano-tubes in intact male Fischer 344 rats, J. Toxicol. Sci. 34:65-76

24. Takagi, A., et al. (2008). Induction of mesothelioma in p53+/- mouse by intraperitoneal application of multi-wall carbon nano-tubes, J. Toxicol. Sci. 33:105-116

25. Hubbs, A., et al. (2009). Persistent Pulmonary Inflammation, Airway Mucous Metaplasia and Migration of Multiwalled Carbon Nano-tubes from the Lung after Subchronic Exposure, Abstract 2193, The Toxicologist CD - An official Journal of the Society of Toxicology, Volume 108, Number S-1, March 2009

26. Lademann, J., et al (2001). Investigation of follicular penetration of topically applied substances. Skin Pharmacol Appl Skin Physiol 14:17-22

27. Aritake, Toshio (2009). Japanese Ministry Issues Guidelines to Reduce Risk from Nanotechnology. The Bureau of National Affairs' Daily Environment Report (http://news.bna.com/deln/DELNWB/split_display.adp?fedfid=11657707&vname=dennotallissues&fcn=6&wsn=499192000&fn=11657707&split=0)

28. Canada Gazette (Feb 21, 2009), Vol. 143, No. 8, Significant New Activity Notice No. 15274

29. European Commission. Nanomaterials in REACH, December 16, 2008, Doc. CA/59/2008 rev. 1 (http://ec.europa.eu/environment/chemicals/reach/pdf/nanomaterials.pdf)

30. Federal Register: October 31, 2008 (Volume 73, Number 212); Toxic Substances Control Act Inventory Status of Carbon Nanotubes (http://www.epa.gov/EPA-TOX/2008/October/Day-31/t26026.htm)

31. Federal Register: November 5, 2008 (Volume 73, Number 215); EPA SNUR for Siloxane modified silica and siloxane modified alumina nanoparticles (http://www.epa.gov/EPA-TOX/2008/November/Day-05/t26409.htm)

32. Federal Register: November 19, 2008 (Volume 73, Number 224); Petition for Rulemaking Requesting EPA Regulate Nano-scale Silver Products as Pesticides; Notice of Availability (http://www.epa.gov/EPA-PEST/2008/November/Day-19/p27204.htm)

33. Federal Register: June 24, 2009 (Volume 74, Number 120); EPA SNUR for Single- and Multi-walled carbon nanotubes (http://www.epa.gov/EPA-TOX/2009/June/Day-24/t14780.htm)

34. Federal Register: November 6, 2009 (Volume 74, Number 214); EPA SNUR for Single- and Multi-walled carbon nanotubes (http://edocket.access.gpo.gov/2009/pdf/E9-26818.pdf)

35. Kreilgaard, M. (2002) Influence of microemulsions on cutaneous drug delivery. Adv Drug Deliv Rev 54:S77-S98

CONTACT INFORMATION

For additional information or discussion, you may contact the following authors:

Ed Wallner
edward.j.wallner@delphi.com

D.H.R. Sarma
d.h.r.sarma@delphi.com

ACKNOWLEDGMENTS

We wish to acknowledge the time and effort of Professor Timothy D. Sands, of the Birck Nanotechnology Center at Purdue University. Due to his technical expertise, Professor Sands has graciously agreed to review and provide input, specifically towards in the area of fundamental physics of nanotechnology.

DEFINITIONS/ABBREVIATIONS

CNT
Carbon Nano-tube

CuO
Cupric oxide

DSC
Differential Scanning Calorimetry

DSL
Domestic Substance list

EPA
Environmental Protection Agency

EU
European Union

FIFRA
Federal Insecticide, Fungicide, Rodenticide Act

IC
Integrated Circuit

iNEMI
International Electronics Manufacturing Initiative

MEMS
Micro Electro Mechanical Sensor

MWCNT
Multi-Walled Carbon Nano-tube

NIOSH
National Institute for Occupational Safety and Health

NSNR
New Substances Notification Regulation

OLED
Organic Light Emitting Diode

PCB
Printed Circuit Board

PMN
Pre Manufacturer Notice

PP
Polypropylene

PPE
Personal Protective Equipment

SMC
Sheet-Molding Compound

SnPb
Tin-Lead

SWCNT
Single-Walled Carbon Nano-tube

SNUR
Significant New Use Rules

TIM
Thermal Interface Materials

TSCA
Toxic Substances Control Act

About the Editor

Dr. Andrew Brown, Jr. is Executive Director & Chief Technologist for Delphi Corporation, and as such he provides leadership on corporate innovation and technology issues to help achieve profitable competitive advantage. Dr. Brown also represents Delphi globally in outside forums on matters of innovation and technology including government and regulatory agencies, customers, alliance partners, vendors, contracting agencies, academia, etc. Prior to this assignment, Dr. Brown had responsibility for common policies, practices, processes and performance across Delphi's 17,000 member technical community globally and its budget of $2.0 billion, including establishing Delphi's global engineering footprint with new centers in Poland, India, China, and Mexico, among others.

In April of 2009, SAE International's Executive Nominating Committee named Dr. Andrew Brown Jr., as its candidate for 2010 SAE International President. He was elected as 2010 SAE International President and Chairman in November of 2009, and was sworn into office in January of 2010. He represents 128,000 members in over 100 countries.

As an NAE member, Dr. Brown was appointed by the National Research Council (NRC) to serve as chair of the Committee on Fuel Economy of Medium and Heavy Duty Vehicles. The report developed by this group was recently referenced by President Obama in his enhanced efforts on fuel economy improvement.

Dr. Brown joined Delphi coming from the GM Research and Development Center in Warren, Michigan, where he was Director - Research, Administration & Strategic Futures. He also served as a Manager of Saturn Car Facilities from 1985 to 1987. At Saturn, he was on the Site Selection Team and responsible for the conceptual design and engineering of this innovative manufacturing facility.

Dr. Brown began his GM career as a Project Engineer at Manufacturing Development in 1973. He progressed in the engineering field as a Senior Project Engineer, Staff Development Engineer, and Manager of R&D for the Manufacturing Staff. During this period, he worked on manufacturing processes and systems with an emphasis on energy systems, productivity improvement and environmental efficiency. Before joining GM, he supervised process development at Allied-Signal Corporation, now Honeywell, Incorporated in Morristown, New Jersey.

Dr. Brown earned a Bachelor of Science Degree in Chemical Engineering from Wayne State University in 1971. He received a Master of Business Administration in Finance and Marketing from Wayne State in 1975 and Master of Science Degree in Mechanical Engineering focused on energy and environmental engineering from the University of Detroit-Mercy in 1978. He completed the Penn State Executive Management Course in 1979. A registered Professional Engineer, Dr. Brown earned a Doctorate of Engineering in September 1992.